Plant Adaptation: Molecular Genetics and Ecology

1

2

3

4

1. Individuals of *Mimulus guttatus* (an outcrosser with large flowers) and *M. alsinoides* (a selfer with small flowers). See Chapter 3. Photo credit: Kermit Ritland.

2. Flowers of *Piriqueta*. See Chapter 9. Photo credit: Mitch Cruzan.

3. *Lasthenia californica* shown growing in a pasture in California. See Chapter 13. Photo credit: Nishanta Rajakaruna.

4. Variation in flower size in individuals of *Collinsia parviflora* grown under uniform conditions. See Chapter 14. Photo credit: Elizabeth Elle.

ubcbotanicalgarden
& centre for plant research
AT THE UNIVERSITY OF BRITISH COLUMBIA

Plant Adaptation: Molecular Genetics and Ecology

Proceedings of an International Workshop sponsored by the UBC Botanical Garden and Centre for Plant Research held December 11–13, 2002 in Vancouver, British Columbia, Canada

Edited by

Q.C.B. Cronk, J. Whitton, R.H. Ree, and I.E.P. Taylor

NRC·CNRC
NRC Research Press
Ottawa 2004

Q.C.B. Cronk, R.H. Ree[1], and I.E.P. Taylor. UBC Botanical Garden and Centre for Plant Research, 6804 SW Marine Drive, Vancouver, BC V6T 1Z4, Canada.

J. Whitton. Department of Botany, University of British Columbia, 3529-6270 University Boulevard, Vancouver, BC V6T 1Z4, Canada

[1]Current address: Field Museum, 1400 S. Lake Shore Dr, Chicago, IL 60605-2496, U.S.A.

Printed in Canada on acid-free paper. ∞

ISBN 0-660-19336-1
NRC No. 46336

National Library of Canada cataloguing in publication

Main entry under title:

Plant adaptation: molecular genetics and ecology

"Proceedings of an International Workshop sponsored by the UBC Botanical
Garden and Centre for Plant Research held December 11–13, 2002 in Vancouver,
British Columbia, Canada"
Includes bibliographical references.
Issued by the National Research Council of Canada.
Issued also on the Internet.
ISBN 0-660-19336-1
Cat. no. NR15-72/2002E

1. Plants — Adaptation — Congresses.
2. Plant ecological genetics — Congresses.
3. Plant genetics — Congresses.
4. Plant ecophysiology — Congresses.
I. Cronk, Quentin C.B.
II National Research Council Canada.

QK912.M64 2004 581.4 C2004-980208-9

Inquiries: Proceedings Program, NRC Research Press, National Research Council of Canada, Ottawa, Ontario K1A 0R6, Canada.

Cover illustration: The cover image was taken on Shell Island, British Columbia and shows a large population of sea blush, *Plectritis congesta* in the foreground. Most *P. congesta* populations are polymorphic for the presence or absence of wings on their fruits. Although the presence of wings is dominant, the polymorphism is hypothesized to be maintained by heterozygote advantage. It is not known whether pleiotropy or linkage contribute to fitness differences among genotypes. Photo credit: Peter Arcese

Correct citation for this publication: Cronk, Q.C.B., Whitton, J., Ree, R.H., and Taylor, I.E.P. (*Editors*). 2004. *Plant Adaptation*: *Molecular Genetics and Ecology*. Proceedings of an International Workshop sponsored by the UBC Botanical Garden and Centre for Plant Research held December 11–13, 2002, in Vancouver, British Columbia, Canada. NRC Research Press, Ottawa, Ontario, Canada. 166 pp.

Contents

Introduction: concepts in plant adaptation . . . 1

Chapter 1. The new science of adaptation: an introduction. . . . 3
Quentin C.B. Cronk

Chapter 2. Discussion report: an intellectual framework for a plant adaptation science . . . 7
Richard Ree

Chapter 3. Pathways to plant population genomics . . . 11
Kermit Ritland

Chapter 4. Evolvability and the raw materials for adaptation . . . 18
Carl D. Schlichting and Courtney J. Murren

Chapter 5. Nucleocytoplasmic incompatibility fosters speciation . . . 30
Donald A. Levin

Chapter 6. Discussion report: answered and unanswered questions in plant adaptation . . . 38
Brian C. Husband

Approaches to the study of plant adaptation . . . 43

Chapter 7. Plant population genomics, linkage disequilibrium mapping and the genetics of adaptation. . . 45
Kenneth M. Olsen and Michael D. Purugganan

Chapter 8. Genomic approaches to identifying quantitative trait loci: lessons from *Arabidopsis thaliana* . . 53
Justin Borevitz

Chapter 9. Experimental analysis of adaptive landscape topographies. . . . 61
Mitchell B. Cruzan and Jennifer M. Rhode

Chapter 10. Terpene synthases and the mediation of plant-insect ecological interactions by terpenoids: a mini-review . . . 70
Dezene P.W. Huber and Jörg Bohlmann

Chapter 11. Adaptation in plant speciation: empirical evidence for the role of selection in the evolution of isolating barriers between plant species. . . . 82
Leonie C. Moyle

Chapter 12. Discussion report: new methods and tools for plant adaptation – what do we need? . . . 94
Barbara K. Mable

Traits, populations, and species: case studies in plant adaptation . . . 101

Chapter 13. Trends in the evolution of edaphic specialists with an example of parallel evolution in the *Lasthenia californica* complex . . . 103
Nishanta Rajakaruna and Jeannette Whitton

Chapter 14. Floral adaptations and biotic and abiotic selection pressures. . . . 111
Elizabeth Elle

Chapter 15. Polyploidy and plant adaptation: a framework for future research . . . 119
Brian C. Husband

Chapter 16. Evolutionary genetics of self-incompatibility in a new "model" plant: *Arabidopsis lyrata* . . 127
Barbara K. Mable

Chapter 17. Natural variation among accessions of *Arabidopsis thaliana*: beyond the flowering date, what morphological traits are relevant to study adaptation? 135
X. Reboud, V. Le Corre, N. Scarcelli, F. Roux, J.L. David, T. Bataillon, C. Camilleri, D. Brunel, and H. McKhann

Chapter 18. The study of ancient adaptation: a case study of a phytochrome gene pair from early-diverging angiosperms 143
Sarah Mathews

Chapter 19. The variable nature of herbivore defense: evidence for a rapidly diverging Kunitz trypsin inhibitor gene in *Populus* 153
Manoela Miranda, Mary Christopher, and C. Peter Constabel

List of participants 159

Index 161

INTRODUCTION:
CONCEPTS IN PLANT ADAPTATION

1
The new science of adaptation: an introduction

Quentin C.B. Cronk

The adaptive recursion

Adaptive recursion is a term borrowed from computer science. A recursive algorithm is a program that runs a new version of itself, by calling itself as it runs. Such recursion can be used, for instance, when searching down a directory tree, or, in the case of adaptive algorithms, to optimize an outcome (Smithies et al. 2004). There are obvious parallels between adaptive algorithms and biological adaptation. Adaptive recursion in computing involves a repeated process, the repeat being dependent upon on the preceding repeat in a particular way. In biological adaptation the "adaptive recursion" is the cycle from gene sequences to their unfolding as functional phenotypes that then interact with the environment leading to gene frequency change in natural populations, and so back to a different starting set of gene sequences. The programme then repeats in a manner that is dependent on the previous iteration. In computing it is important to avoid an infinite loop. In biology however the recursion is potentially infinite, only terminated on extinction.

In most cases the adaptive recursion is only studied by looking at single parts, such as molecular evolution or organism-environment interactions. One study that has been able to examine a substantial chunk of the adaptive recursion is that of Jamaican click beetles (*Pyrophorus plagiophthalamus*) and their evolving luminescence (Stolz et al. 2003). This is an ideal system for studying the adaptive recursion for a number of reasons. First, the beetles are polymorphic for an easily quantified phenotypic character: the colour of their emitted light, which varies from green to yellow-green to orange. Secondly the link from gene to phenotype is both simple and well studied: specific amino acid substitutions in the enzyme luciferase change the colour of the emitted light. Third the environmental function of the light emission is well known: mate attraction. In a highly illuminating piece of work, Stolz and colleagues show that there is a long-term adaptive trend towards orange in Jamaica, driven by natural selection. The click beetle system is an unusually elegant one and it is likely that many other adaptive systems will be much harder to study. This may be particularly true of plants. As plants are prone to phenotypic plasticity it is difficult to quantify the phenotype except in controlled growth chambers. Phenotypic plasticity may also complicate the link from genotype to phenotype. Other problems include the clonal reproduction of plants, which complicates the scoring of allele frequencies.

Quentin C.B. Cronk. UBC Botanical Garden and Centre for Plant Research, 6804 SW Marine Drive, Vancouver, BC V6T 1Z4, Canada. e-mail address: quentin.cronk@ubc.ca
Correct citation: Cronk, Q.C.B. 2004. The new science of adaptation: an introduction. *In* Plant Adaptation: Molecular Genetics and Ecology. Proceedings of an International Workshop held December 11–13, 2002, in Vancouver, British Columbia, Canada. *Edited by* Q.C.B. Cronk, J. Whitton, R.H. Ree, and I.E.P. Taylor. NRC Research Press, Ottawa, Ontario. pp. 3–6.

The Gould objection and neutrality as a null hypothesis

A cursory look at the natural world provides hundreds of apparent instances of the fitting of form to function in plants. However, the uncritical acceptance of ecological determinism prior to the rejection of the appropriate null neutral hypothesis has led uncritically to the conclusion that adaptive evolution is widespread. Naturalists after Darwin chronicled hundreds of instances of form following presumed function as evidence of adaptive evolution (Kerner von Marilaun and Oliver 1902; Lubbock 1905; Fritsch and Salisbury 1953). One example is the prevalence of xeromorphic leaves in dry habitats, apparently conferring resistance to water loss (see Table 1). However, physiological tests of functional efficacy are rare (Givnish 2003; Givnish et al. 2004), and tests of selective advantage rarer still. Thus the idea that these "adaptations" are the result of adaptive evolution has remained speculation only in most cases. However, so plausible has this idea been that it has too rarely been questioned. The main recent challenge has come from Gould and Lewontin, who point out that biologists have a tendency to leap into studies of adaptation without first testing the appropriate null hypothesis — that the variation is neutral.

Gould and Lewontin (1979) point out that even if a trait is currently adaptive, that does not mean that it evolved by adaptive evolution. Functional features of organisms may be merely products of adaptations, rather than adaptations per se. They famously used the example of the spandrels of San Marco, which are triangular not by design but simply because intersecting arches form triangles (Gould and Lewontin 1979). Gould used the term "exaptation" for such non-adaptive features, or features that have evolved adaptively for another function, which are then co-opted (exapted) for a new adaptive function (Gould and Vrba 1982). Thus, although "exaptations" may be adaptive, they have not evolved by adaptive evolution related to their current function. Gould's attack on the adaptationist world-view will stand or fall on the relative prevalence of "exaptations" versus products of conventional adaptive evolution.

Table 1. Some xeromorphic characters assumed to be adaptations to dry habitats arising from selection for drought-tolerance (after Fritsch and Salisbury 1953): examples of adaptive story-telling.

Trait	Example
Thick cuticle	*Vaccinium* spp., *Eryngium*
Stomata in cuticular pits (vestibules)	*Ilex*, *Ulex*
Stomata in grooves	*Nerium oleander*, *Calluna vulgaris*
Stomata protected by hairs	*Stachys* sp.
Stomata protected by rolling of leaves	*Empetrum*, *Festuca*
Massing of leaves in rosettes	*Armeria*, *Saxifraga* spp.
Abscission of leaves in drought	Deciduous trees
Reduction of leaf transpiration surface	Conifers, *Lycopodium*
Mechanical strength of leaves	*Laurus* (less deformation on drying)
Centric organization of the leaf	*Hakea*
Absence or reduction of leaves	Switch plants: *Cytisus*, *Casuarina*
Cladodes (stems as leaves)	*Ruscus*, *Asparagus*
Phyllodes (petioles as leaves)	*Acacia*
Leaf spines	*Ulex*, *Berberis*
Stem spines	*Ulex*
Leaf succulence	*Sedum*, *Sempervivum*
Stem succulence	Cacti
Aqueous hypodermis	*Hakea*, *Streptocarpus saxorum*
Cell mucilage	*Mesembryanthemum*
Vertical leaf angle	*Iris*, *Eucalyptus* spp.
Leaf movements	*Robinia*
High leaf albedo from scales or hairs	*Hippophae*

Molecular tests of neutrality

There is an advantage to starting a study of adaptation at the molecular entry point in the adaptive recursion. This is because robust tools now exist to test the hypothesis of neutrality using gene sequences, thus meeting Gould's objection. Having rejected the hypothesis of neutrality for a given gene, further studies can be reasonably conducted within the context of adaptation, without uncritically assuming ecological determinism. Adaptive explanations can then be sought for the evolution of a phenotype that is controlled by genes known to be under selection, as in the click beetle study.

Most studies of positive selection on genes use comparative data on the proportion of silent substitutions to substitutions that change an amino acid (the dN/dS ratio or omega). These tests can operate on various levels of the comparative hierarchy. Between species methods have been most notably developed by Yang (Anisimova et al. 2001; Yang 1998; Yang and Nielsen 2002), while at the population level various tests, such as the MacDonald-Kreitman test, are used (Purugganan and Suddith 1998). However, positive selection on a gene may be highly localised both in space (to particular amino acids) and time (to particular lineages). A recent study (Ree et al. 2004) coined the term "selectional mosaic" for this phenomenon. This study examined the molecular evolution of two paralogous florally expressed developmental genes in lupins. In one lineage, positive selection (omega > 1) was found to have acted at some codon sites of one paralog. However, on the same lineage the other paralog showed evidence of greater purifying selection (omega < 1) acting at some sites. This heterogeneous selection may be characteristic of the evolution of developmentally important transcription factors.

One significant problem lies in determining the phenotypic effect of a selected amino acid change. This is the essential next step in following the adaptive recursion. In the case of the click beetles the spectral emission from purified luciferase proteins carrying amino acid substitutions can be straightforwardly determined. For most genes however there are no in vitro assays of phenotype, and organismal measures of phenotype are complicated by different genetic backgrounds. Differences in genetic background create a huge amount of noise that complicates efforts to determine whether a given amino acid change is significant. This may be filtered out to a certain extent by large association studies, either under controlled breeding or in large populations. However, tight linkage of other significant mutations is virtually impossible to rule out. Furthermore, phylogenetically deep amino-acid changes, such as changes characteristic of different genera or families, are not amenable to this approach. Instead, the only way to proceed is by transformation of heterologous proteins to complement a knock-out phenotype in a transformable model, such as *Arabidopsis*, and looking for differences in the complementation phenotype. This is slow and is complicated by the introduction of a third genetic background - that of the transformable model. The way forward may come from "soft" reverse genetics techniques such as TILLING (Perry et al. 2003; Till et al. 2003).

Mutations in any amino acid can be created by chemical mutagenesis such as EMS. However, until now it has been impossible to screen large mutagenized populations for amino acid changes in given genes as the phenotypic effects may be small and it may not be clear exactly which phenotypes are candidates for mutations in a particular gene. TILLING solves this problem by pooling the mutagenized population and searching for mutations in the PCR products of a target gene. The enzyme CEL1 cuts only at basepair mismatches so will detect a mutant in a population of pooled PCR products. This technique can be used very efficiently to look for alleles with minor phenotypic effects in a given gene. It can be used in the classical mode on mutagenized populations or to detect single nucleotide polymorphisms in populations, a procedure dubbed ecoTILLING (Comai et al. 2004).

Developmental genes

In a study of the evolution of maize from teosinte under ancient artificial selection, Wang et al. (1999) demonstrated that heavy selection had operated on the cis-regulatory region of the transcription factor teosinte branched 1 (tb1). TB1 has important developmental effects, including effects on maize architecture and branching. What is notable about this study is that the coding region of the genes shows no interesting evolutionary change, instead it is the regulation of the gene that has been significant. Doebley and Lukens (1998) suggested that such changes (in the regulation of regulatory genes) may be of greater significance in evolution than changes in coding regions. Their reasoning is that transcription factors generally have highly specific expression domains and highly targeted effects so changes in these genes can bring about major changes in phenotype without producing deleterious pleiotropic effects that would be the result of evolutionary change in signalling or housekeeping genes. Furthermore, they argue that the evolutionary changes that will be most specific and significant are those that affect the expression of these regulators rather than their mode of action (i.e., changes in cis-regulatory regions).

This "regulation of the regulators" hypothesis of Doebley and Lukens receives some support from traditional evolutionary character analysis. It has long been supposed that many evolutionary changes are either heterochronic or heterotopic, and this consistent with temporal or spatial changes in gene regulation (Cronk 2002). Heterochronic changes, altering the timing or duration of developmental events can lead to different relative sizes and shapes of organs — characters of the sort that commonly differentiate species. Changes in the position of developmental processes can alter the form of organs more radically, such as the conversion of petioles into leaf-like structures in Acacia phyllody and other transferences (Baum and Donoghue 2002; Wang et al. 2004). Such transformations are termed homeosis. There are tremendous opportunities here to integrate developmental genetics, evolution and processes of adaptation (Cronk 2001). However, there are formidable differences too as mutations in cis-regulatory regions are much harder to study than mutations in coding regions. A mutation in a coding region has an easily predictable effect on the protein product and powerful tools such as dN/dS ratio can be brought to bear. In cis-regulatory regions new tools are needed.

The ecomolecular synthesis

Molecular changes, of whatever type, when studied on their own will never give a complete description of the adaptive process. The process of adaptive evolution involves the interaction of genotype with ecology: other organisms (this volume) and climatic (this volume) and soil (Rajakaruna and Whitton, this volume, Chapter 13) factors. Between the genome and the ecosystem is developmental biology, the unfolding of the phenotype, which in plants involves continuous feedback between the environment and the genome. The feedback between environment and development (resulting in phenotypic plasticity) is a particularly notable feature of plants and must be taken into account whenever the adaptive biology of plants is studied (Schlichting and Murren, this volume, Chapter 4). We now have the possibility of combining molecular, developmental and ecological studies into a powerful description of the adaptive process: an ecomolecular synthesis. In some ways the ecomolecular synthesis may be seen as a natural extension of the Darwinian paradigm of natural selection. Darwin was greatly troubled by the then prevailing, and inadequate, genetic theories of blending inheritance and Weissmanism. He did not have access to the modern concept of particulate genetic elements. His theory was organism-centred and involved selection acting on organisms. Fisher, Sewell-Wright, Haldane and others later combined Darwinian selection with the Mendelian notion of the gene. In doing so the focus of selection moved down a level to selection on allele frequencies. Selection on allele frequencies, and the theoretical apparatus surrounding it, is the basis for the "modern synthesis" of evolution, when combined with ecological processes.

However, we now have access to a much more fundamental level of molecular and developmental understanding: the influence of individual nucleotide changes on developmental processes. This information (which is gathering apace from genomics and functional genomics) can then be combined with an ecological understanding of the organism in natural ecosystems. This is the ecomolecular synthesis. The combination of ecology and genomics may shed light on a number of problems that have remained intractable under the pre-genomic "modern synthesis". We have long known that evolution varies strongly in tempo. Some lineages such as *Ginkgo* (maidenhair tree) and *Lingula* (a brachiopod) have varied little (at least in morphology) for hundreds of millions of years. What is not yet clear is the relative contribution of ecological stability versus genomic constraint (developmental canalisation) in maintaining this stasis. Alternatively, evolution, as in rapid island radiations, may proceed at an astonishing rate. Again this fast tempo may be genomic in origin, due to heritable developmental lability for key traits (labile versus constrained development), or ecological because the environment is permissive to phenotypic extremes and may select strongly on these extremes (permissive versus difficult fitness landscapes).

This volume

To address these and other questions raised by the modern study of adaptation in plants, a workshop was held at the Botanical Garden and Centre for Plant Research at the University of British Columbia in Vancouver, Canada from the 11th to 13th December 2002. The chapters of this book are based on presentations made at that workshop, together with three reports of workshop discussions (by Ree, Husband, and Mable).

Participants covered various aspects of the adaptive recursion: (1) molecular evolution (Ritland; Olsen and Purugganan; Matthews), (2) genes, development and the phenotype (Schlichting and Murren; Husband; Levin), (3) phenotype-environment interactions and population processes (Cruzan and Rhode; Elle; Rajakaruna and Whitton; Moyle; Miranda et al.).

Participants also covered adaptation at various levels in the evolutionary hierarchy: (1) adaptation at the infraspecific and population level (Borevitz; Mable; Reboud et al.) (2) speciation and adaptative differences between species (Cruzan and Rhode; Rajakaruna and Whitton) (3) deep adaptation and adaptive differences between major clades, for instance in physiological (phytochrome) and biochemical (terpenoid) traits (Mathews; Huber and Bohlmann).

References

Anisimova, M., Bielawski, J.P., and Yang, Z. 2001. Accuracy and power of the likelihood ratio test in detecting adaptive molecular evolution. Mol. Biol. Evol. **18**: 1585–1592.

Baum, D.A., and Donoghue, M.J. 2002. Transference of function, heterotopy, and the evolution of plant development. *In* Developmental genetics and plant evolution. *Edited by* Q.C.B. Cronk, R.M. Bateman, and J.A. Hawkins. Taylor and Francis, London, pp. 52–69.

Comai, L., Young, K., Till, B.J., Reynolds, S.H., Greene, E.A., Codomo, C.A., Enns, L.C., Johnson, J.E., Burtner, C., Odden, A.R., and Henikoff, S. 2004. Efficient discovery of DNA polymorphisms in natural populations by Ecotilling. Plant. J. **37**: 778–86.

Cronk, Q.C.B. 2001. Plant evolution and development in a post-genomic context. Nat. Rev. Genet. **2**: 607–619.

Cronk, Q.C.B. 2002. Perspectives and paradigms in plant evo-devo. *In* Developmental genetics and plant evolution. *Edited by* Q.C.B. Cronk, R.M. Bateman, and J.A. Hawkins. Taylor & Francis, London, pp. 1–14.

Doebley, J., and Lukens, L. 1998. Transcriptional regulators and the evolution of plant form. Plant Cell **10**: 1075–1082.

Fritsch, F.E., and Salisbury, E. 1953. Plant form and function, 2nd ed. G. Bell. London.

Givnish, T.J. 2003. How a better understanding of adaptations can yield better use of morphology in plant systematics: towards eco-evo-devo. Taxon **53**: 273–296.

Givnish, T.J., Montgomery, R.A., and Goldstein, G. 2004. Adaptive radiation of photosynthetic physiology in the Hawaiian lobeliads: light regimes, static light responses, and whole-plant compensation points. Am. J. Bot. **91**: 228–246.

Gould, S.J., and Lewontin, R. 1979. The spandrels of San Marco and the Panglossian paradigm: a critique of the adaptationist programme. Proc. R. Soc. Lond. B Biol. Sci. **205**: 581–598.

Gould, S.J., and Vrba, E.S. 1982. Exaptation: a missing term in the science of form. Paleobiology **8**: 4–15.

Kerner von Marilaun, A.R., and Oliver, F.W. 1902. The natural history of plants. Blackie and Son, Ltd., London. Lubbock, J. (Lord Avebury). 1905. Notes on the life history of British flowering plants. Macmillan, London.

Perry, J.A., Wang, T.L., Welham, T.J., Gardner, S., Pike, J.M., Yoshida, S., and Parniske, M. 2003. A TILLING reverse genetics tool and a web-accessible collection of mutants of the legume *Lotus japonicus*. Plant Physiol. **131**: 866–871.

Purugganan M.D., and Suddith J.I. 1998. Molecular population genetics of the *Arabidopsis* CAULIFLOWER regulatory gene: nonneutral evolution and naturally occurring variation in floral homeotic function. Proc. Natl. Acad. Sci. U.S.A. **95**: 8130–8134.

Ree, R.H., Citerne, H.L., Lavin, M., and Cronk, Q.C.B. 2004. Heterogeneous selection on LEGCYC paralogs in relation to flower morphology and the phylogeny of *Lupinus* (Leguminosae). Mol. Biol. Evol. **21**: 321–331.

Smithies, R., Salhi, S., and Queen, N. 2004 Adaptive hybrid learning for neural networks. Neural Comput. **16**: 139–157.

Stolz, U., Velez, S., Wood, K.V., Wood, M., and Feder, J.L. 2003. Darwinian natural selection for orange bioluminescent color in a Jamaican click beetle. Proc. Natl. Acad. Sci. U.S.A. **100**: 14955–14959.

Till, B.J., Reynolds, S.H., Greene, E.A., Codomo, C.A., Enns, L.C., Johnson, J.E., Burtner, C., Odden, A.R., Young, K., Taylor, N.E., Henikoff, J.G., Comai, L., and Henikoff, S. 2003. Large-scale discovery of induced point mutations with high-throughput TILLING. Genome Res. **13**: 524–530.

Wang, C.-N., Moller, M., and Cronk, Q.C.B. 2004. Altered expression of GFLO, the Gesneriaceae homologue of FLORICAULA/LEAFY, is associated with the transition to bulbil formation in *Titanotrichum oldhamii*. Dev. Genes Evol. **214**: 122–127.

Wang, R.-L., Stec, A., Lukens, L., Hey, J., and Doebley, J. 1999. The limits of selection during maize domestication. Nature **398**: 263–269.

Yang, Z. 1998. Likelihood ratio tests for detecting positive selection and application to primate lysozyme evolution. Mol. Biol. Evol. **15**: 568–573.

Yang, Z., and Nielsen, R. 2002. Codon-substitution models for detecting molecular adaptation at individual sites along specific lineages. Mol. Biol. Evol. **19**: 908–917.

2
Discussion report: an intellectual framework for a plant adaptation science

Richard Ree

Introduction

The first day of the workshop concluded with a discussion on "the intellectual framework for a plant adaptation science", chaired by Quentin Cronk (UBC). The topic was particularly appropriate, as the two sessions that day were organized around the theme of abiotic evolutionary challenges faced by plants (adaptation to the edaphic environment and light climate, respectively), and the talks presented varied widely in methodology and organismal scope. In general the discussion revolved around the idea of an "eco-molecular synthesis" in studies of plant adaptation: what this comprises, why it is a worthwhile goal, and what is needed to achieve it. The main participants in the discussion were Justin Borevitz (Salk Institute), Toby Bradshaw (U. Washington), Quentin Cronk (UBC), Daniel Fulop (Harvard), Tom Givnish (U. Wisconsin), Barbara Mable (U. Guelph), Michael Purugganan (NCSU), Carl Schlichting (U. Connecticut), and Jeannette Whitton (UBC).

Here I provide an overview of the salient points made during the ensuing forty-five minute brainstorming session. They fall into three major threads. (1) Approaches to studying plant adaptation — what are their similarities and differences? How might their incompatibilities and the barriers between their respective research communities be overcome? (2) An "eco-molecular synthesis" — what would the theoretical structure of such a thing look like? Are there general principles or paradigmatic empirical expectations that could help unify studies of plant adaptation? (3) Adaptation — is a unified concept necessary for progress to be made?

An intellectual framework for studying plant adaptation

Approaches

The talks presented earlier in the day were demonstrative of the progress being made in understanding the adaptive value of ecologically relevant traits. Adaptation studies generally fall into three categories: molecular genetic, ecological, and phylogenetic, each having strong methodological traditions and active research communities. The approaches correspond to hierarchical levels in genetic organization, namely at the level of the individual, population, and species, respectively. All were represented in the day's presentations. Molecular genetic studies focus on the evolution of genes or gene networks known or thought to be functionally important in organismal development, as illustrated by Michael Purugganan's studies of genes involved in *Arabidopsis* inflorescence development. Ecological studies analyze environmental correlates of variation with genotype and phenotype in the context of populations of a single species or its close relatives, as shown by Nishanta Rajakaruna's study of parallel edaphic specialization of *Lasthenia californica* populations. Phylogenetic studies use comparative methods to study trait evolution in lineages separated by deep divergences. Suhua Shi (Zhongshan University) presented examples of convergence in morphological and life-history traits across disparate lineages of mangrove plants. Sarah Mathews (U. Missouri) presented an analysis of molecular evolution in phytochrome A sequences in angiosperms, providing evidence of positive selection at that locus.

There is currently little overlap or continuity between these research groups. John Doebley's work on maize was mentioned as an exemplar of synthetic interdisciplinary research. Participants expressed general optimism that in the future it will be increasingly possible to link molecular, ecological, and phylogenetic approaches together and make strides in discovering the "natural history" of genes in species outside the current set of models. It was suggested that a rewarding line of investigation would follow a top-down research cascade. This would begin with the demonstration of a trait's adaptive value in an ecological context, followed by the identification and mapping of the QTLs underlying the trait. The significant genes at those loci would then be cloned and functionally classified. Finally, phylogenetic analysis would reveal the genealogies and patterns of molecular evolution of those genes. Another proposal advocated adopting a strategy of reciprocal illumination that alternates focus between the species of interest and a model system like *Arabidopsis*. In this scenario, one would generate evolutionary hypotheses from studies of the species of interest, then test those hypotheses using the superior molecular tools

Richard Ree.[1] UBC Botanical Garden and Centre for Plant Research, 6804 SW Marine Drive, Vancouver, BC V6T 1Z4, Canada. e-mail address: rree@fieldmuseum.org

Correct citation: Ree, R. 2004. Discussion report: an intellectual framework for a plant adaptation science. *In* Plant Adaptation: Molecular Genetics and Ecology. *Proceedings of an International Workshop held December 11–13, 2002, in Vancouver, British Columbia, Canada. Edited by* Q.C.B. Cronk, J. Whitton, R.H. Ree, and I.E.P. Taylor. NRC Research Press, Ottawa, Ontario. pp. 7–10.

[1]Current address: Department of Botany, Field Museum of Natural History, 1400 S. Lake Shore Drive, Chicago, IL 60605, U.S.A.

and knowledge base available for the model species, e.g. by transforming candidate genes from the species of interest into *Arabidopsis* and measuring their function. These results could then be used to refine the original hypotheses for further studies of the species of interest.

The point was made that this latter approach will probably be effective only if the chosen model system is sufficiently closely related to the species of interest. More importantly, while it may be successful in showing *conservation* of function, it is not well suited for studies of adaptive *differentiation*. The general problem is that if the gene of interest is shown to have a different function in *Arabidopsis*, one cannot tell whether it is because the gene has actually diverged in function, or whether the difference is due to the genetic background of *Arabidopsis*. Regulatory genes in particular can affect multiple pathways depending on the context in which they are expressed. Such outcomes are not without value, however, as interaction with background is an inherently interesting aspect of a gene's natural history.

Differences between approaches

The preceding highlights a significant point of conflict between strictly molecular genetic studies and the primary goal of discovering the genetic basis of adaptive evolution. In functional studies of genes at the molecular level, e.g. in studies of "evo-devo", it is always easier to show conservation of function: if knocking out a candidate gene produces the same developmental effects in different genetic backgrounds, that is positive evidence that its function is the same. However from an evolutionary standpoint it is generally more interesting to study differences in gene function, not similarities, because the functional diversity of genetic variation is most likely to have fitness consequences.

This follows from what seems to be a general disconnect between approaches in their inferential scope. At the population level it is possible to show experimentally that particular traits are adaptive in terms of the fitness consequences of natural variation, but in general little is known about the genetic basis of such traits. In contrast, at the molecular level we are equipped to determine that a gene is undergoing adaptive evolution. We know how to recognize and detect the signature of positive selection in multiple sequence alignments and data sets of allele frequencies, for example, but we generally don't know *why* such selection occurred or what its significance is in the context of phenotype or fitness. Fitting these two inferential toolkits together, i.e., linking cause and effect is currently extremely difficult. This difficulty may be partly due to the fact that cutting-edge molecular approaches have only made substantial inroads at the intraspecific level. More generally, the difficulty may lie in how we atomize traits in an ecological setting (e.g. specific leaf area in relation to shade) versus their developmental genetic basis.

Taxonomic focus is another obvious difference between molecular genetic studies and ecological or phylogenetic studies. The former tend to focus on model species, the most prominent in plants being *Arabidopsis thaliana*, while the latter tend to focus on phylogenetically more diverse assemblages of species that are genetically poorly characterized. One of the main advantages of using model species is that they are the subject of genome sequencing efforts and have vast amounts of genomic data available in public repositories. Other benefits include ease of propagation, optimized protocols, mail-order seed bank and ecotype stocks, etc. While it may be unrealistic to expect the same level of convenience in studies of non-model species, genomic tools that are applicable to such species are needed nonetheless. These include cDNA libraries, microarrays, knockout mutant collections, and so on. Genome sequences of judiciously chosen non-model species would certainly increase the prospects of developing genomic tools and methods that could be used in a truly comparative framework. Due to the high cost of genome sequencing projects, high-throughput sequencing centers will need guidance from all corners of the plant adaptation community, ideally including input from researchers in developmental genetics, ecology, and phylogenetics. The community could also collectively shoulder the burden of developing genomic tools like cDNA libraries. Perhaps the difficulty lies in selecting which species will receive the funding and attention necessary to develop them. A desirable outcome for the meeting would be to propose a set of candidate species for coordinated, detailed study at a variety of levels. One prime candidate for this set is *Mimulus*, given the considerable amount of genetic, natural history, and experimental data that has been gathered by Bradshaw and others.

An "eco-molecular" synthesis

The importance of "evo-devo" — evolutionary developmental biology — in understanding the role of genetics in morphological development (i.e., the genotype-phenotype map considered by many to be the Holy Grail of evolutionary biology) has become abundantly clear over the past several years. But it is equally apparent that this is only part of a fully synthetic understanding of plant adaptation. What is really needed is an "eco-evo-devo" framework that integrates the study of all the most important aspects of adaptive phenotypic evolution, such as developmental pathways, ecologically determined selection pressures, and phylogeny. An "eco-molecular" synthesis, as proposed by Quentin Cronk in his introductory remarks to the conference, would tie together molecular development, ecology, and molecular evolution, allowing more elegant, satisfying studies than are currently possible.

It was noted that such a synthesis may have particular import for plants. Compared to animals, plants are generally more modular and prone to plasticity. Deep ties clearly exist between plant morphology and the ecological context in which morphological phenotypes develop and evolve. In Liliales, for example, species that occur in shady habitats commonly possess a convergent suite of traits (net venation, rhizomatous roots, inconspicuous flowers, and fleshy fruits).

An eco-molecular synthesis will require conceptual paradigms to organize and focus research efforts. The need for this is underscored by the concern that empirical studies of the evolution of development are merely anecdotal in nature, i.e., discrete snippets of knowledge that are non-applicable outside a specific context, and which therefore lack the power necessary to achieve a general understanding of the developmental process. If this is true, it begs the question of whether there is in fact a general process to discover and understand at all. For example, what are the general principles underlying the evolution of genetic pathways and regulatory networks? Do we have reason to believe that generalities exist, or can they all be simply characterized by idiosyncrasies? Certain types of pathways, e.g., glycolytic or metabolic pathways, can use metabolic control theory to generate evolutionary predictions; for other types, such as developmental pathways, what are our expectations regarding their constraint or evolutionary opportunities? Are certain nodes more prone to respond to selection than others are? The general sentiment of the group was that at this point in time there is little basis for generalities to be drawn, or for inferring any rules that universally apply.

The utility of a unified conceptual framework is most clearly seen in experimental design, in terms of selecting the most appropriate hypotheses and study systems. At what phylogenetic level — recent or ancient divergences — should we focus studies of comparative evo-devo in order to discover adaptive changes? The current intellectual climate is most receptive to studies targeted at comparisons across deep splits, e.g. annelids versus molluscs, or angiosperms versus gymnosperms. But we do not yet have enough information to predict where the most interesting changes are to be found; in fact, there are good reasons to expect to find them at multiple levels. In angiosperms, for example, fleshy fruits have evolved multiple times in different tissues. In most cases, we have no idea as to the developmental basis of those evolutionary transformations. The observation that a character changes frequently in a clade can be interpreted as suggestive of a simple, common genetic mechanism that can account for some fundamental aspect of the change. Alternatively, such observations could indicate many different ways in which the character has evolved, particularly in cases where the end products (fleshy fruits) are easily identified as convergent homoplasy and not synapomorphy. The only way to differentiate these possibilities is to do the requisite comparative genomic studies.

Convergence is a common observation in agriculture. The literature on crop domestication provides many examples of how strong directional selection produces the same phenotypic change in different species. In general, we expect convergence as long as the same underlying developmental pathways are in place. Conversely, differences in genetic architecture are more likely to lead to different adaptive outcomes from the same selective regime. For example, within a group like Asteraceae we might expect to find that separate instances of edaphic specialization involve the same or very similar mutations in the same sets of genes, while between more distantly related lineages that are more divergent in terms of regulatory canalization, the genetics of edaphic specialization might be very different, involving different pathways. In other words, we expect any particular developmental architecture to be canalized, allowing certain mutations but excluding others.

Hidden synapomorphies

Fleshy fruits in angiosperms are examples of homoplasy, the evolutionary recurrence of similar traits. Over the past decade numerous plant molecular systematic studies have concluded that traits of all kinds that were thought to be uniquely derived and diagnostic of natural groups have in fact evolved independently multiple times. Moreover, it is often the case that homoplasy is confined to a single clade, suggesting an "underlying synapomorphy" or "latent homology" that predisposes the lineage to converge. Could the search for hidden synapomorphies underlying phylogenetically clustered instances of convergent evolution ("apomorphic tendencies") become a general or unifying principle in the study of adaptation? It would certainly be exciting to genetically characterize a "pre-adaptive" trait that has enabled independent adaptive changes in descendant lineages, especially if the pre-adaptive and adaptive traits are functionally unlinked.

The question arose that if we were to discover a hidden synapomorphy and were able to determine its genetic basis, would we expect to find evidence of positive selection in the genes involved? In other words, if underlying synapomorphies are real and pervasive in plant evolution, to what extent are they themselves adaptive? A cautionary point was made that the particular ecological context and circumstances in which novelties arise may render conditions unsuitable for selection to occur. For example the trait could have arisen in the absence of competitors, as is commonly the case following island colonization. More fundamentally, however, the answer to this question may depend on one's particular views on the overall importance of adaptation and selection in the evolutionary process.

The potential for discovering latent homology is also exemplified by nodulation, a symbiosis that exhibits a striking phylogenetic distribution: it is has arisen independently around ten times only within the Rosid I clade of angiosperms. Rhizobial nodules are unique to Fabaceae, in which they have arisen three times. That such a complex association is so phylogenetically constrained is strongly suggestive of a common underlying mechanism. How would one go about developing nodulation as a model system for discovering the genetic basis of apomorphic tendencies? Much baseline data, such as the genome sequence of *Sinorhizobium meliloti*, are available. Genomic studies of nodulation may provide an opportunity to get at the fundamental genes that activate downstream pathways.

A possible start down this path would be to sequence the transcripts of all the genes known to be expressed in nodules across a range of species, and then screen for the expression of those genes in related species that do not exhibit nodulation. Finding that those genes are down regulated or turned off entirely would provide positive evidence for their common function. However a potentially confounding result would be that some, possibly many, of the putative nodulation genes are still being expressed in plants lacking nodulation, but have been co-opted for different functions. This is especially true for regulatory genes that may interact with multiple pathways that are not related to nodulation. An alternative strategy would be to assess the complementarity of nodulation genetic systems in different species by crossing experiments, or using transgenesis techniques.

Targeting the natural history of exemplar genes

Research on non-model species is hampered by the lack of genomic data and tools that are available for model species. The observation that certain genes such as phytochromes are found repeatedly at the root of QTLs in *Arabidopsis* suggests that these genes are particularly important. One suspects the reason is that adaptation to light climate is pervasive throughout the evolutionary history of plants. This advances a case for making a concerted effort to study the natural history of that gene. In other systems, the important genes responsible for adaptation may not be so obvious. For example, soil specialization in Asteraceae occurs on a wide range of soil types (gypsum, saline, etc.) and different genes may be involved in each case.

To take this approach one would need to make the requisite libraries to pull out the genes of interest from a range of different plants, and then perform functional studies. It is important to keep in mind, however, that the functional differences among the genes themselves may be far removed from the adaptation being studied, if the genes are high up in a developmental cascade. The same signal-sensing apparatus may exist in all the species being studied, but the direction of regulation by the gene of interest may differ dramatically, e.g., with the same phytochrome genes, certain plants respond positively to a long day and others to a short day.

The challenge of the exemplar genes approach is to determine which classes of genes (e.g., cis-regulatory elements or other transcription factors, phytochromes, hormones, etc.) will yield the greatest rewards in linking adaptation from genotype to phenotype. The discovery of animal homeobox genes, for example, raised high hopes of solving the major questions of developmental biology, until it became apparent that depending on the location and context in which the genes are expressed, completely unexpected and different developmental outcomes, such as ectopic eye development in *Drosophila*, were possible.

The nature of adaptation

As commonly happens during discussions of conceptual issues, there came a time when definitions of terms needed to be clarified. In this case, the nature of adaptation itself came under scrutiny. There was general agreement that adaptations are highly context-specific, both ecologically (extrinsically) and organismally (intrinsically). A trait that confers a fitness advantage in one ecological community or set of environmental conditions may be maladaptive under difference circumstances. Similarly, it was suggested that an adaptation should only be interpreted in the context of the biology of the whole organism possessing it.

But what is meant by the term "adaptation"? Two opposing views were put forth in the discussion, differing largely along the lines of adaptation being an active *process* versus a passive *pattern*. From the process perspective, Tom Givnish (U. Wisconsin) defined adaptation as "phenotypic variation that enhances, on average, the fitness of the bearers of that variation, relative to other members of the population". From this quantitative view it follows that measuring adaptation requires there to be adaptive genetic variation, i.e., variation that can be shown to confer a fitness differential across the population. This also means that traits that are fixed cannot be considered to be adaptive.

Quentin Cronk advanced the pattern view of adaptation, suggesting a "weaker but still useful" definition: "demonstrated utility". Thus the beak of a bird can be demonstrated to be a suitable size and shape for eating food of a particular size, but it is another matter entirely to demonstrate the fitness consequences of this trait variation through selection experiments. The utility definition therefore has pragmatic value, and under this view, an adaptation can be considered to be a testable hypothesis, nothing more. Skeptics, however, suggested that a more appropriate label for this is "hypothetical adaptation" or, more bluntly, "variation of unknown significance". It should be noted however that many, if not most, of the traits of organisms that are commonly considered to be adaptations have never been formally tested.

Conclusions

The discussion, which ended the first day of the conference, drew to the forefront fundamental issues in adaptation research. The notion of an "eco-molecular synthesis", while regarded as a worthwhile objective, was nonetheless acknowledged to face challenges requiring time, funding, and (perhaps most importantly) a community-based effort to overcome, particularly with regard to the development of genomic tools for non-model species. There was a clear desire to move beyond the "anecdotal" nature of findings from studies of molecular genetics and development of model systems, and link these together with an ecological and phylogenetic context, forming a more comprehensive understanding of plant adaptation. Conceptually, the most promising unifying principle raised during the discussion was that of hidden synapomorphies: lineage-specific transformations that release constraints and allow repeated instances of adaptation to occur in closely related species.

3
Pathways to plant population genomics

Kermit Ritland

Abstract: Population genomics is a field in its infancy. It is inherently a multilocus activity, involving simultaneous comparison of many variants at different gene loci. Here, some approaches to plant population genomics are given, with examples taken from several plant species. First, studies are just beginning to discover the extent of linkage disequilibrium and its implications for natural selection and genetic load. Second, the correlation of gene diversity can be studied in the same manner as linkage disequilibrium, with an example from lodgepole pine "diversity mapping". Third, strategies for SNP detection are discussed with an emphasis on finding SNPs of adaptive value. Fourth, a simultaneous comparison of entire suites of genes (in this case, the chloroplast) can infer longer-term patterns of population change, in this case, the relative rates of deleterious substitutions. The special features of plants that distinguish them from animals include: greater population differentiation, greater inbreeding and clonality, the chloroplast genome, and the gametophyte.

Introduction

Lewontin's 1974 book, *The Genetic Basis of Evolutionary Change*, heralded the dramatic change in population genetics brought on by the discovery of abundant molecular genetic variation in natural populations. It was much larger than expected, as compared to morphology. More importantly, it could be compared among species, and this variation enabled parameters of many population genetic models to be estimated for the first time. Since then, incremental advances involving other types of markers have improved the resolution and scope of population genetics. With the advent of many complete genome sequences, is it possible that the paradigm will shift? Can knowing the genome, at least in model organisms, provide new directions in population genetics? What types of genomic approaches can be used in non-model organisms to provide insights into the genome processes that operate in natural populations?

Population genomics is a field in its infancy. So far, it has been mainly involved with activities as the analysis of human mitochondrial DNA (Hedges 2000), or for intensive genotyping of individuals for neutral genetic markers (Pletcher and Stumpf 2002) with the usual goal of identifying genes underlying human disease. In plants, the word is not currently found. In a broad sense, population genomics should be a multilocus activity (Goldstein and Weale 2001), involving simultaneous comparison of many variants at different gene loci. Inferring the roles of genes, the roles of particular mutations within these genes, and the co-dependent evolutionary pathways that allelic variants of these genes follow, is the pathway that population genomics can follow. While current genomics activities can provide some premonition of what population genomics might become, in fact, a good amount of existing population genetics theory is applicable to genomics data.

Here, I present some pathways of research in this emerging field. Classically, population genetics had data from just a few genes, and as such, was limited to inferences about inter-population processes. With data from many genes, and from genes with adaptive allelic variation, the focus shifts to the roles of genes, their adaptation and coadaptation. In addition, the dimensions of gene diversity increase: the pattern of variation along chromosomes adds a dimension to the diversity of a population, and the patterns of divergence for a population of genes add another dimension to the diversity of species. The clonal and sedentary nature of plants, the extent of gene duplication and polyploidy in plants, and the ease of their experimental manipulation, should allow plants to contribute much to this field over the next few years.

The space of variation

With the advent of whole genome sequences, we now know the approximate numbers of genes in a genome, and we can more accurately guess the numbers of possible genotypes in a population. While this depends upon the level of nucleotide polymorphism and the role of non-coding regions, with ca. 30,000 genes (*Arabidopsis*) and 10 polymorphisms per gene (1000 base pairs per gene, 0.01% polymorphism per pair), the total number of genotypes is astronomical: $2^{300,000} \cong 10^{100,000}$, a number far greater than the number of atoms in the universe. There must be a subset of this "space of variation" that populations occupy. Many genotypes would be inherently impossible because of intrinsically low fitness: even if the population was infinite, the number of habitats might be small with each habitat having an optimal adapted genotype (the ecological limit concept), or alternatively, fitnesses of alleles at alternative loci may be highly dependent, placing constraints on accessibility to the genomic landscape (the epistatic limit concept, c.f. Lewontin and White 1960).

Kermit Ritland. Department of Forest Sciences, University of British Columbia, Vancouver, BC V6T 1Z4, Canada.
e-mail address: Kermit.Ritland@ubc.ca
Correct citation: Ritland, K. 2004. Pathways to plant population genomics. *In* Plant Adaptation: Molecular Genetics and Ecology. Proceedings of an International Workshop held December 11–13, 2002, in Vancouver, British Columbia, Canada. *Edited by* Q.C.B. Cronk, J. Whitton, R.H. Ree, and I.E.P. Taylor. NRC Research Press, Ottawa, Ontario. pp. 11–17.

Indeed, recent findings about multilocus structure indicate the potential space of variation can actually be much smaller. In humans, Reich et al. (2001) found extensive linkage disequilibrium (LD), or non-random associations of alleles; across 160 kilobases in each of 19 genomic regions, and found normalized values of D to be above 0.5 between sites separated by up to 60 kb (however, in other regions, LD extended only over a few kilobases).

LD much reduces the space of variation, as unit of variation is a combination of alleles among loci, or the "haplotype". The reduction in attainable space incurred by haplotype structure is illustrated by Fig. 1, which depicts 4 regions of LD. Within each, a few haplotypes are descended from an ancestor as indicated by the genealogy; mutations, as indicated by hash marks on branches leading to a site, generate new haplotypes; the number of haplotypes equals the number of sites with mutational changes. At each nucleotide site, there is a SNP, and if we disregard allelic associations, the number of SNP combinations is $2^{13} = 8{,}192$. In contrast, the number of haplotype combinations is $3 \times 4 \times 4 \times 2 = 96$, a reduction of 99% in the number of possible genotypes!

Extensive haplotype structure would have several implications for the adaptive evolution of plant genomes, mainly that (*i*) hitchhiking effects would be more prevalent than expected, and (*ii*) haplotypes containing coadapted substitutions ("superalleles") can evolve. Models for the evolution of genetic load and for coadapted gene complexes might recognize underlying functional mechanisms through which adaptive genetic variation, as well as deleterious mutations that cause genetic load, spread and are maintained in populations of plant species. Of particular interest would be the role of gene families and gene conversion in generating haplotype structure.

Empirically, in plants, while it is well documented that plants species differ widely in their levels of gene diversity, significant studies of LD have been done only with maize and *Arabidopsis* (http://www.arabidopsis.org/cereon/). In a survey of six genes, Remington et al. (2001) found that intragenic LD generally declined rapidly with distance (normalized $D < 0.1$ within 1500 bp), but rates of decline were highly variable among genes. Ching et al. (2002) examined 18 maize genes in 36 maize inbreds, and found each gene to contain a small number (2–8) of distinct and highly diverse haplotypes, and consequently, significant LD within genes.

Two documented factors that cause variation of LD are (*i*) GC composition (72-fold lower recombination frequency in the GC-poor isochores compared to GC-rich isochors, Eisenbarth et al. 2001) and (*ii*) recombination hotspots, which creates "blocks" of LD. However, genetic drift alone can generate block-like patterns of LD similar to those observed in human populations (Zhang et al. 2003). There is a desperate need for studies of patterns of LD in other species, particularly comparative studies in relation to genome size, levels of variability and inbreeding, clonality, and other aspects of life history.

Fig. 1. Phylogeny of haplotypes in a chromosomal region with four blocks of linkage disequilibrium.

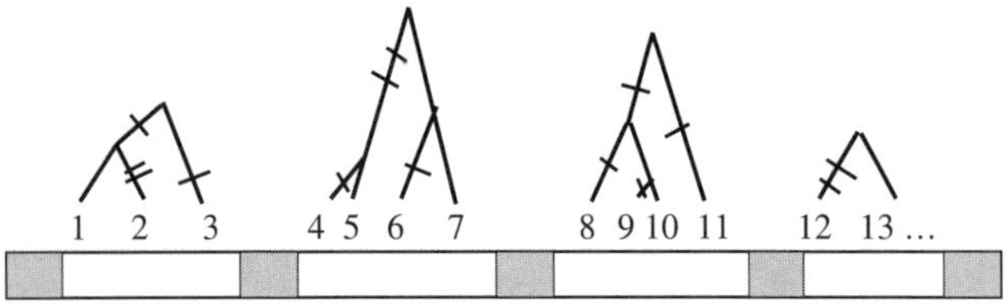

The landscape of chromosomal diversity

These findings of haplotype structure, and the general importance of multilocus models in genomics, place new emphasis on linkage disequilibria, a topic born from the Lewontin and Karlin schools of the 1970's. Historically, two-locus models have received much attention, as they are the minimalist description. Three two-locus inbreeding coefficients are needed to model the evolution of two locus systems (Strobeck and Morgan 1978). As diagrammed in Fig. 2, these respectively involve gene identities (triple lines) of two loci (*i*) among two gametes (Φ_{AB}), (*ii*) among three gametes (Γ_{AB}), and (*iii*) among four gametes (Δ_{AB}). Recursions of these quantities, as a function of mutation rate and population size, are given by Stobeck and Morgan (1978). These quantities can be related to genotypic frequencies as

[1] $$\Phi_{AB} = \sum_{i,j} p_{ij}^2 = J^2 + J(1-J)\,[R_{AB}^{D} + 2R_{AB}^{H,D} + R_{AB}^{H}]$$

$$\Gamma_{AB} = \sum_{i,j} p_{ij}p_i p_j = J^2 + J(1-J)\,[R_{AB}^{H,D} + R_{AB}^{H}]$$

$$\Delta_{AB} = \sum_{i,j} p_i^2 p_j^2 = J^2 + J(1-J)R_{AB}^{H}$$

where i indexes alleles at locus A and j indexes alleles at locus B; p_i is the frequency of allele A_i, p_j is the frequency of allele B_j, and p_{ij} is the frequency of gamete A_iB_j. J is the expected homozygosity (sum of squared gene frequency) at loci chosen at random from the genome (e.g., outside "reference loci" are needed; hence Equation [1] should be computed across many pairs of loci a given map distance apart). R_{AB}^{D} is equilivalent the squared normalized linkage disequilibrium between all alleles, and is estimated as $R_{AB}^{D} = (\Phi_{AB} - 2\Gamma_{AB} + \Delta_{AB})/J(1-J)$.

R_{AB}^{H}, the correlation of diversity between loci, is a quantity that has received no formal treatment, nor much empirical attention at least as an estimable parameter of genome structure. "Hitchhiking", or the change of gene frequency at neutral loci linked to selectively favored alleles (Maynard Smith and Haigh 1974), results in an increase of linkage disequilibrium but also a reduction in diversity around the selected locus. While LD (R_{AB}^{D}) declines rapidly with time, by contrast, the reduction in diversity, or equivalently, the correlation of diversity between linked loci (R_{AB}^{H}), decays much slower, as

Fig. 2. The three configurations of gene identity-by-descent between two loci, involving two, three or four gametes, respectively.

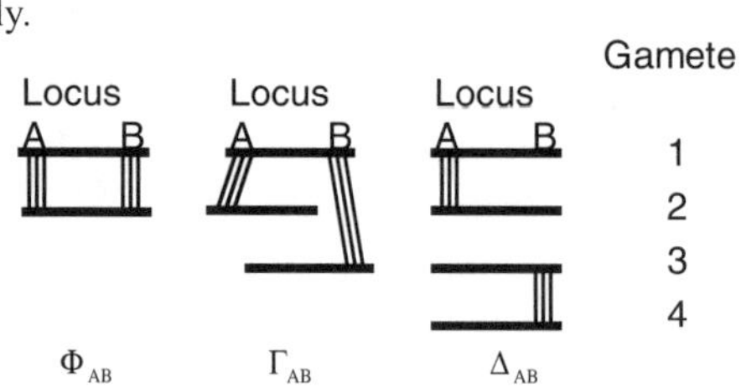

only new mutations restore diversity, and mutation is rarer than recombination.

We are currently implementing a procedure termed "diversity mapping", which maps diversity along linkage groups with progeny arrays, and are implementing it in lodgepole pine, to the end of detecting selective sweeps and/or enhanced drift due to migrational bottlenecks. *Pinus contorta* (lodgepole pine) is a conifer distributed throughout western North America, and a close relative to *Pinus taeda* (loblolly pine), the conifer that has received the most genome work to date. Subspecies *P. contorta latifolia* is an aggressive pioneer, often occurring in small isolated populations and, since deglaciation, it has migrated progressively northward for thousands of kilometers; the most northerly populations only a few hundred years old, as shown by pollen sediment dating. Its genome has likely seen several bottleneck events and selective sweeps by genes adapted for dispersal or for northern habitats.

As part of a larger study, conducted by Cherdsak Liewlaksaneeyanawin (PhD candidate, Dept. Forest Sciences, UBC), 20 progeny from each of 20 parents were assayed for a total of 434 AFLP loci. A novel procedure was used to jointly infer parent genotypes and pairwise recombination rates (Hu and Ritland, submitted) using methods analogous to progeny array analysis of mating systems (Ritland 1986), in that parental genotypes are inferred in probability. Fig. 3 plots all estimates of pairwise recombination rates. It shows a significant correlation of diversity, with more closely linked loci showing higher similarity of diversity; closely linked loci $r <$ 0.05) show a 9% correlation, and this correlation declines in an approximate linear manner out to $r = 0.25$, where it reaches zero. This correlation of diversity seems to extend much further than the few thousand nucleotides predicted by equilibrium drift theory, and suggests that bottlenecks and migrational events are strong in this tree species.

A more complete analysis will identify specific regions of reduced diversity. Such linked clusters of loci with especially low heterozygosity would also be skewed toward rare alleles, due to recent natural selection, possibly in the context of adaptation to new environments. These areas of localized reduced diversity can ultimately be surveyed for candidate genes for adaptation.

Statistical aspects of detecting SNPs of adaptive importance

Over six million SNPs have been documented in the human genome. While plant genomes can be smaller (mainly with less repetitive DNA), levels of nucleotide variability in plants are ca. 10 times higher. Besides providing ample markers for studies of population genetic parameters, SNPs may be employed in association studies to understand genome evolution and adaptive variation.

A picture of SNP evolution is shown in Fig. 4. Excluding natural selection, there are two important features of SNP evolution. First, as SNPs undergo genetic drift over longer periods of time, those still surviving tend to be of higher frequency. Second, all SNPs initially arise as new mutations, and are in complete linkage disequilibrium with whatever variants occur on the same chromosome as the original mutation, but this disequilibrium declines through time.

To give us quantitative predictions about the expected association between SNPs and other variants, this process can be quantified as follows. In a population of size N, an allele of frequency p has an expected age of $t = [(-4Np)/(1-p)]\ln(p)$ (Kimura and Ohta 1973). For example, for $N = 10^4$, then for $p = 0.01$ the average age is 0.093N, while for $p = 0.5$, the average age is 1.38N. The second factor, linkage disequilibrium (LD), declines geometrically with time; at generation t it is $(1-c)^t$, for c the crossover rate. As new mutations are in complete disequilibrium, the expected LD as a function of N, c and p is $D' = (1-c)^{[(-4Np)/(1-p)]\ln(p)} \cong 1 + 4Nc\,p\ln(p)$, which is small except when c, p and N are all small.

From this, a rough idea of the range over which disequilibrium is expected can be obtained as follows. Typically, in plants (such as *Populus*) there are ca. 10^5 nucleotide sites per centimorgan; if effective population size is $N = 10^4$ (effective population sizes are always much less than actual), then $D' > 0.5$ occurs within about 25 bp of a SNP of 0.01 fre-

Fig. 3. The correlation of gene diversity between pairs of AFLP loci in lodgepole pine, plotted against estimated recombination rate between pairs of loci.

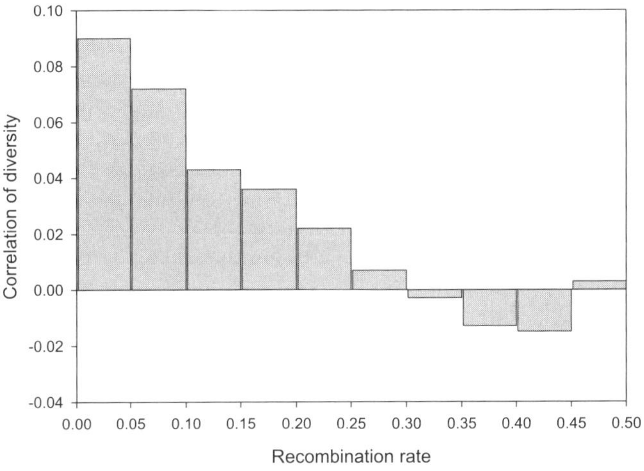

Fig. 4. Example of genetic drift, showing how older alleles tend to have higher frequency, and the speed of adaptive substitution.

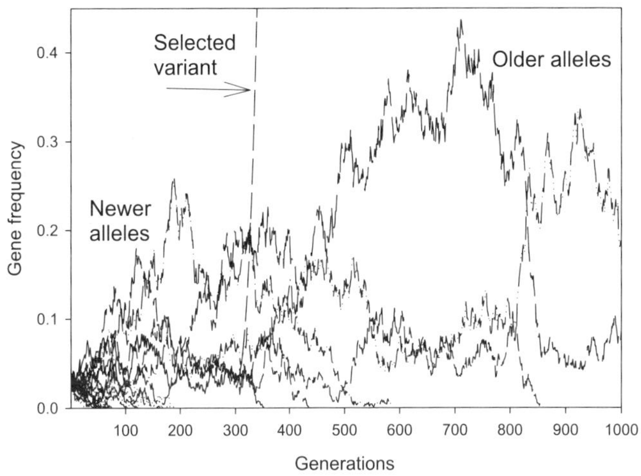

quency, and within about 175 bp of a SNP of 0.001 frequency. While this is a rather short distance, it should be noted that variation in the rate of recombination and the stochastic nature of LD creates a larger window of opportunity for LD. As well, the human results (above) suggests that much higher LD is possible.

This places a high importance upon finding rare SNPs. How can we detect rare SNPs? A common method to detect SNPs is the "*in silico*" examination of EST databases for polymorphisms. While, the numbers of different genetic individuals in these databases can be small, a common rule of thumb is that only a few (3–6) individuals capture sufficient polymorphism. To give a prediction about what spectrum of alleles we can recover from such examinations, we can conduct a doubly "*in-silico*" investigation, applying population genetics theory to DNA databases.

In a population of constant size N, a segment of DNA with mutation rate u, the expected distribution of allele frequency is proportional to $\exp\left(\int_0^x \frac{2\mu N}{y(1-y)}dy\right)$. A computer program was written to estimate the sample distribution of allele frequency estimates. Fig. 5 shows the sample distributions for samples of N = 1, 2..6 individuals (as we assume diploidy, the copy number is twice this). This figure clearly shows that increasing the number of sampled individuals from 1 to 6 doesn't substantially increase the detection of rare SNPs, as the number of rarer SNPs less than doubles from 1 to 6, and this suggests that even a sample of a single individual (as many databases have) is practical enough. At least, .the fraction of SNPs increases well with number of individuals used (inset of Fig. 5).

An alternative to "*in silico*" is the old-fashioned wet lab. Here, pooling of DNA samples from several individuals would best detect rarer SNPs. Ritland (2002) showed that pooling DNA from different individuals increases the efficiency of allele frequency estimation. In that design, several

Fig. 5. The distribution of SNP gene frequency as a function of number of individuals used (screened) for in-silico SNP finding, showing that the rare allele class is very difficult to sample. At least, the fraction of SNPs increases with number of individuals used (inset).

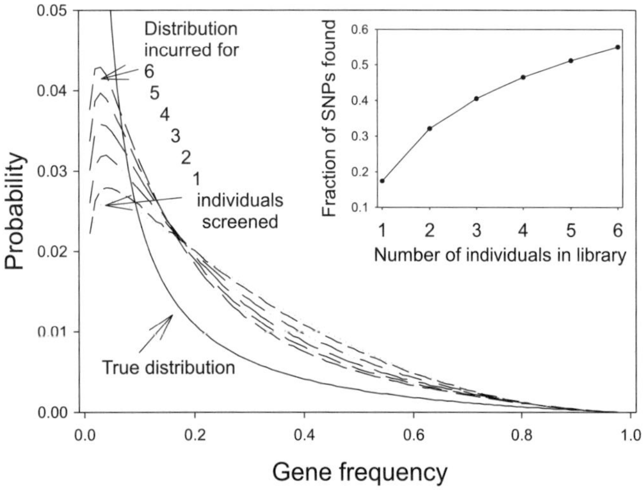

pools each of size n are genotyped for presence/absence of bands. If a locus is diallelic, the optimal pool size is about the inverse of its gene frequency, e.g., for estimating frequencies of alleles of ca. 0.01 frequency, pools of 100 individuals are best. The optimal pool size for merely detecting alleles has not been studied. For this, an "array pool", where individuals are placed in an array, and pools taken across rows and across columns, giving $2n$ pools for an array of n^2 individuals, warrants attention.

Superimposed upon the landscape of genetic drift, in the genome, a few alleles arise, and are immediately favored by natural selection. The pathway of such a "selected variant" is also illustrated in Fig. 4. It shows the rapid nature of adaptive substitutions, compared to alleles subject to pure drift; their life as polymorphic variants are short, and the stronger the selective substitution, the less likely we will observe it as a polymorphism in a species. This suggests that fixed differences found between species are more likely due to adaptive substitutions. Even more interestingly, this suggests that mapping of quantitative trait loci that differentiate recently diverged species might be a powerful method to detect adaptive substitutions.

Demonstrating adaptive SNPs with patterns of variation within populations

Neutral polymorphic loci are commonly used to study patterns of genetic differentiation among populations, and to estimate parameters of population structure such as effective population size, gene flow and selfing rates. Natural selection acting on allelic variants at a locus can reshape its diversity in a locus specific manner, and hence is often regarded as a nuisance. Conversely, such "outlier loci" can be used to infer the presence of natural selection and adaptive variation (reviewed in Schlötterer 2002). Lewontin and Krakauer (1973) proposed two tests for detecting deviations from neutrality, both based upon F_{st} (the standardized variance of gene frequency among populations) for specific alleles. However, genetic drift has a large evolutionary variance, e.g. F_{st} can vary among loci due to chance along (value), so it takes rather extreme values of F_{st} to ascertain selection. In addition, the variance of F_{st} is quite sensitive to population history. This problem can be reduced by conducting pairwise analyses of populations (Vitalis et al. 2001)

Now, if allelic variation among populations correlates with environmental factors, particularly clines such as day length or temperature, the likelihood of natural selection is much greater. An example is the currently EU-funded project TREESNIPS, where SNP variation in populations throughout Europe will be examined at loci that might be related to the steep latitudinal cline in timing of growth in Scots pine (Outi Savolainen, personal communication). This might be termed "population covariance analysis" as the covariance of gene frequency with another factor is evaluated across populations.

One class of "factors" is the other alleles in the genome. Normally, allelic variation is used to infer relatedness between populations using a population × allele matrix. An example of this is shown in the top of Fig. 6, which is based upon a subset of data on isozyme variation in the mountain monkeyflower, from the study of Ritland (1989). We can

Fig. 6. (*a*) Phylogeny of populations, contrasted with (*b*) associations of alleles among these same populations. Alleles that cluster together tend to co-vary among populations, suggesting functional relationship.

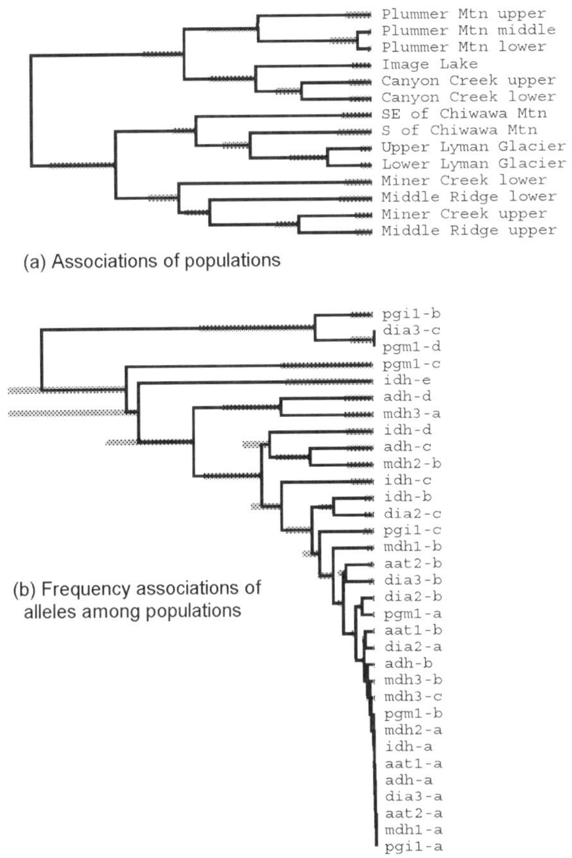

transpose this matrix to an allele × population matrix, and cluster alleles based upon their correlation of gene frequency across populations (Fig. 6, bottom). Of course, monomorphic loci cannot be examined in this type of covariance analysis, and at diallelic loci, the second allele is perfectly negatively correlated with the first allele. Ideally, we should look for the clustering of alleles from different loci across populations. Fig. 6 suggests there are some associations between alleles among loci, suggesting functional relationships; for example, the clustering of pgi1-b and dia3-c and pgm1-d (Fig. 6). However, the statistical tests assume populations are independently evolving; covariance of relatedness among populations needs to be taken into account, otherwise the statistical error is underestimated.

Another way to increase power is to focus on a few, candidate SNPs. Several factors increase the likelihood that a SNP is of functional importance: (*i*) occurrence in region of probable functional importance such as a membrane spanning site, (*ii*) analogous studies in other organisms demonstrating the role of the gene or allelic variant in adaptation. Our discovery of the gene for the "Kermode" bear, a white coat colored variant of the black bear, provides a good example of the power of the candidate gene approach. Besides showing a perfect correlation between SNP genotypes at the mc1r locus and coat color, mutations at other sites in related species show similar coat color changes (Ritland et al. 2001). The allele for white coat color also seems to cluster geographically with an observed level of F_{st} higher than 70 of 77 microsatellite alleles (Marshall and Ritland 2002), and also interestingly, there is strong evidence for assortative mating at this locus (Ritland et al. 2001).

Evolutionary comparisons of SNPs and their implications for population processes

In species comparisons, individual genes do not provide sufficient numbers of SNPs to allow inferences about adaptive processes. Among many SNPs, genome-wide adaptive processes were revealed by DNA sequence variation. Between species, evolutionary comparisons of populations of genes can give information about population-level processes: sequence comparisons among taxa allow estimates of the deleterious mutation rate (Eyre-Walker and Keightley 1999), and when combined with knowledge of nucleotide heterozygosities within taxa, allow inferences about adaptive genetic divergence (Smith and Eyre-Walker 2002).

The chloroplast genome of plants is a very convenient model for "whole" genome analysis. While genes do move from chloroplasts to the nucleus, or are lost, the vast majority of the ca. 120 genes are found in most plant taxa, and they encode ribosomal RNAs and proteins, tRNAs, and proteins involved in photosynthesis. Every gene on this organelle genome can be identified, and sequence conservation allows easy alignment and comparison between species, and confusion between paralogy and homology is unlikely.

Currently, there are 17 available chloroplast genome sequences from green plants. Five of the angiosperm sequences (*Arabidopsis*, *Lotus*, corn, spinach and poplar) are from species quite mutually unrelated. For our example, a total of 34 genes containing ca. 4,750 codons were compiled. Genes used were: atpe, atph, atpi, ndhc, ndhe, ndhg, ndhh, peta, psaa, psab, psac, psba, psbb, psbc, psbd, psbe, psbf, psbh, psbi, psbl, psbm, psbn, rbcl, rpl14, rpl20, rpoa, rpob, rpoc2, rps11, rps14, rps2, rps3, rps4, and rps8. While there are normally over 100 chloroplast genes, this subset of 34 genes could be unambiguously aligned and were present in all five taxa.

An unrooted phylogeny of these species (based upon the Neighbor Joining method applied to the matrix of nonsynonymous differences) gives a nearly perfect "star phylogeny" within the angiosperms (Fig. 7). The star allows easy inference of the changes along each evolutionary pathway from the common ancestor.

Fig. 7. Unrooted phylogeny of the five species used for chloroplast sequence analysis, based upon nucleotide similarity. A near perfect star phylogeny is evident.

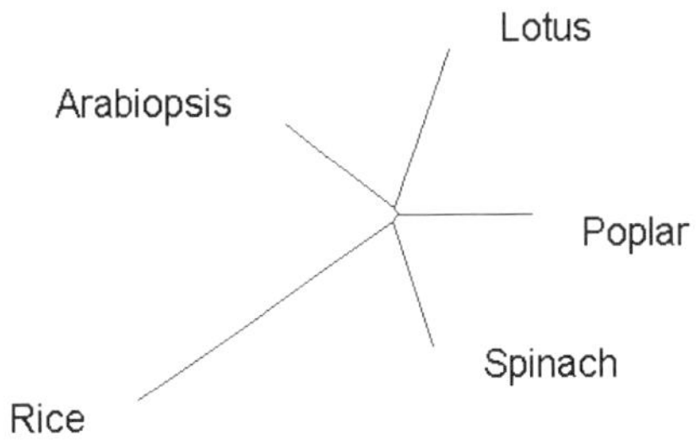

Table 1. Estimates of the proportion expected number of non-synonymous changes (K_{ne}) along each lineage of an angiosperm phylogeny, compared to observed (K_n).

Lineage	K_{ne}	K_n	K_n/K_{ne}	$K_{s,ts}$	$K_{s,tv}$	$p_{n,ts}$	$p_{n,tv}$	# codons
Arabidopsis	677.7	443	0.65	629	317	0.661	0.827	4753
Lotus	731.2	520	0.71	698	325	0.662	0.828	4751
Spinach	680.3	410	0.60	677	282	0.661	0.826	4746
Rice	920.2	944	1.03	912	385	0.660	0.827	4730
Poplar	558.2	400	0.72	611	187	0.661	0.826	4753

One comparative activity is to infer the rate of deleterious mutation. Eyre-Walker and Keightley (1999) adopted a variation of a molecular method (Kondrashov and Crow 1993) to estimate the genomic mutation rate in humans. They assumed that (*i*) the synonymous substitution rate represents the neutral mutation rate, (*ii*) the non-synonymous substitution rate is reduced due to rejection of deleterious mutations by natural selection, and (*iii*) therefore, that the rate of deleterious mutation is the difference between these two rates. While it is difficult to attach a "generation time" to chloroplasts, we can at least compare the relative numbers of deleterious mutations among the lineages that lead to the five species.

Using the observed number of synonymous changes, the expected number of non-synonymous (amino acid changing) substitutions, K_{ne}, along a lineage can be written as

[2] $$K_{ne} = K_{s,ts}p_{n,ts} + K_{s,tv}p_{n,tv},$$

where $K_{s,ts}$ and $K_{s,tv}$ are the observed numbers of synonymous transitions and synonymous transversions, respectively, and $p_{n,ts}$ and $p_{n,pv}$ are the probability that transitions and transversions, respectively, result in a non-synonymous change. This formula is subdivided into transition and tranversion changes because their rates differ significantly. The p's are found by taking the observed contemporary sequence, and calculating what proportions of hypothetical nucleotide substitutions cause amino acid changes, separately for transitions and transversions.

Table 1 gives estimates of these quantities for each branch, assuming a star phylogeny. For all species except rice, the proportion of non-synonymous changes that were accepted were about 60–70%, while in rice, essentially all non-synonymous changes were accepted. This implies that in rice, more deleterious mutations have accumulated.

The special features of plants

While the inferential procedures discussed in this paper are as applicable to animals as they are to plants, there are certain features of plants that lend an advantage to their use. First and foremost, plants are sedentary and "wait to be counted", allowing easy collection of demographic data (Harper 1977). Second, many plant species are colonizing species, and experience bottlenecks and migration, with consequential accentuated population subdivision (Baker and Stebbins 1965). Third, inbreeding and self-fertilization is common in plants (Jain 1976), further favoring allele associations among loci. Fourth, plants are often clonal (Kroon and van Groenendael 1997), allowing easy replication of genotypes and studies of somatic mutation rates. Fifth, the chloroplast genome provides as a small-scale model for genome evolution, as exemplified above. Sixth, in conifers, the gametophyte, a copy of the maternal haploid contribution to the embryo, can be assayed. Appreciable amounts of DNA (10 μg) can be obtained from many *Pinus* species, while lesser amounts (20–50 ng) can be obtained from spruce and cedar. Use of this tissue allows (a) linkage phase and linkage disequilibrium to be directly inferred, (b) direct observation of haploid meiotic products, enabling rapid genetic mapping, and (c) easy verification of SNP homology, via assay of putative heterozygotes. While "single-cell PCR" (Hahn et al. 2000) might be applied to pollen to obtain the equivalent information in angiosperms, there are as yet no known successful applications of such a haplotyping technique to plants.

Acknowledgments

I thank my graduate student, Cherdsak Liewlaksaneeyanawin, for his unpublished data, Gerald Tuskin for providing the poplar chloroplast sequence, and the organizers of this symposium, particularly Quentin Cronk, for their time and efforts.

References

Baker, H.G., and Stebbins, G.L. (*Editors*). 1965. The genetics of colonizing species. Academic Press, New York.

Ching A., Caldwell, K.S., Jung, M., Dolan, M., Smith, O.S., Tingey, S., Morgante, M., and Rafalski, A.J. 2002. SNP frequency, haplotype structure and linkage disequilibrium in elite maize inbred lines. BMC Genet. **3**: 19.

Eisenbarth, I., Striebel, A.M., Moschgath, E., Vogel, W., and Assum, G. 2001. Long-range sequence composition mirrors linkage disequilibrium pattern in a 1.13 Mb region of human chromosome 22. Hum. Mol. Genet. **10**: 2833–2839.

Eyre-Walker, A., and Keightley, P.D. 1999. High genomic deleterious mutation rate in hominids. Nature **397**: 344–347.

Goldstein, D.B., and Weale, M.E. 2001. Population genomics: Linkage disequilibrium holds the key. Curr. Biol. **11**: 576–579.

Hahn S., Zhong, X.Y., Troeger, C,. Burgemeister, R., Gloning, K., and Holzgreve, W. 2000. Current applications of single-cell PCR. Cell. Mol. Life Sci. **57**: 96–105.

Harper, J.L. 1977. Population biology of plants. Academic Press, New York.

Hedges, S.B. 2000. Human evolution: A start for population genomics. Nature **408**: 652–653.

Jain, S.K. 1976. The evolution of inbreeding in plants. Annu. Rev. Ecol. Syst. **7**: 469–495.

Kimura, M., and Ohta, T. 1973. The age of a neutral mutant persisting in a finite population. Genetics **75**: 199–212.

Kondrashov, A.S., and Crow, J.F. 1993. A molecular approach to estimating the human deleterious mutation rate. Hum. Mutat. **2**: 229–234.

Kroon, H. de, and van Groenendael, J. (*Editors*). 1997. The Ecology and evolution of clonal plants. Backhuys Publishers, Leiden, The Netherlands

Lewontin, R.C., and White, M.J.D. 1960. Interaction between inversion polymorphisms of two chromosome pairs in the grasshopper, *Moraba scurra*. Evolution **14**: 116–129.

Lewontin, R.C., and Krakauer, J. 1973. Distribution of gene frequency as a test of the theory of the selective neutrality of polymorphisms. Genetics **74**: 175–195.

Marshall, H.D., and Ritland, K. 2002. Genetic diversity and differentiation of Kermode bear populations. Mol. Ecol. **11**: 685–697.

Maynard Smith, J. and Haigh, J. 1974. The hitch-hiking effect of a favorable gene. Genet. Res. **23**: 23–35.

Pletcher, S.D., and Stumpf, M.P. 2002. Population genomics: ageing by association. Curr. Biol. **12**: 328–30.

Reich, D.E., Cargill, M., Bolk, M., Irelan, J., Sabeti, P.C., Richter, D.J., Lavery, T., Kouyoumjian, R., Farhadian, F., Ward, R., and Lander, E.S. 2001. Linkage disequilibrium in the human genome. Nature **411**: 199–204.

Remington, D., Thornsberry, J.M., Matsuoka, Y., Wilson, L.M., Whitt, S.R., Doebley, J., Kresovich, S., Goodman, M.M. and Buckler, E.S. IV. 2001. Structure of linkage disequilibrium and phenotypic associations in the maize genome Proc. Natl. Acad. Sci. U.S.A. **98**: 11479–11484.

Ritland, K. 1986. Joint maximum likelihood estimation of genetic and mating structure using open-pollinated progenies. Biometrics **42**: 25–43.

Ritland, K. 1989. Genetic differentiation, diversity and inbreeding in the mountain monkeyflower (*Mimulus caespitosus*) of the Washington Cascades. Can. J. Bot. **67**: 2017–2024.

Ritland, K. 2002. Estimation of gene frequency and heterozygosity from pooled samples. Mol. Ecol. Notes 78. **2**: 370–372.

Ritland, K., Newton, C., and Marshall, D. 2001. Inheritance and population structure of the white-phased "Kermode" black bear. Curr. Biol. **11**: 1468–1472.

Schlötterer, C. 2002. Towards a molecular characterization of adaptation in local populations Curr. Opin. Genet. Dev. **12**: 683–687.

Smith, N., and Eyre-Walker, A. 2002. Adaptive protein evolution in *Drosophila*. Nature **415**: 1022–1024.

Strobeck, C., and Morgan, K. 1978. The effect of intragenic recombination on the number of alleles in a finite population. Genetics **88**: 829–844

Vitalis, R., Dawson, K., and Boursot, P. 2001. Interpretation of variation across marker loci as evidence of selection. Genetics **158**: 1811–1823.

Zhang K, Akey, J.M., Wang, N., Xiong, M., Chakraborty, R., and Jin, L. 2003. Randomly distributed crossovers may generate block-like patterns of linkage disequilibrium: an act of genetic drift. Hum. Genet. **113**: 51–59.

4 Evolvability and the raw materials for adaptation

Carl D. Schlichting and Courtney J. Murren

Abstract: The concept of evolvability has received increasing attention as a means of encapsulating perceptions that lineages of organisms vary not only in diversity, but also in their ability to generate such diversity. However, there has been scant attention to what the underlying components of evolvability may be. Because the raw materials for any evolution are phenotypic and genetic variation, we propose that the sources of evolvability can be found in two distinct phases in the production of the phenotype: (1) the generation of genetic variability *per se*, and (2) the interpretation of that variability via the suite of processes comprising the genotype to phenotype map. Key components of evolvability may include the ability of a genotype/lineage to generate genetic variability, or perhaps to maintain phenotypic variation over evolutionary time, enabling the pursuit of diverse evolutionary trajectories. Here we outline an approach to studying evolvability, with an expanded view of the possible sources of variation available to evolutionary processes that highlights the role of environmental variability in producing and storing variation.

Introduction

The question of evolutionary success has intrigued biologists for a long time – what are the characteristics of genotypes or lineages that propel them to prosperity? Is success due to the form of the genetic architecture, particular components of the genetic system, the buffering capacity of the internal environment, or behavioral flexibility? Or is the success of a lineage derived from some aspect of the way in which genes interact with the environment?

Several lines of inquiry have been developed over the years in relation to the genesis or radiation of groups of organisms – the origin of novelty (e.g., Mayr 1960; Nitecki 1990*a*; Sommer 1997), key innovations (Liem 1973; Nitecki 1990*b*; Galis and Drucker 1996; Bond and Opell 1998; Hunter 1998), and versatility (Vermeij 1974). Recently a number of authors have begun to use the term *evolvability* as another perspective on this complex issue. The term, however, has been used in different contexts or with distinctly different intentions, even though most usages refer to aspects of lineage diversification (Braterman 1988; Dawkins 1988; Arnold 1989; Alberch 1991).

We will briefly review several definitions, but our specific intent here is twofold: first, to provide a deconstruction of evolvability in terms of the processes that can promote (or impede) evolutionary responsiveness, and second, to identify the components of evolvability, i.e., the features of organisms that represent the raw material for evolutionary change. We would like to move evolvability from a purely descriptive, retrospective concept to a dynamic one amenable to prediction and hypothesis testing.

Evolvability

von Dassow and Munro (1999) stated that the central problem of evolutionary developmental biology "is to understand how the architecture of development confers evolvability". In its simplest form, evolvability is the ability to evolve, but this simple tautology is *not* simply interpreted. Is the phenotypic diversity of arthropods, for example, due to enhanced evolvability? Do organisms possess specific developmental and genetic properties that allow us to distinguish them as evolvable? What is the appropriate hierarchical level at which to examine evolvability: a lineage, species, or population?

We employ Wagner and Altenberg's (1996) distinction between *variability*, the "propensity to vary", and *variation,* the current differences between individuals at some level of a hierarchy. We refine this distinction as follows: variability refers to the processes potentially generating *genetic* diversity, whereas variation represents *phenotypic* diversity. In between lies the *genotype to phenotype mapping function,* referring to those developmental (epigenetic) processes that carry out or modify the expression of a given genetic sequence en route to the final phenotype (Scharloo 1987; Altenberg 1995; Wagner and Altenberg 1996).

Most variation-based views of evolvability are implicitly backwards looking: inductive, lineage-based, comparative approaches. Definitions from this perspective are limited to descriptions of currently observable phenotypic differences. The designation of a lineage as "evolvable" sometimes merely indicates different amplitudes of morphological diversity between clades — a more evolvable clade has a vast array of phenotypic variation (or a large number of taxa). This descriptive approach has its value and may enable us to pose specific questions within a well-defined phylogenetic context about causes and consequences of the breadth of phenotypic variation. However, it also has its limitations: the genetic and selective bases for the described phenotypic

C. D. Schlichting and Courtney J. Murren[1]. Department of Ecology and Evolutionary Biology, University of Connecticut, Storrs, CT 06269-0043, U.S.A.
Correct citation: Schlichting, C.D., and Murren, C.J. 2004. Evolvability and the raw materials for adaptation. *In* Plant Adaptation: Molecular Genetics and Ecology. Proceedings of an International Workshop held December 11–13, 2002, in Vancouver, British Columbia, Canada. *Edited by* Q.C.B. Cronk, J. Whitton, R.H. Ree, and I.E.P. Taylor. NRC Research Press, Ottawa, Ontario. pp. 18–29.

[1]Current address: Department of Biology, H.J. Patterson Hall, University of Maryland, College Park, MD 20742, U.S.A.

variation are rarely investigated, and the link between current phenotypic variation and a clade's past "evolvability" is usually unspecified.

The identification of key innovations is also variation based, but is viewed backwards as through a funnel – at the narrow point of the funnel, *something* must have happened to make it broaden, spurring a search for a synapomorphy that can be interpreted as a trigger for diversification (Cracraft 1990). Key innovations are often seen as enhancers of adaptability, increasing an organism's invasive capability or expanding its niche (e.g., the evolution of subdigital toe-pads in *Anolis*: Warheit et al. 1999). Alternatively, they may be precursors of characters that facilitate speciation (e.g., the colored wing scales of Lepidoptera, Galant et al. 1998). However, key innovations must also instigate, or promote, diversification, and little is understood about this role.

An examination of the presence of key innovations (Lauder 1981; Arnold 1989; Nijhout and Emlen 1998) can begin to bridge the gap between variation- and variability-based definitions of evolvability, *if* the role of the innovations in generating phenotypic diversity can be assessed. The columbines (*Aquilegia*) are an instructive example. Hodges and Arnold (1994, 1995) suggested that the nectar spur of columbine flowers represents a key innovation for this group: relatively small differences in spur size or morphology can favor distinct pollinator faunas, leading to rapid reproductive isolation. Interestingly, although *Aquilegia* is speciose, divergence among species is largely restricted to floral traits; in other related widespread genera, there is divergence for a variety of morphological and floral characters. This raises the issue of which characteristics will be most informative as measures of variation.

Definitions

Most of the confusion in recent discussions of evolvability resides in the variability and G ➔ P mapping components. Authors have variously emphasized phenotypic, genetic or developmental perspectives in their definitions. Arnold et al.'s (1989) definition (Table 1) is the most phylogenetic in context, describing evolvability in terms of lineage diversification, but is rather vague about the basis for diversification. Kauffman (1993) implicitly refers to variability, with new mutants "searching" phenotypic space. Other authors allude to the genotype to phenotype (G ➔ P) map in their definition: e.g., Kirschner and Gerhart (1998) link the generation of variability to selection on the phenotype, and Wagner and Altenberg (1996) explicitly discuss variability, the genetic system, and adaptation in terms of the G ➔ P map. Although conceptually broad, this array of definitions provides few testable notions, and leaves other issues unexplored.

Table 1. Definitions of evolvability.

1. **Arnold** (1989): a "special class of novelties or watershed events that presage … a proliferation of lineages and a diversification of morphology"
2. **Houle** (1992): "the ability of a population to respond to natural or artificial selection"
3. **Kauffman** (1993): "the capacity to search a reasonable fraction of space, which may be optimized when landscape structure, mutation rate, and population size are adjusted"
4. **Wagner and Altenberg** (1996): "the genome's ability to produce adaptive variants when acted upon by the genetic system"
5. **Kirschner and Gerhart** (1998): "an organism's capacity to generate heritable (selectable) phenotypic variation"

We define evolvability as: *the tendency of a genotype or lineage to generate genetic variability and produce or maintain phenotypic variation over evolutionary time, enabling it to pursue diverse evolutionary trajectories*. Defined in this way, we can recognize three facets of evolvability: (1) processes that generate mutations and new gene arrangements, (2) processes that interpret genetic information, and (3) the resulting phenotypic variants (Fig. 1). Evolvability then has its roots in facets 1 and 2: the production of genetic variability *per se*, and the epigenetic processes (the G ➔ P map) that translate genetic variability into phenotypic variation. Various components of each phase are quantifiable, and thus questions of interest can be focused on generating and testing hypotheses: e.g., do evolutionarily successful lineages share certain features of one or both phases, or is such success idiosyncratic?

Apart from our focus on quantification and testability, there are several other distinctions between our perspective and earlier ones. First, we explicitly refer to the *maintenance* of variation as a possible component of evolvability (Rutherford and Lindquist 1998; Rasmuson 2002). Second, *contra* Wagner and Altenberg (1996), we do not restrict evolvability to include only "adaptive variants". We view this as both an onerous and an unnecessary restriction: onerous, given the difficulty of demonstrating the adaptive value of *any* trait, and unnecessary, given that a significant fraction of evolutionary diversification may be non-adaptive. Third, we underscore the potentially significant role of environmental variation in altering components of evolvability by explicitly including phenotypic plasticity among the factors affecting variability, the G ➔ P map, and variation.

We use this framework to discuss a set of parameters whose potential relationship with evolvability may be evaluated. The list of candidates is extensive, perhaps precluding any exhaustive inspection. However, in order to provide insights into the importance of evolvability for evolution, and on the evolution of evolvability itself, empirical tests of the relevance of, and the relationships among, these factors will ultimately be more fruitful than retrospective, pattern based analyses.

Fig. 1.

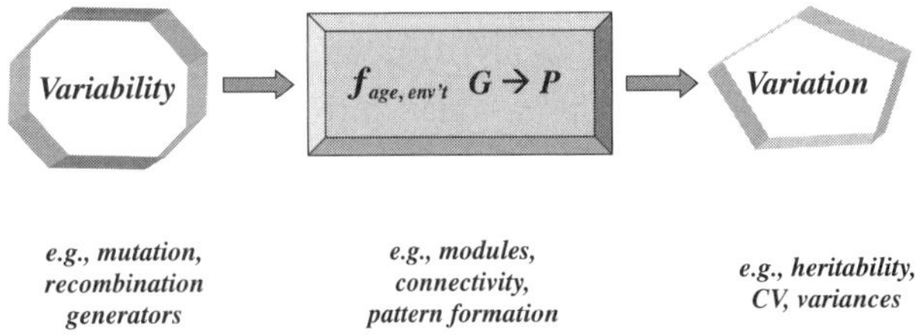

The raw materials for adaptive (and non-adaptive) evolution

Evolution cannot occur without variation: even genetic drift requires at a minimum genetic variation, while natural selection requires both phenotypic and genetic variation. Such variation is produced at the individual level, but can accumulate within and among populations (Ancel Meyers and Bull 2002).

Mechanisms generating genetic variability

Mutation and Recombination

Differences among species and populations in the "propensity to vary", referring to the processes which produce novel genetic variants, represent a potential basis for differences in evolvability (Caporale 2000; Tenaillon et al. 2001; Betancourt and Presgraves 2002; Rice 2002). Mutation and recombination rates are the obvious variability generating sources (Table 2). There is evidence that mutation and recombination rates vary among species and populations (Taddei et al. 1997; Naylor and Gerstein 2000; Sanchez-Moran et al. 2002). Analyses of quantitative traits indicate that mutations appear to arise at fairly high rates (see Drake et al. 1998; Fry et al. 1999; Schultz et al. 1999). Lynch et al. (1998) argue that these data are consistent with the idea that the vast majority of standing genetic variance for life-history characters may result from recurrent mutation (see also Houle et al. 1994). Finally, Ofria et al. (2003) have suggested that it is likely that there has been selection at the level of the genetic code for tolerance to mutation; this will tend to increase the proportion of mutations that are neutral, thus contributing to standing variation.

Table 2. Components of variability. These measures are of the features that are responsible for generating genetic variability, and may be surveyed for individuals, populations or higher taxa.

1. **Mutation rates**: measures of the rate of mutations — nucleotide, amino acid, protein
2. **Mutational range**: what is the fraction and frequency of possible mutations that are observed?
3. **Recombination rates**: measures of the rate of recombination frequency
4. **Genomic rearrangement**: inversions, translocations, transposons — events that can change the genetic milieu
5. **Phenotypic plasticity**: are any of the above altered by environmental conditions?

A variety of studies have now examined the sizes and distributions of mutational effects (Shaw et al. 2000; Azevedo et al. 2002; Caballero et al. 2002), providing insights into variability generating processes. Analyses of the effects of mutations suggest that at least some mutations are of relatively large effect (Keightley and Bataillon 2000; Caballero et al. 2002), and although most studies find that the majority of mutations are deleterious, a study on *Arabidopsis* revealed a symmetrical distribution of positive and negative effects (Shaw et al. 2000). Together, the increasing data on mutation rates and on the nature of mutational effects is leading towards a more general theory of the role of mutation in evolution.

Are mutational parameters related to evolvability? ***Mutation rates*:** de Visser et al. (1999) found that the rate of adaptation in *Escherichia coli* was proportional to its mutation supply rate, but only in small or initially well-adapted populations. Vulic et al. (1999) described how mutations in the DNA mismatch repair system can increase mutation and recombination rates and thus influence speciation dynamics. ***Types of mutation:*** Despite evidence for variation in mutation rates within a genome (Denver et al. 2000; Kovalchuk et al. 2000), there is scant information to address whether particular mutational events may be prohibited or enhanced (Yampolsky and Stoltzfus 2001; Ito-Harashima et al. 2002). Grogan et al. (2001) reported that the overall mutation rate of the thermoacidophilic bacteria *Sulfolobus acidocaldarius* was not greater than that of other bacteria; however, it had a relative scarcity of base pair substitutions compared to other organisms. Moore et al. (2000) studied compensatory adaptation (an increase in fitness even though the initially deleterious mutation is retained) over 200 generations in *Escherichia coli*, and found that certain types of mutations were more likely to produce compensation. Mutations of large effect were substituted more often, and rates of compensatory adaptation were faster for synergistic than multiplicative pairs of mutations (see also Harada 1995). Cowell et al. (1999) showed an intensification of mutability in the complementarity-determining regions of immunoglobulin genes; mutations in these regions have a greater chance of being advantageous.

Recombination rates also exhibit both hot and cold spots across a genome (Nachman 2002; Reich et al. 2002). Lercher and Hurst (2002) have found that higher mutation rates are associated with regions of high recombination. Map-based investigations into this relationship may lead to predictions of the relative mutability of different phenotypic characters.

Although the effects of transposons on eukaryotic fitness are generally small, there are several reports of relationships with mutation and recombination rates. The most interesting from our perspective is the study of Martusewitsch et al. (2000) who found that rates of spontaneous mutation in the *pyrE* gene of the archaebacterium *S. solfataricus* were increased by an order of magnitude by insertion of *IS* transposable elements. Studies on *Drosophila melanogaster* (Duret et al. 2000) and *Caenorhabditis elegans* (Rizzon et al. 2002) both found negative associations between density of DNA transposons and recombination rates (the pattern expected if selection operates against transposons), but only the *Drosophila* study found a negative association also for the RNA (retro) transposons.

Environmental influences

A number of studies have now found that both mutation and recombination rates are environmentally sensitive (Radman 1999; Caporale 2000; McKenzie and Rosenberg 2001; Ancel Meyers and Bull 2002; Bedau and Packard 2003). Although the phenomenon of 'adaptive' or 'hyper'- mutation has generated considerable discussion (e.g., Sniegowski et al. 2000; de Visser 2002), the generality of such mechanisms is becoming clear with reports from a variety of organisms: bacteria (Foster 2000; McKenzie et al. 2000; Giraud et al.

2001), green algae (Goho and Bell 2000), fungi (Marini et al. 1999), plants (Lucht et al. 2002), insects (Gorodetskii et al. 1991; Fry and Heinsohn 2002), and vertebrates (Newell and Heddle 2002). For example, Lamb et al. (1998) document both evolutionary and environmental differences in mutation rates of the fungus *Sordaria fimicola*: lab strains derived from the "harsher" habitat maintained their higher rates. Lucht et al. (2002) found that *Arabidopsis* plants attacked by the oomycete *Peronospora parasitica* had a significant increase in recombination rates. Other results indicate that the effects of mutation are themselves environment-dependent (Fry et al. 1996; Fernandez and Lopez-Fanjul 1997; You and Yin 2002), or age-dependent (Pletcher et al. 1998; Mack et al. 2000).

Sniegowski et al. (2000) suggested that, because of the higher likelihood of deleterious mutations, selection will in general favor the reduction of mutation rate to the minimum possible. However, as these authors also point out, the evolution of environmentally-dependent mutation rates gets around this problem: selection may have favored lineages with higher rates triggered by unfavorable environments (Travis and Travis 2002), a scenario that fits perfectly with the general model of the evolution of adaptive plasticity (Schlichting and Pigliucci 1998). Along these lines, Koren et al. (2002) proposed a model that suggests that environmental factors may have an important role in initiating recombination.

Although we are ultimately interested in those mutations that confer fitness advantages, we note that the continuing production of neutral or even slightly deleterious variants may also prove to enhance evolvability. The role of such mutants will especially come into play when they are expressed in new environments (e.g., sickle-cell anemia) or new genetic backgrounds (Elena and Lenski 2001).

Translating variability into variation: the G ➔ P map

The concept of the genotype to phenotype map has arisen as shorthand for the processes that transform the molecules generated from the genetic instructions into phenotypic features. What aspects of the genotype to phenotype map account for differences in evolvability? Recent discussions have emphasized several components: modularity, connectivity, and gene duplication/divergence. However, there are other components of the epigenetic system that merit attention as well: phenotypic plasticity, canalization and homeostasis.

Modularity and connectivity

Modularity has become a unifying theme for evolutionists (Bonner 1998; Raff 1996; von Dassow and Munro 1999; Schlosser 2002), despite disparity in the use of the term (Winther 2001). Raff (1996) proposed that modules have unique portions of the genome turned on at specific times during development, are hierarchically organized, yet interact with other modules. Wagner and Altenberg (1996) viewed modules as functionally distinct character complexes, and pleiotropy would occur most frequently among genes for traits within a given complex. These semi-autonomous units would be critical for phenotype construction. These ideas were extended by Wagner and Schwenk (2000) with the concept of evolutionary stable configurations. Kirschner and Gerhart (1998) proposed the term compartmentalization for the partially independent subsets of expressed genes that build structures at the cellular level. Ptashne and Gunn (1998) proposed that locally responsive gene regulation would generate complexity and evolvability (see also, Thieffry and Romero 1999). Despite differences in emphasis, these authors agree on the importance of modular construction for evolvability (but see Hansen 2003, for a somewhat different perspective).

An essential feature of modules *vis a vis* evolvability is that they must be capable of being deconstructed and their components re-used as building blocks: deconstruction allows for further evolution of modularity (e.g., Gatesy and Middleton 1997). The degree of connectivity (coupling/decoupling; association/dissociation; integration/parcellation) within and among modules must be malleable as well. Connectivity has good news/bad news features: connected traits can function cooperatively and be co-regulated, but connectivity also may restrict the independent evolution of its separate components (due to genetic correlation, but see Beldade et al. 2002). Modules can be dissociated (Minelli 1998) through changes in gene regulation (e.g., via heterochrony – changes in timing, and heterotopy – changes in location of expression), or through environmentally mediated changes in interactions between modules (see below). Rockman and Wray (2002) have documented considerable variation in *cis*-regulatory function in the human genome, and suggest that the consequences of such variation for epistasis and genotype-environment interaction are enormous.

A variety of studies have been made of the modularity/connectivity elements (Table 3) of the G ➔ P map. Raff (1996) cited perturbation studies as evidence for dissociability of developmental processes: for example, the dissection of the complex interactions among organs during development of amphibian eyes has revealed a multitude of interconnections between modules. Nijhout et al. (2003) examined the evolutionary dynamics of the mitogen-activated protein kinase (MAPK) cascade, and found that different sub-components differed in their ability to limit the evolvability of the pathway as a whole.

Table 3. Components of the G ➔ P Map. These are measures of the modularity or connectivity of the organismal 'architecture', and of the capacity for plasticity of development. They can be recorded for both genetic and phenotypic (in the quantitative genetics sense) entities.

6. **Number of modules**: this could be determined in morphological terms as the number of separate functional systems, or by examining correlations (genetic or phenotypic) among traits.
7. **Absolute value of correlations and variance of correlation values**: summary statistics for the degree of connectedness between measured traits.
8. **Developmental connectivity**: patterns of connection (measured as in #5 and #6) may change through ontogeny — e.g., for holometabolous insects.
9. **Phenotypic plasticity**: (*a*) Do developmental pathways shift in response to environmental changes? (*b*) Are measures of modularity/connectivity altered by environmental conditions?

Several studies have provided evidence of modularity of genetic architecture (Nemeschkal 1999; Mezey et al. 2000; Raff and Sly 2000; von Dassow et al. 2000). Modularity of architecture can also be represented as genetic/phenotypic variances and covariances, and differences in these quantities have been detected among populations or species (e.g., Steppan 1997; Badyaev and Hill 2000; Carroll et al. 2001; Murren et al. 2002). Another study examined the effects of mutation on correlations (the M matrix, Camara and Pigliucci 1999).

Similarities between phenotypic and genetic modularity have also been reported. Nemeschkal (1999) showed that certain patterns of correlation among bird skeletal elements were related to expression of developmental 'control' genes. Stock (2001) identified similar types of genetic and phenotypic modules in the dentition of vertebrates, although he concluded that it is too early to claim exact correspondences.

Studies of modularity of sets of genes, proteins or cells are accumulating rapidly (Rives and Galitski 2003; Tanay et al. 2004), fueled by the explosion of gene expression data (e.g., Klebes et al. 2002; Horvath et al. 2003; Stuart et al. 2003). These studies should quickly give us broad insights into the degree of modularity and connectivity for various organisms. For example, Wang et al. (2002), distilled a fairly detailed picture of regulatory networks from microarray studies of *Saccharomyces cerevisiae*. They identified particular signal transduction pathways and their associated transcription modules (the transcription factors and their target genes). Finally, for 21 different assay conditions, they examined the environmental dependence of the activity of the transcription modules, as well as the interactions between these modules.

Both connectivity and modularity (represented by correlation matrices) have also been shown to be environmentally sensitive (reviewed in Schlichting and Pigliucci 1998; Badyaev and Foresman 2000; Kause and Morin 2001; Pigliucci and Hayden 2001). For example, Nicotra et al. (1997) explored modularity in two species of *Piper*, finding evidence for distinct modules of vegetative and photosynthetic traits; the relationships within modules were altered between high and low light levels. Such environmental sensitivity of modularity is likely to be extremely important for enabling adaptive plasticity, and perhaps for allowing persistence through changing climatic conditions.

Plasticity of gene expression

Adaptive plastic responses are executed by changes in gene regulation and expression (Smith 1990; Pigliucci 2001; Schlichting and Smith 2002). Studies documenting plastic changes in gene expression are numerous (e.g., Pigliucci 1996; Schlichting and Pigliucci 1998; Lachke et al. 2000; Schenk et al. 2000). For example, Chen et al. (2002) used microarrays to assay expression of 402 transcriptional regulators in *Arabidopsis*. Their results indicated a diversity of patterns of gene expression in response to various treatments (e.g., salinity, cold, virus, fungi, wounding). Some patterns of response were similar (salt and fungus), while others (e.g., cold and wounding) were quite divergent (Table 4).

Table 4. Correlations of expression levels of transcription factors between different experimental treatments (n = 402).

	Salt	Wound	Fungus	Virus
Cold	0.73	0.45	0.82	0.52
Salt		0.71	0.87	0.71
Wound			0.60	0.67
Fungus				0.65

Note: Calculated from data in Table 1, Chen et al. 2002: *Cold*: 4°C @ 27 h (expt. 14); *Salt*: NaCl 100 µmol @ 27 h (expt. 16); *Wounding* @ 6 h (expt. 23); *Fungus*: *Botrytis* @ 36 h (expt. 40); *Virus*: Cucumber mosaic virus @ 48 h (expt. 8). All correlations are significantly >0, and significantly <1.

Although we may infer that developmental pathways are altered when we observe plastic responses of gene expression or phenotypes, to date there has been little investigation of the explicit links between changes in gene expression, modularity or phenotypes. One example is provided by Zhang and co-workers (Zhang and Forde 1998; Zhang et al. 1999) who identified a gene, *ANR1,* in *Arabidopsis* whose expression is rapidly induced by nitrate. *ANR1* expression is limited to roots, inducing changes in development of root architecture, perhaps by increasing the rate of lateral root elongation. Other excellent examples are provided by documentation of the differing patterns of gene expression in insects that produce discrete phenotypes from a common genotype, e.g., castes (Evans and Wheeler 2001*a*, *b*; Abouheif and Wray 2002) or polyphenisms (Zhao and Zera 2002; Nijhout 2003). As West-Eberhard (1989) has pointed out, such alternative developmental pathways may represent important elements for subsequent evolutionary change.

Canalization and hidden reaction norms

Evolvability lies not just in the generation of genetic variability but in its maintenance as well (Ohta 2002; Schlichting 2004). Environmental modifiability of the G → P map may be an important component of both. Phenotypic plasticity in response to environmental change represents not only the production of variation, but also a potential means of preserving it (Schlichting 2004). The key link is canalization of development (Meiklejohn and Hartl 2002; Nijhout 2002): a developmental program or adaptive plastic response that produces the optimum for a particular trait will effectively hide the genetic architecture of that trait, because the intensity of selection on the trait has been reduced to zero. Stabilizing (canalizing) selection that favors particular norms of reaction or phenotypes thus can "hide" among-genotype variation (e.g., Wright 1931; Levin 1988; Bruno and Edmunds 1997). Evidence for adaptive convergence supports this hypothesis: when it is the final phenotype that is under selection, the details of how to get there do not necessarily have to be the same (Gilchrist and Partridge 1999; Riley et al. 2001; Cuevas et al. 2002). However, when environmental conditions transcend the zone of canalized expression, then those deviations may be expressed. We refer to this phenomenon as a *hidden reaction norm* because the variation is elicited as a plastic response only beyond the normal environmental range (Schlichting and Pigliucci 1998).

Besides observations of novel phenotypes when animals and plants are grown in new environments, there are a variety of other phenomena that fall within the scope of hidden reaction norms. The classic phenocopies of Goldschmidt

(1938) and Waddington (1956) are environmentally induced 'versions' of known mutations. The documentation of "latent genetic potential" in novel environments (Stern 1958; Stebbins Jr. and Hartl 1988; Silva and Dykhuizen 1993) is evidence for canalization of phenotypes in the normal environmental range.

The release of hidden or cryptic variation following mutation (Rutherford and Lindquist 1998; Queitsch et al. 2002), or changes in gene expression in different genetic backgrounds (e.g., Harland 1936; Van Delden et al. 1978; Dean 1995; Lauter and Doebley 2002) also implicate canalization; in these cases it is the alteration of internal environmental conditions that elicits the novel phenotypes (Schlichting 2003). Genetic redundancy also acts to canalize phenotypes and may release variation with environmental change (Schlichting and Pigliucci 1998; Wagner 1999; Gu et al. 2003).

Gene duplication and divergence

Gene duplication remains a fundamental topic related to evolvability and modularity because of its ability to create *de novo* the raw material for new genes, new modules or for new patterns of regulation (Ohno 1970; Smith 1990; Ohta 1991; Bird 1995; Long 2001). The processes leading to new functions for duplicated genes are also receiving increasing attention (Averof et al. 1996; Lynch and Force 2000; Wagner 2001; Ohta 2003).

Patterns of variation

Studies that document aspects of variation (Table 5) have been the backbone of traditional evolutionary investigation. Hoffmann and colleagues have done a series of studies on *Drosophila melanogaster* representative of recent approaches (review in Hoffmann and Merila 1999). For wing traits, but not bristle traits, heritabilities were relatively lower in the stressful environment, largely due to increases in the environmental variance under stress. However, levels of additive genetic variance were relatively constant, and coefficients of additive genetic variation (i.e., Houle's CV_A) were largely similar between environments (Hoffmann and Schiffer 1998). Both heritability and CV_A estimates were higher in a combined stress treatment, reflecting an apparent increase in the additive genetic variance (Schlichting and Pigliucci 1998). Flies exposed to 14°C had higher heritability (compared to 28°C) for fecundity, but lower heritability and CV_A for development time (Sgrò and Hoffmann 1998). A study of variation in QTLs among *Arabidopsis* accessions (Pérez-Péreza et al. 2002) revealed a large amount of naturally occurring variation for leaf architecture. Both classic quantitative genetic parameters measured across environments, in combination with new approaches such as QTL or microarrays will lead to novel insights into patterns of variation and may elucidate which components are important contributors to evolvability.

Table 5. Components of variation. These include standard and not-so-standard measures of variation in traits in populations of organisms.

10. **Ranges, variances** of traits
11. **Houle's CV_A**: standardized measure of genetic variance relative to trait mean
12. **Morphospace estimates**: how is the diversity of actual phenotypes related to possible phenotypes?
13. **Phenotypic plasticity and hidden reaction norms**: are any of the above (10–12) altered by environmental conditions?
14. **Gene flow/dispersal rates**: are there characteristics of lineages that may have promoted macroevolutionary trends (e.g., species/lineage selection)?

Evolvability: experimental approaches

Leroi (2000; see also Cracraft 1990) has pointed out that the hypothesized effects of evolvability are typically untestable. To save evolvability from the fate suffered by other such intuitive concepts (e.g., the over-application of 'constraint'), care must be taken to provide, at the least, alternative hypotheses, and preferably the tests themselves. The evidence for evolvability has most often been adduced from species diversity, but with experimental approaches other measures such as rates of diversification may be appropriate. Here we will discuss published analyses which begin to scratch the surface as tests of evolvability, and suggest further experimental approaches which might provide fruitful data.

There are two ways that the roles played by factors in the above lists can be evaluated. The first is strictly correlative — phylogenetic comparisons of characters with standing (or fossil) diversity (with replicates or independent contrasts for statistical evaluation). For example, mutation rates could be estimated for a number of different Dipteran lineages — we might hypothesize that the Hawaiian *Drosophila* have a particularly high rate — or the number of morphological modules of extinct taxa might be estimated and correlated with subsequent evolutionary diversification. The correlation of a phenotypic character innovation and a radiation of a clade may be a starting point to examine evolvability, especially the components of the variability of the genetic system and the G → P map that underlie the phenotypic variation observed. An example of this approach is the study by de Queiroz (1999), who examined the suggestion that the origin of image-forming eyes may promote evolutionary diversification, contrasting species diversity of 12 pairs of clades with and without eyes. There was no relationship between species diversity and the presence of eyes, nor did taxa with eyes appear to have higher locomotory activity than those without. Although this study is exemplary, it is burdened by a variety of shortcomings, as de Queiroz points out. The species diversities are based on small sample sizes and extant taxa only – he notes especially the current paucity of cephalopods relative to the gastropods, despite contrary patterns in the fossil record. The characteristic "image-forming eyes" is not homologous, nor is the difference between 'haves' and 'have-nots' the same for all pairs of taxa. Even if an analysis can include fossil taxa, the completeness of the fossil record remains an issue. Despite these drawbacks, such approaches may aid in focusing on particular clades or characters that may provide appropriate test systems (de Queiroz 2002).

Hunter and Jernvall (1995) analyzed the relationship between taxonomic diversity and the possession of a hypocone, an added cusp on the upper molar teeth of therian mammals. Phylogenetic evidence indicates that hypocones are 'easy' to evolve, with more than 20 independent origins. The major radiations of mammals on each continent were of groups with a hypocone, and the many mammalian higher taxa without hypocones show no significant increases in diversity. These results were interpreted to "strongly support recognition of the hypocone as a key innovation that has allowed invasion of, and diversification within, herbivorous adaptive zones." (for further discussion see Leroi 2000). Yang (2001) presents a test for a relationship between modularity and evolvability (again measured as diversification). He predicted that Holometabola, with modular (juvenile vs. adult) life stages, should exhibit higher rates of diversification than the less modular Hemimetabola. He found that Holometabola had diversified significantly more than both the Hemimetabola, and its subclade Eumetabola (the putative sister group to Holometabola). These data are suggestive of the potential role of modularity in diversification.

Several studies have also investigated predictions related to modularity not directly tied to diversification. For example, a correlated prediction of modularity is that characters in more modular clades will exhibit greater levels of variation due to their independence (Lauder 1981; Yang 2001). Schaefer and Lauder (1996) predicted that decoupling of structures (an increase in module number) should be accompanied by an increase in variation in the decoupled components. They examined patterns of variation in Loricarioid catfish, whose jaw components have been significantly decoupled during the course of evolution (e.g., upper jaw uncoupled from cranium, and lower jaw uncoupled from the opercular series), and found increased morphological diversity in the derived clades.

Such correlative analyses may be most successfully employed for recent large radiations — e.g., Hawaiian silverswords (Baldwin and Wessa 2000; Barrier et al. 2001) and *Drosophila* (Kambysellis et al. 1995; Baker and DeSalle 1997), rift lake cichlid fishes (Danley and Kocher 2001; Takahashi et al. 2001), or even Darwin's finches, for which detailed phylogenetic hypotheses exist. However, such correlative analyses are only a first step towards understanding the mechanisms generating phenotypic variation.

Whether any of the suggested components are responsible for true evolvability can only be determined by the second method — via specific empirical tests. Jernvall (2000) provides an example of this method with a study of the population variation in tooth cusps of seals. Extant variation covers the range from 3–5 cusps. Examination of developmental variation suggested that those cusps that are not always present are those that develop last. Jernvall concluded from this developmental analysis that teeth should be recognized "as highly evolvable because only small developmental changes are needed to produce large changes in size and number of small cusps".

In an experimental approach, modularity could be dissected into its component parts, which might be subjected to manipulation or selection.. The number of modules, the level of connection among components of the module, the integration among modules (Schlichting 1989; Gatesy and Middleton 1997), and the changes in modularity through ontogeny, are all measurable entities, and thus a set of quantitative measurements of the G ➔ P map can be made. Testing for modularity can begin by identifying a set of interacting phenotypic traits and working backwards to identify the set of genes which interact during development to produce this character. Some approaches include QTL mapping (e.g., Mezey et al. 2000; Murren and Kover 2004) and defining evolutionary stable configurations via correlations (Wagner and Schwenk 2000). How gene expression changes in time and space through development, and as environments change, is the next requisite step for understanding modular construction. When we are able to obtain this information rapidly for a number of clades we should be able to examine the evolutionary-developmental-genetic mechanisms underlying phenotypic variation.

Selection experiments, however, represent the best test of relationships between measures that may represent the potential for diversification, and any diversification that ensues. The experiments envisioned would, for example, contrast two or more taxa for which putative indicators of evolvability could be measured. Replicate populations of each of these would then be exposed to selection, either for new trait values or trait combinations, or in novel environmental conditions. Responses to selection would then be compared for the different taxa. An interesting system in this regard is that of the heat shock protein brought to light by Rutherford and Lindquist (1998) — when *Hsp90* expression was inhibited, significant, and otherwise hidden, variation was expressed. Such a system could be exploited to generate lines that differ in their intrinsic "variation" (see Wagner et al. 1999), and then subjected to various selection regimes to assay their evolvability.

Selection experiments will largely be limited to organisms with rapid generation times. Bacteria, unicellular protistans (e.g., *Chlamydomonas*), and yeasts represent excellent experimental systems, albeit with limited morphological phenotypes. Multicellular organisms have more complex phenotypes but correspondingly longer life cycles — however, current model organisms (e.g., *Caenorhabditis, Arabidopsis,* and *Drosophila*) as well as other short generation beasties can be experimentally manipulated, and tens of generations may provide insights into more complex systems.

Although many of the entities in our tables are not easily measured at present, most will become feasible as current techniques mature or new ones are developed. We see special promise for microarray techniques that are already providing unique insights into gene expression and development of model organisms. As mentioned above, these will be quite useful in delimiting functional modules, and identifying changes in module membership through ontogeny and across environments. Another emerging technique is that of RNA interference (also called post-transcriptional gene silencing): the insertion of strands of mRNA molecules of a gene has been shown to silence the gene expression (Hannon 2002). This promises the ability to very specifically control the expression of particular genes to examine their precise role in development (Hughes and Kaufman 2000; Kidner and Martienssen 2004).

The evolution of evolvability

Discussions of evolvability have spanned many levels of biological organization. In our opinion, the often-imprecise language has led to some confusion and may have hindered explicit quantification of the mechanisms generating genetic variability and phenotypic variation. Wagner and Altenberg (1996) alternately refer to individuals, to populations, and to species in their discussion of the topic. Ganfornina and Sanchez (1999) suggested that evolvability is one step of a lineage towards speciation. Arnold et al. (1989) take a lineage-based view: more prolific lineages with greater morphological diversity are more evolvable. Other authors explicitly state that for evolvability itself to evolve, selection must occur above the individual level (Conrad 1990; West-Eberhard 1998). In contrast, Leroi (2000) vehemently argues against the proposal that various features of metazoan development have conferred clade-level evolvability.

We concur with West-Eberhard (1998) that differential evolvability among clades may be an indication of selection above the individual level, and the evaluation of how phenotypic variation is partitioned among clades is a valuable starting point for understanding evolvability. However, we also emphasize the investigation of mechanisms behind how *current* phenotypic variation is maintained or generated. Several of the quantifiable components we discuss are parameters in traditional population genetics models. As is evident from our definition, we consider these measures to be properties of genotypes *and* lineages (*sensu lato* – including populations, species etc.); each of these individual components (e.g., mutation rates, developmental connectivity, dispersal characteristics, Tables 2, 3, 5) can evolve. We favor a comparative approach towards understanding evolvability: specifying the level of organization and examining differentiation among some of the components that may be related to evolvability.

The framework we have proposed differs from that of Raff (1996) in our increased emphasis on the importance of environmental conditions on the interconnection of genes, gene products and phenotypes. In addition, we re-emphasize that the components of the G → P map are themselves under genetic control and therefore able to evolve (Conrad 1990; Wagner and Altenberg 1996). In contrast to Wagner and Altenberg's definition, we do not limit evolvability to the production of only adaptive variants; we also recognize the potential for a significant role for drift (Ohta 2002). Although the evolvability implied by the existence of adaptive radiations is of special interest to evolutionists, there are instances of broad diversification that may lack any strong adaptive element. In such cases the production of nearly neutral variation could be a key component.

The approach we have taken to disentangling the components of evolvability will allow further testable hypotheses to be generated. Of course, the components of evolvability need not be the same across groups — in fact, it would be surprising if they were (Poole et al. 2003). If we can make the appropriate measurements, we need not wonder whether "certain kinds of embryology may be pre-disposed to spawn rich evolutionary radiations" (Arnold et al. 1989). We can put such surmises directly to the test.

Acknowledgments

We thank Kitty Engelmann for extensive discussion of these ideas. This work has been supported by the University of Connecticut Research Foundation.

References

Abouheif, E., and Wray, G.A. 2002. Evolution of the gene network underlying wing polyphenism in ants. Science (Washington, D.C.) **297**: 249–252.

Alberch, P. 1991. From genes to phenotypes: dynamical systems and evolvability. Genetica **84**: 5–11.

Altenberg, L. 1995. Genome growth and the evolution of the genotype-phenotype map. *In* Evolution and biocomputation: computational models of evolution. *Edited by* W. Banzhaf and F.H. Eeckman. SpringerVerlag, Berlin. pp. 205–259.

Ancel Meyers, L., and Bull, J.J. 2002. Fighting change with change: adaptive variation in an uncertain world. Trends Ecol. Evol. **17**: 551–557.

Arnold, S.J. 1989. Group report: How do complex organisms evolve? *In* Complex organismal functions: integration and evolution in vertebrates. *Edited by* D.B. Wake and G. Roth. John Wiley & Sons, Chichester, UK. pp. 403–433.

Averof, M., Dawes, R., and Ferrier, D. 1996. Diversification of arthropod Hox genes as a paradigm for the evolution of gene functions. Semin. Cell Dev. Biol. **7**: 539–551.

Azevedo, R.B.R., Keightley P.D., Laurén-Määttä, C., Vassilieva, L.L., Lynch, M., and Leroi. A.M. 2002. Spontaneous mutational variation for body size in *Caenorhabditis elegans*. Genetics **162**: 755–765.

Badyaev, A.V., and Foresman, K.R. 2000. Extreme environmental change and evolution: stress-induced morphological variation is strongly concordant with patterns of evolutionary divergence in shrew mandibles. Proc. Roy. Soc. London Biol. Sci. **267**: 371–378.

Badyaev, A.V., and Hill, G.E. 2000. The evolution of sexual dimorphism in the house finch. I. Population divergence in morphological covariance structure. Evolution **54**: 1784–1794.

Baker, R.H., and DeSalle, R. 1997. Multiple sources of character information and the phylogeny of Hawaiian Drosophilids. Syst. Biol. **46**: 654–673.

Baldwin, B.G., and Wessa, B.L.. 2000. Origin and relationships of the tarweed-silversword lineage (Compositae-Madiinae). Am. J. Bot. **87**: 1890–1908.

Barrier, M., Robichaux, R.H., and Purugganan M.D. 2001. Accelerated regulatory gene evolution in an adaptive radiation. Proc. Natl. Acad. Sci. U.S.A. **98**: 10208–10213.

Bedau, M.A., and Packard, N.H. 2003. Evolution of evolvability via adaptation of mutation rates. Biosystems **69**: 143–162.

Beldade, P., Koops, K., and Brakefield, P.M. 2002. Modularity, individuality, and evo-devo in butterfly wings. Proc. Natl. Acad. Sci. U.S.A. **99**: 14262–14267.

Betancourt, A.J., and Presgraves, D.C. 2002. Linkage limits the power of natural selection in *Drosophila*. Proc. Natl. Acad. Sci. U.S.A. **99**: 13616–13620.

Bird, A.P. 1995. Gene number, noise reduction and biological complexity. Trends Genet. **11**: 94–100.

Bond, J.E., and Opell, B.D. 1998. Testing adaptive radiation and key innovation hypotheses in spiders. Evolution **52**: 403–414.

Bonner, J.T. 1988. The evolution of complexity by means of natural selection. Princeton University Press, Princeton, NJ.

Braterman, P.S. 1988. The evolution of evolvability: the pedigree principle. *In* Clay minerals and the origin of life. *Edited by* A.G. Cairns-Smith and H. Hartman. Cambridge University Press, Cambridge, UK. pp. 152–153.

Bruno, J.F., and Edmunds, P.J. 1997. Clonal variation for phenotypic plasticity in the coral *Madracis mirabilis*. Ecology **78**: 2177–2190.

Caballero, A., Cusi, E., Garcia, C., and Garcia-Dorado, A. 2002. Accumulation of deleterious mutations: additional *Drosophila melanogaster* estimates and a simulation of the effects of selection. Evolution **56**: 1150–1159.

Camara, M.D., and Pigliucci, M. 1999. Mutational contributions to genetic variance-covariance matrices: an experimental approach using induced mutations in *Arabidopsis thaliana*. Evolution **53**: 1692–1703.

Caporale, L.H. 2000. Mutation is modulated: implications for evolution. BioEssays **22**: 388–395.

Carroll, S.P., Dingle, H., Famula, T.H., and Fox, C.W. 2001. Genetic architecture of adaptive differentiation in evolving host races of the soapberry bug, *Jadera haematoloma*. Genetica **112–113**: 257–272.

Chen, W., Provart, N.J., Glazebrook, J., Katagiri, F., Chang, H.-S., Eulgem, T., Mauch, F., Luan, S., Zou, G., Whitham, S.A., Budworth, P.R., Tao, Y., Xie, Z., Chen, X., Lam, S., Kreps, J.A., Harper, J.F., Si-Ammour, A., Mauch-Mani, B., Heinlein, M., Kobayashi, K., Hohn, T., Dangl, J.L., Wang, X., and Zhu. T., 2002. Expression profile matrix of *Arabidopsis* transcription factor genes suggests their putative functions in response to environmental stresses. Plant Cell **14**: 559–574.

Conrad, M. 1990. The geometry of evolution. Biosystems **24**: 61–81.

Cowell, L.G., Kim, H.-J., Humaljoki, T., Berek, C., and Kepler T.B. 1999. Enhanced evolvability in immunoglobulin V genes under somatic hypermutation. J. Mol. Evol. **49**: 23–26.

Cracraft, J. 1990. The origin of evolutionary novelties: pattern and process at different hierarchical levels. *In* Evolutionary innovations. *Edited by* M. H. Nitecki. University of Chicago Press, Chicago, IL. pp. 21–44.

Cuevas, J.M., Elena, S.F., and Moya, A. 2002. Molecular basis of adaptive convergence in experimental populations of RNA viruses. Genetics **162**: 533–542.

Danley, P.D., and Kocher, T.D. 2001. Speciation in rapidly diverging systems: lessons from Lake Malawi. Mol. Ecol. **10**: 1075–1086.

Dawkins, R. 1988. The evolution of evolvability. *In* Artificial life. *Edited by* C.G. Langton. Addison-Wesley, Redwood City, CA. pp. 201–220.

de Queiroz, A. 1999. Do image-forming eyes promote evolutionary diversification? Evolution **53**: 1654–1664.

de Queiroz, A. 2002. Contingent predictability in evolution: key traits and diversification. Syst. Biol. **51**: 917–929.

de Visser, J.A.G.M. 2002. The fate of microbial mutators. Microbiology **148**: 1247–1252.

de Visser, J.A.G.M., Zeyl, C.W., Gerrish, P.J., Blanchard, J.L., and Lenski R.E. 1999. Diminishing returns from mutation supply rate in asexual populations. Science (Washington, D.C.) **283**: 404–406.

Dean, A.M. 1995. A molecular investigation of genotype by environment interactions. Genetics **139**: 19–33.

Denver, D.R., Morris, K., Lynch, M., Vassilieva, L.L., and Thomas, W.K. 2000. High direct estimate of the mutation rate in the mitochondrial genome of *Caenorhabditis elegans*. Science (Washington, D.C.) **289**: 2342–2344.

Drake, J.W., Charlesworth, B., Charlesworth, D., and Crow. J.F. 1998. Rates of spontaneous mutation. Genetics **148**: 1667–1686.

Duret, L., Marais, G., and Biemont C. 2000. Transposons but not retrotransposons are located preferentially in regions of high recombination rate in *Caenorhabditis elegans*. Genetics **156**: 1661–1669.

Elena, S.F., and Lenski, R.E. 2001. Epistasis between new mutations and genetic background and a test of genetic canalization. Evolution **55**: 1746–1752.

Evans, J.D., and Wheeler, D.E. 2001*a*. Expression profiles during honeybee caste determination. Genome Biol. **2**: r1–r6.

Evans, J. D., and Wheeler, D.E. 2001*b*. Gene expression and the evolution of insect polyphenisms. BioEssays **23**: 62–68.

Fernandez, A., and Lopez-Fanjul, C. 1997. Spontaneous mutational genotype-environment interaction for fitness-realted traits in *Drosophila melanogaster*. Evolution **51**: 856–864.

Foster, P.L. 2000. Adaptive mutation: implications for evolution. BioEssays **22**: 1067–1074.

Fry, J.D., and Heinsohn, S.L. 2002. Environment dependence of mutational parameters for viability in *Drosophila melanogaster*. Genetics **161**: 1155–1167.

Fry, J.D., Heinsohn, S L., and Mackay, T.F.C. 1996. The contribution of new mutations to genotype-environment interaction for fitness in *Drosophila melanogaster*. Evolution **50**: 2316–2327.

Fry, J.D.,. Keightley, P.D., Heinsohn, S.L., and Nuzhdin, S.V. 1999. New estimates of the rates and effects of mildly deleterious mutation in *Drosophila melanogaster*. Proc. Natl. Acad. Sci. U.S.A. **96**: 574–579.

Galant, R., Skeath, J.B., Paddock, S., Lewis, D.L., and Carroll, S.B. 1998. Expression pattern of a butterfly achaete-scute homolog reveals the homology of butterfly wing scales and insect sensory bristles. Curr. Biol. **8**: 807–813.

Galis, F., and Drucker, E.G. 1996. Pharyngeal biting mechanics in centrarchid and cichlid fishes: insights into a key evolutionary innovation. J. Evol. Biol. **9**: 641–670.

Ganfornina, M.D., and Sanchez, D. 1999. Generation of evolutionary novelty by functional shift. BioEssays **21**: 432–439.

Gatesy, S.M., and Middleton, K.M. 1997. Bipedalism, flight, and the evolution of theropod locomotor diversity. J. Vert. Paleontol. **17**: 308–329.

Gilchrist, A.S., and Partridge, L. 1999. A comparison of the genetic basis of wing size divergence in three parallel body size clines of *Drosophila melanogaster*. Genetics **153**: 1775–1787.

Giraud, A., Matic, I., Tenaillon, O., Clara, A., Radman, M., Fons, M., and Taddei, F. 2001. Costs and benefits of high mutation rates: adaptive evolution of bacteria in the mouse gut. Science (Washington, D.C.) **291**: 2602–2608.

Goho, S., and Bell, G. 2000. Mild environmental stress elicits mutations affecting fitness in *Chlamydomonas*. Proc. R. Soc. Lond. B Biol. Sci. **267**: 123–130.

Goldschmidt, R.B. 1938. Physiological genetics. McGraw-Hill, New Haven, CT.

Gorodetskii, V.P., Zhuchenko, A.A., and Korol, A.B. 1991. Efficiency of feedback selection for recombination in *Drosophila*. Soviet Genet. **26**: 1261–1269.

Grogan, D.W., Carver, G.T., and Drake J.W. 2001. Genetic fidelity under harsh conditions: analysis of spontaneous mutation in the thermoacidophilic archaeon *Sulfolobus acidocaldarius*. Proc. Natl. Acad. Sci. U.S.A. **98**: 7928–7933.

Gu, Z., Steinmetz, L.M., Gu, X., Scharfe, C., Davis, R.W., and Li W.-H.. 2003. Role of duplicate genes in genetic robustness against null mutations. Nature **421**: 63–66.

Hannon, G.J. 2002. RNA interference. Nature **418**: 244–251.

Hansen, T.F. 2003. Is modularity necessary for evolvability? Remarks on the relationship between pleiotropy and evolvability. Biosystems **69**: 83–94.

Harada, K. 1995. A quantitative analysis of modifier mutations which occur in mutation accumulation lines in *Drosophila melanogaster*. Heredity **75**: 589–598.

Harland, S.C. 1936. The genetical conception of the species. Biol. Rev. Camb. Philos. Soc. **11**: 83–112.

Hodges, S.A., and Arnold, M.L. 1994. Columbines: a geographically widespread species flock. Proc. Natl. Acad. Sci. U.S.A. **91**: 5129–5132.

Hodges, S.A., and Arnold, M.L. 1995. Spurring plant diversification: are floral nectar spurs a key innovation? Proc. R. Soc. Lond. B Biol. Sci. **262**: 343–348.

Hoffmann, A.A., and Merila, J. 1999. Heritable variation and evolution under favourable and unfavourable conditions. Trends Ecol. Evol. **14**: 96–101.

Hoffmann, A.A., and Schiffer, M. 1998. Changes in the heritability of five morphological traits under combined environmental stresses in *Drosophila melanogaster*. Evolution **52**: 1207–1212.

Horvath, D.P., Schaffer, R., West, M., and Wisman, E. 2003. *Arabidopsis* microarrays identify conserved and differentially expressed genes involved in shoot growth and development from distantly related plant species. Plant J. **34**: 125–134.

Houle, D. 1992. Comparing evolvability and variability of quantitative traits. Genetics **130**: 195–204.

Houle, D., Hughes, K.A., Hoffmaster, D.K., Ihara, J., Assimacopoulos, S., Canada, D., and Charlesworth, B. 1994. The effects of spontaneous mutation on quantitative traits. I. Variances and covariances of life history traits. Genetics **138**: 773–785.

Hughes, C.L., and Kaufman, T.C. 2000. RNAi analysis of *Deformed, proboscipedia* and *Sex combs reduced* in the milkweed bug *Oncopeltus fasciatus*: novel roles for Hox genes in the hemipteran head. Development **127**: 3683–3694.

Hunter, J.P. 1998. Key innovations and the ecology of macroevolution. Trends Ecol. Evol. **13**: 31–36.

Hunter, J.P., and Jernvall, J. 1995. The hypocone as a key innovation in mammalian evolution. Proc. Natl. Acad. Sci. U.S.A. **92**: 10718–10722.

Ito-Harashima, S., Hartzog, P.E., Sinha, H., and McCusker, J.H. 2002. The tRNA-Tyr gene family of *Saccharomyces cerevisiae*: agents of phenotypic variation and position effects on mutation frequency. Genetics **161**: 1395–1410.

Jernvall, J. 2000. Linking development with generation of novelty in mammalian teeth. Proc. Natl. Acad. Sci. U.S.A. **97**: 2641–2645.

Kambysellis, M.P., Ho, K.F., Craddock, E.M., Piano, F., Parisi, M., and Cohen, J. 1995. Pattern of ecological shifts in the diversification of Hawaiian *Drosophila* inferred from a molecular phylogeny. Curr. Biol. **5**: 1129–1139.

Kauffman, S.A. 1993. The origins of order: self-organization and selection in evolution. Oxford University Press, New York, NY.

Kause, A., and Morin, J.P. 2001. Seasonality and genetic architecture of development time and body size of the birch feeding sawfly *Priophorus pallipes*. Genet. Res. **78**: 31–40.

Keightley, P.D., and Bataillon, T.M. 2000. Multigeneration maximum-likelihood analysis applied to mutation-accumulation experiments in *Caenorhabditis elegans*. Genetics **154**: 1193–1201.

Kidner, C.A., and Martienssen, R.A. 2004. Spatially restricted microRNA directs leaf polarity through *ARGONAUTE1*. Nature **428**: 81–84.

Kirschner, M., and Gerhart, J. 1998. Evolvability. Proc. Natl. Acad. Sci. U.S.A. **95**: 8420–8427.

Klebes, A., Biehs, B., Cifuentes, F., and Kornberg, T.B. 2002. Expression profiling of *Drosophila* imaginal discs. Genome Biol. **3**: research0038.1–0038.16.

Koren, A., Ben-Aroya, S., and Kupiec, M. 2002. Control of meiotic recombination initiation: a role for the environment? Curr. Genet. **42**: 129–139.

Kovalchuk, I., Kovalchuk, O., and Hohn, B. 2000. Genome-wide variation of the somatic mutation frequency in transgenic plants. EMBO J. **19**: 4431–4438.

Lachke, S.A., Srikantha, T., Tsai, L.K., Daniels, K., and Soll, D.R. 2000. Phenotypic switching in *Candida glabrata* involves phase-specific regulation of the metallothionein gene *MT-II* and the newly discovered hemolysin gene *HLP*. Infect. Immunol. **68**: 884–895.

Lamb, B.C., Saleem, M., Scott, W., Thapa, N., and Nevo E. 1998. Inherited and environmentally induced differences in mutation frequencies between wild strains of *Sordaria fimicola* from "Evolution Canyon". Genetics **149**: 87–99.

Lauder, G.V. 1981. Form and function: structural analysis in evolutionary morphology. Paleobiology **7**: 430–442.

Lauter, N., and Doebley, J. 2002. Genetic variation for phenotypically invariant traits detected in teosinte: implications for the evolution of novel forms. Genetics **160**: 332–342.

Lercher, M.J., and Hurst L.D. 2002. Human SNP variability and mutation rate are higher in regions of high recombination. Trends Genet. **18**: 337–340.

Leroi, A.M. 2000. The scale independence of evolution. Evol. Dev. **2**: 67–77.

Levin, D.A. 1988. Plasticity, canalization and evolutionary stasis in plants. *In* Plant population biology. *Edited by* A.J. Davy, M.J. Hutchings and A. Watkinson. Blackwell, Oxford, UK. pp. 35–45.

Liem, K F. 1973. Evolutionary strategies and morphological innovations: cichlid pharyngeal jaws. Syst. Zool. **22**: 425–441.

Long, M. 2001. Evolution of novel genes. Curr. Opin. Genet. Dev. **11**: 673–680.

Lucht, J.M., Mauch-Mani, B., Steiner, H.-Y., Metraux, J.-P., Ryals, J., and Hohn, B. 2002. Pathogen stress increases somatic recombination frequency in *Arabidopsis*. Nature Genet. **30**: 311–314.

Lynch, M., and Force, A. 2000. The probability of duplicate gene preservation by subfunctionalization. Genetics **154**: 459–473.

Lynch, M., Latta, L., Hicks, J., and Giorgianni, M. 1998. Mutation, selection, and the maintenance of life-history variation in a natural population. Evolution **52**: 727–733.

Mack, P.D., Lester, V.K., and Promislow, D.E.L. 2000. Age-specific effects of novel mutations in *Drosophila melanogaster*. II. Fecundity and male mating ability. Genetica **110**: 31–41.

Marini, A., Matmati, N. and Morpurgo, G. 1999. Starvation in yeast increases non-adaptive mutation. Curr. Genet. **35**: 77–81.

Martusewitsch, E., Sensen, C.W., and Schleper, C. 2000. High spontaneous mutation rate in the hyperthermophilic archaeon *Sulfolobus solfataricus* is mediated by transposable elements. J. Bacteriol. **182**: 2574–2581.

Mayr, E. 1960. The emergence of evolutionary novelties. *In* Evolution after Darwin: the evolution of life. *Edited by* S. Tax. University of Chicago Press, Chicago, IL. pp. 349–380.

McKenzie, G.J., Harris, R.S., Lee, P.L., and Rosenberg S.M.. 2000. The SOS response regulates adaptive mutation. Proc. Natl. Acad. Sci. U.S.A. **97**: 6646–6651.

McKenzie, G.J., and Rosenberg, S.M. 2001. Adaptive mutations, mutator DNA polymerases and genetic change strategies of pathogens. Curr. Opin. Microbiol. **4**: 586–594.

Meiklejohn, C.D., and Hartl, D.L. 2002. A single mode of canalization. Trends Ecol. Evol. **17**: 468–473.

Mezey, J.G., Cheverud, J.M., and Wagner, G.P. 2000. Is the genotype-phenotype map modular?: a statistical approach using mouse quantitative trait locus data. Genetics **156**: 305–311.

Minelli, A. 1998. Molecules, developmental modules, and phenotypes: A combinatorial approach to homology. Mol Phylogenet. Evol. **9**: 340–347.

Moore, F.B.-G., Rozen, D.E., and Lenski R.E. 2000. Pervasive compensatory adaptation in *Escherichia coli*. Proc. R. Soc. Lond. B Biol. Sci. **267**: 515–522.

Murren, C.J., and Kover, P.X. 2004. QTL mapping: a first step toward an understanding of molecular genetic mechanisms behind phenotypic complexity/integration. *In* Phenotypic integration: studying the ecology and evolution of complex phenotypes. *Edited by* M. Pigliucci and K.A. Preston. Oxford University Press, Oxford, UK. pp. 195–212.

Murren, C.J., Pendleton, N. and Pigliucci, M. 2002. Evolution of phenotypic integration in *Brassica* (Brassicaceae). Am. J. Bot. **89**: 655–663.

Nachman, M.W. 2002. Variation in recombination rate across the genome: evidence and implications. Curr. Opin. Genet. Dev. **12**: 657–663.

Naylor, G.J.P., and Gerstein, M. 2000. Measuring shifts in function and evolutionary opportunity using variability profiles: a case study of the globins. J. Mol. Evol. **51**: 223–233.

Nemeschkal, H.L. 1999. Morphometric correlation patterns of adult birds (Fringillidae: Passeriformes and Columbiformes) mirror the expression of developmental control genes. Evolution **53**: 899–918.

Newell, L.E., and Heddle, J.A. 2002. The effect of dietary restriction during development *in utero* on the frequency of spontaneous somatic mutations. Mutagenesis **17**: 289–292.

Nicotra, A.B., Chazdon, R.L., and Schlichting, C.D. 1997. Patterns of genotypic variation and phenotypic plasticity of light response in two tropical *Piper* species. Am. J. Bot. **84**: 1542–1552.

Nijhout, H.F. 2002. The nature of robustness in development. Bioessays **24**: 553–563.

Nijhout, H.F. 2003. Development and evolution of adaptive polyphenisms. Evol. Dev. **5**: 9–18.

Nijhout, H.F., Berg, A.M., and Gibson, W.T. 2003. A mechanistic study of evolvability using the mitogen-activated protein kinase cascade. Evol. Dev. **5**: 281–294.

Nijhout, H.F., and Emlen, D.J. 1998. Competition among body parts in the development and evolution of insect morphology. Proc. Natl. Acad. Sci. U.S.A. **95**: 3685–3689.

Nitecki, M. H. 1990*a*. Evolutionary innovations. University of Chicago Press, Chicago, IL.

Nitecki, M. H. 1990*b*. The plurality of evolutionary innovations. *In* Evolutionary innovations. *Edited by* M. H. Nitecki. University of Chicago Press, Chicago, IL. pp. 3–18.

Ofria, C., Adami, C., and Collier, T.C. 2003. Selective pressures on genomes in molecular evolution. J. Theor. Biol. **222**: 477–483.

Ohno, S. 1970. Evolution by gene duplication. Springer-Verlag, New York, NY.

Ohta, T. 1991. Multigene families and the evolution of complexity. J. Mol. Evol. **33**: 34–41.

Ohta, T. 2002. Near-neutrality in evolution of genes and gene regulation. Proc. Natl. Acad. Sci. U.S.A. **99**: 16134–16137.

Ohta, T. 2003. Evolution by gene duplication revisited: differentiation of regulatory elements versus proteins. Genetica **118**: 209–216.

Pérez-Péreza, J.M., Serrano-Cartagena, J., and Micol, J.L. 2002. Genetic analysis of natural variations in the architecture of *Arabidopsis thaliana* vegetative leaves. Genetics **162**: 893–915.

Pigliucci, M. 1996. How organisms respond to environmental change: from phenotypes to molecules (and vice versa). Trends Ecol. Evol. **11**: 168–173.

Pigliucci, M. 2001. Phenotypic plasticity: beyond nature and nurture. Johns Hopkins University Press, Baltimore, MD.

Pigliucci, M., and Hayden, K.P. 2001. Phenotypic plasticity is the major determinant of changes in phenotypic integration in *Arabidopsis*. New Phytol. **152**: 419–430.

Pletcher, S.D., Houle, D., and Curtsinger, J.W. 1998. Age-specific properties of spontaneous mutations affecting mortality in *Drosophila melanogaster*. Genetics **148**: 287–303.

Poole, A.M., Phillips M.J., and Penny D. 2003. Prokaryote and eukaryote evolvability. Biosystems **69**: 163–185.

Ptashne, M., and Gann, A. 1998. Imposing specificity by localization: mechanism and evolvability. Curr. Biol. **8**: 812–822.

Queitsch, C., Sangster, T.A., and Lindquist, S.L. 2002. Hsp90 as a capacitor of phenotypic variation. Nature **417**: 618–624.

Radman, M. 1999. Enzymes of evolutionary change. Nature **401**: 866–868.

Raff, R.A. 1996. The shape of life: genes, development, and the evolution of animal form. University of Chicago Press, Chicago, IL.

Raff, R.A., and Sly, B.J. 2000. Modularity and dissociation in the evolution of gene expression territories in development. Evol. Dev. **2**: 102–113.

Rasmuson, M. 2002. The genotype-phenotype link. Hereditas **136**: 1–6.

Reich, D.E., Schaffner, S.F., Daly, M.J., McVean, G., Mullikin, J.C., Higgins, J.M., Richter, D.J., Lander, E.S., and Altshuler, D. 2002. Human genome sequence variation and the influence of gene history, mutation and recombination. Nature Genet. **32**: 135–142.

Rice, W.R. 2002. Experimental tests of the adaptive significance of sexual recombination. Nat. Rev. Genet. **3**: 241–251.

Riley, M.S., Cooper, V.S., Lenski, R.E., Forney, L.J., and Marsh T.L. 2001. Rapid phenotypic change and diversification of a soil bacterium during 1000 generations of experimental evolution. Microbiology **147**: 995–1006.

Rives, A.W., and Galitski, T. 2003. Modular organization of cellular networks. Proc. Natl. Acad. Sci. U.S.A. **100**: 1128–1133.

Rizzon, C., Marais, G., Gouy, M., and Biemont C. 2002. Recombination rate and the distribution of transposable elements in the *Drosophila melanogaster* genome. Genome Res. **12**: 400–407.

Rockman, M.V., and Wray G.A. 2002. Abundant raw material for *cis*-regulatory evolution in humans. Mol. Biol. Evol. **19**: 1991–2004.

Rutherford, S.L., and Lindquist, S. 1998. Hsp90 as a capacitor for morphological evolution. Nature **396**: 336–342.

Sanchez-Moran, E., Armstrong, S.J., Santos, J.L., Franklin, F.C.H., and Jones G.H.. 2002. Variation in chiasma frequency among eight accessions of *Arabidopsis thaliana*. Genetics **162**: 1415–1422.

Schaefer, S.A., and Lauder, G.V. 1996. Testing historical hypotheses of morphological change: biomechanical decoupling in Loricarioid catfishes. Evolution **50**: 1661–1675.

Scharloo, W. 1987. Constraints in selection response. *In* Genetic constraints on adaptive evolution. *Edited by* V. Loeschcke. Springer-Verlag, Berlin. pp. 125–149.

Schenk, P.M., Kazan, K., Wilson, I., Anderson, J.P., Richmond, T., Somerville, S.C., and Manners, J.M. 2000. Coordinated plant defense responses in *Arabidopsis* revealed by microarray analysis. Proc. Natl. Acad. Sci. U.S.A. **97**: 11655–11660.

Schlichting, C.D. 1989. Phenotypic integration and environmental change. BioScience **39**: 460–464.

Schlichting, C.D. 2003. The origins of differentiation via phenotypic plasticity. Evol. Dev. **5**: 98–105.

Schlichting, C.D. 2004. The role of phenotypic plasticity in diversification. *In* Phenotypic plasticity: functional and conceptual approaches. *Edited by* T.J. DeWitt and S.M. Scheiner, Oxford University Press, Oxford, UK. pp. 191–200.

Schlichting, C.D., and Pigliucci, M. 1998. Phenotypic evolution: a reaction norm perspective. Sinauer Associates, Sunderland, MA.

Schlichting, C.D., and H. Smith. 2002. Phenotypic plasticity: linking molecular mechanisms with evolutionary outcomes. Evol. Ecol. **16**: 189–211.

Schlosser, G. 2002. Modularity and the units of evolution. Theor. Biosci. **121**: 1–80.

Schultz, S.T., Lynch, M., and Willis. J.H. 1999. Spontaneous deleterious mutation in *Arabidopsis thaliana*. Proc. Natl. Acad. Sci. U.S.A. **96**: 11393–11398.

Sgrò, C.M., and Hoffmann, A.A. 1998. Effects of temperature extremes on genetic variances for life history traits in *Drosophila melanogaster* as determined from parent-offspring comparisons. J. Evol. Biol. **11**: 1–20.

Shaw, R.G., Byers, D.L., and Darmo, E. 2000. Spontaneous mutational effects on reproductive traits of *Arabidopsis thaliana*. Genetics **155**: 369–378.

Silva, P.J.N., and Dykhuizen, D.E. 1993. The increased potential for selection of the *lac* operon of *Escherichia coli*. Evolution **47**: 741–749.

Smith, H. 1990. Signal perception, differential expression within multigene families and the molecular basis of phenotypic plasticity. Plant Cell Environ. **13**: 585–594.

Sniegowski, P.D., Gerrish, P.J., Johnson, T., and Shaver A. 2000. The evolution of mutation rates: separating causes from consequences. BioEssays **22**: 1057–1066.

Sommer, R.J. 1997. Evolutionary changes of developmental mechanisms in the absence of cell lineage alterations during vulva formation in the Diplogastridae (Nematoda). Development **124**: 243–251.

Stebbins Jr., G.L., and Hartl, D.L. 1988. Comparative evolution: latent potentials for anagenetic advance. Proc. Natl. Acad. Sci. U.S.A. **85**: 5141–5145.

Steppan, S.J. 1997. Phylogenetic analysis of phenotypic covariance structure. I. Contrasting results from matrix correlation and common principal component analyses. Evolution **51**: 571–586.

Stern, C. 1958. Selection for subthreshold differences and the origin of pseudoexogenous adaptations. Am. Nat. **92**: 313–316.

Stock, D.W. 2001. The genetic basis of modularity in the development and evolution of the vertebrate dentition. Phil. Trans. Roy. Soc. Lond. B Biol. Sci. **356**: 1633–1653.

Stuart, J.M., Segal, E., Koller, D., and Kim, S.K. 2003. A genecoexpression network for global discovery of conserved genetic modules. Science (Washington, D.C.) **302**: 249–255.

Taddei, F., Radman, M., Maynard-Smith, J., Toupance, J., Gouyon, P.H., and Godelle B. 1997. Role of mutator alleles in adaptive evolution. Nature **387**: 700–702.

Takahashi, K., Nishida, M., Yuma, M., and Okada, N. 2001. Retroposition of the AFC family of SINEs (short interspersed repetitive elements) before and during the adaptive radiation of cichlid fishes in Lake Malawi and related inferences about phylogeny. J. Mol. Evol. **53**: 496–507.

Tanay, A., Sharan, R., Kupiec, M., and Shamir, R.. 2004. Revealing modularity and organization in the yeast molecular network by integrated analysis of highly heterogeneous genomewide data. Proc. Natl. Acad. Sci. U.S.A. **101**: 2981–2986.

Tenaillon, O., Taddei, F., Radman, M., and Matic, I., 2001. Second-order selection in bacterial evolution: selection acting on mutation and recombination rates in the course of adaptation. Res. Microbiol. **152**: 11–16.

Thieffry, D., and Romero, D. 1999. The modularity of biological regulatory networks. BioSystems **50**: 49–59.

Travis, J.M.J., and Travis, E.R. 2002. Mutator dynamics in fluctuating environments. Proc. R. Soc. Lond. B Biol. Sci. **269**: 591–597.

Van Delden, W., Boerema, A.C., and Kamping, A. 1978. The alcohol dehydrogenase polymorphism in populations of *Drosophila melanogaster* I. Selection in different environments. Genetics **90**: 161–191.

Vermeij, G.J. 1974. Adaptation, versatility, and evolution. Syst. Zool. **22**: 466–477.

von Dassow, G., Meir, E., Munro, E.M., and Odell G.M. 2000. The segment polarity network is a robust developmental module. Nature **406**: 188–192.

von Dassow, G., and Munro, E. 1999. Modularity in animal development and evolution: elements of a conceptual framework for EvoDevo. J. Exp. Zool. (Mol. Dev. Evol.) **285**: 307–325.

Vulic, M., Lenski, R.E., and Radman, M. 1999. Mutation, recombination, and incipient speciation of bacteria in the laboratory. Proc. Natl. Acad. Sci. U.S.A. **96**: 7348–7351.

Waddington, C.H. 1956. Genetic assimilation of the *bithorax* phenotype. Evolution **10**: 1–13.

Wagner, A. 1999. Redundant gene functions and natural selection. J. Evol. Biol. **12**: 1–16.

Wagner, A. 2001. The yeast protein interaction network evolves rapidly and contains few redundant duplicate genes. Mol. Biol. Evol. **18**: 1283–1292.

Wagner, G.P., and Altenberg, L. 1996. Perspective: complex adaptations and the evolution of evolvability. Evolution **50**: 967–976.

Wagner, G.P., Chiu, C.-H., and Hansen T.F. 1999. Is Hsp90 a regulator of evolvability? Mol. Dev. Evol. **285**: 116–118.

Wagner, G.P., and Schwenk, K. 2000. Evolutionary stable configurations: functional integration and the evolution of phenotypic stability. Evol. Biol. **31**: 155–217.

Wang, W., Cherry, J.M., Botstein, D., and Li, H. 2002. A systematic approach to reconstructing transcription networks in *Saccharomyces cerevisiae*. Proc. Natl. Acad. Sci. U.S.A. **99**: 16893–16898.

Warheit, K.I., Forman, J.D., Losos, J.B., and Miles D.B. 1999. Morphological diversification and adaptive radiation: A comparison of two diverse lizard clades. Evolution **53**: 1226–1234.

West-Eberhard, M.J. 1989. Phenotypic plasticity and the origins of diversity. Annu. Rev. Ecolo. .Syst. **20**: 249–278.

West-Eberhard, M. J. 1998. Evolution in the light of developmental and cell biology, and *vice versa*. Proc. Natl. Acad. Sci. U.S.A. **95**: 8417–8419.

Winther, R.G. 2001. Varieties of modules: kinds, levels, origins, and behaviors. J. Exp. Zool. (Mol. Dev. Evol.) **291**: 116–129.

Wright, S. 1931. Evolution in Mendelian populations. Genetics **16**: 97–159.

Yampolsky, L.Y., and Stoltzfus, A. 2001. Bias in the introduction of variation as an orienting factor in evolution. Evol. Dev. **3**: 73–83.

Yang, A.S. 2001. Modularity, evolvability, and adaptive radiations: a comparison of the hemi- and holometabolous insects. Evol. Dev. **3**: 59–72.

You, L.C., and Yin, J. 2002. Dependence of epistasis on environment and mutation severity as revealed by *in silico* mutagenesis of Phage T7. Genetics **160**: 1273–1281.

Zhang, H., and Forde, B.G.. 1998. An *Arabidopsis* MADS box gene that controls nutrient-induced changes in root architecture. Science (Washington, D.C.) **279**: 407–409.

Zhang, H., Jennings, A., Barlow, P.W., and Forde B.G. 1999. Dual pathways for regulation of root branching by nitrate. Proc. Natl. Acad. Sci. U.S.A. **96**: 6529–6534.

Zhao, Z., and Zera A.J. 2002. Differential lipid biosynthesis underlies a tradeoff between reproduction and flight capability in a wing-polymorphic cricket. Proc. Natl. Acad. Sci. U.S.A. **99**: 16829–16834.

5 Nucleocytoplasmic incompatibility fosters speciation

Donald A. Levin

Abstract: The role of nucleocytoplasmic interactions in the genesis of post-zygotic isolation has been given little attention by plant evolutionists. I present evidence from reciprocal crosses, cytoplasmic substitution lines, and cell fusion lines that hybrid weakness and sterility often arise from interactions between the nuclear genome and the chloroplast and mitochondrial genomes. These interactions may be important in the origin and isolation of species. The strength of the post-zygotic barriers tends to be a function of cytoplasmic divergence. Nucleocytoplasmic incompatibility may stand in the way of chloroplast capture, which often is offered as the explanation for the incongruence between nuclear and chloroplast gene trees. However, nucleocytoplasmic heterosis is known in some species combinations. The properties and evolutionary potential of allopolyploids and diploid hybrid derivatives may be influenced by cytoplasmic factors. Cytoplasmic factors thus may contribute to the novelty of lineages as well as to their isolation.

Introduction

By most accounts, the emergence of reproductive barriers to gene exchange between plant lineages signals that the lineages have diverged to the level of species (Levin 2000). The character of these barriers has been studied intensively in a very large number of genera. Thus we know a great deal about (1) differences in the ecological tolerances and phenology of species, (2) the penchant and ability of pollinators to effectively cross-pollinate related species, and (3) the incompatibility of divergent genomes as expressed in interspecific cross-incompatibility, and hybrid inviability, weakness, and sterility.

The genetic basis for differences in ecological amplitude and phenology, and in floral traits affecting the foraging behavior of pollinators is understood in a few cases (Levin 2000). Similarly, the genetic causes of cross-incompatibility and of hybrid inviability, weakness, and sterility are understood in a few instances. We also know that in addition to nuclear genetic factors, hybrid sterility may be due to the failure of chromosomes to pair and/or abnormal pairing and abnormal disjunction (Levin 2002).

In contrast to nuclear genic and chromosomal factors, the role of chloroplast and mitochondrial genes in the post-zygotic isolation of species has been largely ignored. This is somewhat surprising, because for a long time systematists have been using organelle genome differences (which often are not subtle) to establish species' relationships.

The only time that the role of organelle genes in post-zygotic isolation has been raised is when the first generation hybrids between species differ in fitness, depending on which species served as the maternal parent (Levin 2000). This asymmetry arises from the compatibility of the nuclear genome of species A and the cytoplasmic genome of species B on the one hand, and the incompatibility of the nuclear genome of species B and cytoplasmic genome of species A on the other hand.

The tacit assumption that pervades the literature is that the lack of major reciprocal differences in species hybrids means that cytoplasmic factors have little role in species' isolation. But why can't weakness in hybrids from crossing combinations A × B and B × A also involve cytoplasmic factors? There is no rule that cytoplasmic effects must be asymmetrical. Indeed, symmetry is more likely to be the case, because there is little reason to assume that the nuclear genome of species A will work better with the cytoplasmic genome of species B than the nuclear of B will work with the cytoplasm of species A. The alien cytoplasmic genomes are equally divergent from the resident cytoplasmic genomes.

The tacit assumption that cytoplasmic factors have little role in species' isolation stands in face of an abundant literature on chloroplast and mitochondrial divergence between congeneric species. This divergence is important because it is the basis for nucleocytoplasmic incompatibility in interspecific hybrids. This incompatibility arises because normal cell function requires the well-regulated interplay between nuclear and cytoplasmic genes, and this process is upset when the nuclear genome of one species must function in the presence of alien chloroplast and mitochondrial genomes. The cell must coordinate the expression of nuclear-encoded genes, which are present in one or a few copies, with the expression of cytoplasmic genes, which are present in several hundred or thousands of copies (Leon et al. 1998). For example, while nuclear genes are pre-eminent in chloroplast biosynthesis, the expression of these genes encoding proteins related to photosynthesis is regulated by different signal pathways from the plastids (Gray et al. 2003).

The cytoplasmic genome of plants contains 95–100 genes in the chloroplast (Sugiura 1992) and 120–140 genes in the

Donald A. Levin. Section of Integrative Biology, University of Texas, Austin, TX 78712, U.S.A.
Correct citation: Levin, D.A. 2004. Nucleocytoplasmic incompatibility fosters speciation. *In* Plant Adaptation: Molecular Genetics and Ecology. Proceedings of an International Workshop held December 11–13, 2002, in Vancouver, British Columbia, Canada. *Edited by* Q.C.B. Cronk, J. Whitton, R.H. Ree, and I.E.P. Taylor. NRC Research Press, Ottawa, Ontario. pp. 30–37.

mitochondrion (Schuster and Bennicke 1994). These genes must work in concert with nuclear genes, because over evolutionary time, a majority of the genes required for organelle structure and function were transferred to the nucleus. The regulatory interactions between the organelle genomes and the nuclear genome have been synchronized over evolutionary time, so that these genomes are functionally co-adapted partners. We may say that positive intergenomic epistasis is required for normal cell and plant function. As sets of co-adapted organellar and nuclear genomes diverge, the intergenomic interactions between sets become increasingly negative. Post-zygotic barriers arise from these negative interactions, and thus organellar divergence promotes the process of speciation.

Some insights into the cause of nucleocytoplasmic disharmony recently were obtained by Schmitz-Linneweber et al. (2002). They compared the base sequences of the plastid genomes of *Atropa belladonna* and *Nicotiana tabacum*, both members of the Solanaceae. It appears that regulatory elements and genes (including introns) are well conserved. However, there are significant differences in the RNA editotypes of the two species. The chloroplast RNA editing (that takes place in about 30 sites) changes the information content of messenger RNA. This editing involves nuclear-encoded editing factors that are species-specific. Accordingly, when a nucleus is coupled with alien plastids, chloroplast RNA editing may go awry in this and other species' combinations, the result being reduced rates of photosynthesis.

There are three types of investigations that shed light on whether nucleocytoplasmic interactions have a bearing on post-zygotic isolation. One type involves the production of reciprocal, first-generation hybrids. These hybrids contain half of the nuclear genomes of two species and the cytoplasmic genome of one. A second type of investigation is the formulation of cytoplasmic substitution lines. These are lines that contain the nucleus of one species and the organelles of another. They are made by successive generations of backcrossing to one of the species, always using that (the recurrent) species as the pollen parent. These lines are referred to alloplasmic, whereas the lines containing the nucleus and cytoplasm of the same species are referred to as euplasmic.

The third approach to evaluating the compatibility of genomes from different species is to fuse cells of two species and make cybrids. Cybrids contain the nucleus and chloroplasts of one species and the mitochondria of another. They have the same chromosome number as does the nuclear "parent" of cybrids. Reciprocal cybrids could be produced, but this is a rare occurrence. As with alloplasmic lines, investigators are interested in improving the performance of specific crops rather than understanding the broader issue of nucleocytoplasmic compatibility. The production of reciprocal cybrids and reciprocal substitution lines is essential for a better understanding of the cytoplasmic factor in post-zygotic barrier formation, and thus speciation.

The primary purpose of this paper is to show that nucleocytoplasmic incompatibility may produce strong post-zygotic barriers to interspecific gene exchange. Evidence is presented from reciprocal crosses, cytoplasmic substitution lines, and cybrids. The second purpose of the paper is to show that cytoplasmic constitution of an allopolyploid may influence many of its characteristics. An allopolyploid with the cytoplasm of one parental species may have different phenotypic or physiological expressions than an allopolyploid with the same nuclear genomes but the cytoplasm of the other parental species. All cytoplasmic effects are not negative. The third purpose of this paper is to consider the possibility of cytoplasmic capture in light of nucleocytoplasmic interactions. Capture is unlikely when the presence of an alien cytoplasm compromises female fitness. However, there are species combinations in which alien cytoplasm confers a fitness advantage.

The cytoplasm as a factor in postzygotic isolation

The following information on the cytoplasmic factor in the genesis of post-zygotic barriers is a by-product of attempts to improve the agronomic attributes of plants. There have been no systematic investigations of the topic *per se*. Accordingly, most of the studies on nucleocytoplasmic interactions are incomplete from an evolutionary perspective. Nevertheless, plant scientists provide us with incontrovertible evidence that nucleocytoplasmic interactions typically are negative at the species level. The data they provide is far more informative than that from reciprocal crosses, which were made infrequently by plant evolutionists for most of the 20th century.

The most extensive data set on cytoplasmic substitution comes from various studies that replace the cytoplasm of breadwheat (*Triticum aestivum*) with cytoplasms from other *Triticum* species and species of *Aegilops* and *Secale*.

Male sterility is a common outcome of this substitution (Wilson and Driscoll 1983; Jones et al. 1998). Breadwheat with the cytoplasm of several *Aegilops* species also have reduced vigor and delayed sterility. Some of these alloplasmic lines suffer from increased susceptibility to fungal pathogens such as wheat leaf rust (*Puccinia recondite*; Washington and Mann 1974) and loose smut (Dhitaphichit et al. 1989).

The physiological abnormalities that underlie hybrid weakness are known for a few *Triticum* alloplasmic lines. Ikeda et al. (1994) compared the cytochrome oxidase activities of the mitochondria in euplasmic breadwheat with alloplasmic lines containing the cytoplasms of *Aegilops columnaris*, *Ae. crassa*, *Ae. mutica*, and *Ae. triuncialis*. The activity of this enzyme was impaired, and this resulted in a severe depression in respiratory activity in these plants.

Developmental abnormalities affecting fertility, growth, and phenology have been described in the alloplasmic lines of several other domesticated species. One of the more interesting studies involved substituting the cytoplasm of the *Helianthus annuus* with the cytoplasms of some perennial congeners (*H. angustifolius*, *H. divaricatus*, *H. grosseserratus*, *H. maximilliani*, and *H. mollis* (Jan 1992)). The alloplasmic lines were weak, and had delayed maturity and reduced seed production. The stunted growth was due to the interaction of a single nuclear gene with the alien cytoplasms. All of the perennial species had the same gene.

Male sterility in alloplasmic lines of *Triticum aestivum*, *Helianthus annuus* and numerous other species is due to nuclear-mitochondrial interaction. Male-sterile plants have different mitochondrial gene sequences than male-fertile plants

(Hanson 1991; Saumitou-Laprade et. al. 1994). Mitochondrial change or the lack thereof also is at the core of reciprocal differences in male function when *Epilobium watsonii* is crossed with three other species (Schmitz 1988; Schmitz and Michaelis 1988).

Hybrid weakness in alloplasmic lines is attributable to nuclear-chloroplast interactions when the level of photosynthesis is depressed or when chloroplast development is abnormal (Wilson and Driscoll 1983; Jones et al. 1998). Otherwise, the cytological basis for weakness is poorly understood. That the mitochondria might be involved was demonstrated when the nucleus and chloroplasts of *Nicotiana tabacum* were combined with the mitochondria of *N. debneyi* in a cell fusion product (Asahi et al. 1988; Pollak 1991).

Through its effect on viability, nucleocytoplasmic incompatibility also may result in atypical segregation ratios in second-generation hybrids. A prime illustration comes from a pair of F_2 populations derived from crosses between *Iris brevicaulis* and *I. fulva* (Burke et al. 1998*b*). One population had the cytoplasm of the former, and the other the cytoplasm of the latter. In the population with the *I. fulva* cytoplasm, there was a significant excess of *I. fulva* genotypes and a deficiency of *I. brevicaulis* genotypes at one of three loci studied. Conversely, the segregation pattern was normal on the *I. brevicaulis* background.

Distorted segregation ratios in the direction of the cytoplasmic donor also are evident in second generation hybrids between *Mimulus guttatus* and *M. nasutus*. Fishman et al. (2001) speculate that this result may be due to dissonant nucleocytoplamic interactions.

Isolation as a function of cytoplasmic divergence

Thus far we have seen that when nuclei of species are combined with cytoplasmic genomes of others plant performance suffers. Given the coevolution of nuclear and organelle genomes within species, we would expect that as alien organelles became more divergent from the native organelle the level of abnormality in substitution lines or in cybrids would increase. This expectation tends to be met in the small number of genera where genetic distances between chloroplast genomes have been measured and where substitution lines or cybrids have been produced.

Triticum and its allies is the prime arena for viewing how organelle divergence affects plant performance. The vigor of breadwheat alloplasmic lines can be viewed with reference to the genetic distances between breadwheat cytoplasms and those of other species based on nucleotide sequence differences for both mitochondrial and chloroplast genomes (Wang et al. 1997). In general, the vigor of alloplasmic lines decreases as organelle genome distance increases (Wilson and Driscoll 1983).

The effect of increasing nucleocytoplasmic dissonance in breadwheat substitution lines is apparent early in development. Plant biomass in three-week old seedlings is inversely correlated with cytoplasmic genetic distance (Nakamura et al. 1991).

Maize is another crop where numerous alloplasmic lines have been generated and where cytoplasmic distances have been determined. The lines contain the cytoplasm of *Zea mays* ssp. *mexicana*, *Z. mays* ssp. *parviflora* and *Z. perennis*. The cytoplasm of the taxon with the greatest genetic distance, *Z. perennis*, produced the greatest depression in vigor relative to euplasmic maize (Doebley 1990*a*, *b*; Edwards and Coors 1996).

Studies in the Solanaceae provide another opportunity to see how chloroplast divergence impacts plant vigor. This family has been fertile ground for the production of cybrids, with the potato (*Solanum tuberosum*) being a pivotal species in this regard. Its nucleus and mitochondria have been combined with the chloroplasts of 14 other species. Oddly enough, plant growth was normal even when the chloroplasts were imported from the most distantly related species (Perl et al. 1991). However, the cybrids containing the genetically distant chloroplasts were male sterile.

Intergeneric solanaceous cybrids have been made in various combinations. Now we see nucleocytoplasmic incompatibility in action. Cybrids containing the tobacco nucleus and mitochondrial genomes and the chloroplast genome of *Hyocyamus niger* have retarded growth in part as a result of a chlorophyll deficiency (Zubko et al. 2001). Cybrids with the nucleus of tobacco and chloroplasts of *Lycium barbarium*, *Scopolia carniolica*, *Nolina paradoxa*, and *Physochlaine officinalis* have reduced growth rates because of chlorophyll deficiencies (Babiychuk et al. 1995). When tobacco chloroplasts replaced those of *Atropa belladonna*, the plantlets lacked chlorophyll (Kushnir et al. 1991). Many attempts to make wide solanaceous cybrids have been unsuccessful. This is because "captured" chloroplasts may not develop or function normally (Medgyesy 1994). It is not clear how the failure rate is related to genetic distances between the native and alien chloroplast genomes.

In summary, there is an abundance of evidence that the nuclear genomes of one species and the organellar genomes of another species usually are quite incompatible. And this is under good culture conditions. If plants were compared under more stressful (natural) conditions, it is likely that the negative nucleocytoplasmic effects would be magnified, just as inbreeding depression is greater under stressful conditions (e.g., buckwheat, Komaki 1982; redwoods, Libby et al. 1981; *Crepis sancta*, Cheptou et al. 2000).

If nucleocytoplasmic incompatibility increases with the increasing genetic distance of chloroplast genomes and if nucleocytoplasmic incompatibility promotes genetic isolation, then we may speculate that lineages with relatively high rates of chloroplast gene substitution should have higher rates of speciation than those with relatively slow rates. In order to assess the merits of this supposition, there must be heterogeneity in the rates of sequence divergence among lineages; and indeed there is. Using the *rbcL* (ribulose-1,5-bisphosphate carboxylase) gene as a proxy for the chloroplast genome as a whole, Bousquet et al. (1992) found that the fastest rates of substitution occurred in the Asteridae and Poaceae, and the slowest rates occurred in the perennial dicots and early diverging monocots. Substantial rate heterogeneity also is evident among grass subfamilies (Gaut et al. 1997). The rates of mitochondrial genome evolution also vary among angiosperm families (Palmer et al. 2000; Adams et al. 2002).

In order to determine whether there was a relationship between rates of molecular evolution and speciation rates, Barraclough and Savolainen (2001) contrasted species numbers and rates of *rbcL* evolution for sister families of angiosperms. They found a statistically significant positive correlation between the two variables. They also found a positive correlation between species numbers and rates of nuclear gene evolution. Rates of molecular evolution were not associated with rates of adaptive phenotypic change. The aforementioned relationships suggest that molecular divergence may be involved speciation, and that cytoplasmic divergence may play a role in such.

Chloroplast capture

The chloroplast genomes of congeneric species often are well differentiated from one another; and they have been useful in the establishing species' relationships. However, the phylogenetic trees based on chloroplast markers are not always in accord with trees based on nuclear markers (Wendel and Doyle 1998). In many instances the position of species on chloroplast trees is much closer than their positions on nuclear gene trees. One explanation for this incongruence is chloroplast capture, the replacement of the native chloroplast genome by the genome of another species. Because the position of species on chloroplast trees is often much closer than their positions on nuclear gene trees, it is assumed that chloroplast capture occurred long after the species had diverged from a common ancestor. Put another way, chloroplast introgression had occurred between distantly related species.

For capture to occur, the alien chloroplast must at least have a neutral effect on seed production. In the most plausible scenarios for angioperms, the carriers of alien chloroplasts must have greater seed set than carriers of native chloroplasts (Tsitrone et al. 2003). Female reproductive superiority is the key to chloroplast substitution in plants where chloroplast transmission is through the egg. This usually is the case in angiosperms (Reboud and Zeyl 1994).

In most flowering plants, replacement of the chloroplasts occurs concurrently with the replacement of mitochondria, because both typically undergo maternal inheritance. With strict co-inheritance, each chloroplast–mitochondrial genome pair acts as a single clonal lineage with a shared evolutionary history. Accordingly, when populations have multiple chloroplast and multiple mitochondrial haplotypes (variants), haplotypes of the two are in strong linkage disequilibrium, as found in *Beta vulgaris* (Desplanque et al. 2000) and *Silene vulgaris* (Olsen and McCauley 2000). At the phylogenetic level, the signature of strict co-inheritance is complete congruence between chloroplast and mitochondrial gene trees. This result has been found in many genera (e.g., *Quercus*, Dumolin-Lapègue et al. 1998) and *Silene* (Olsen and McCauley 2000).

The concurrent inheritance of both chloroplasts and mitochondria means that if either organelle genome is detrimental to the female function then chloroplast capture will be opposed. Accordingly, if alien chloroplasts had no negative effect on the recipient but alien mitochondria did, then chloroplast capture would not occur. This assumes that capture is selective. Incidental capture is not likely in the face of negative nucleocytoplasmic effects.

The information presented earlier stressed that the interaction between a resident nucleus and alien cytoplasm usually is neutral, or more negative one. The operative term here is usually. Occasionally the cytoplasm of one species enhances the performance of another, as has been reported in *Avena*. Lines with 87% of the *A. sativa* nuclear genome and 13% of the *A. sterilis* nuclear genome were formulated with cytoplasms from each species (Beavis and Frey 1987). Two of the 38 isopopulations with the *A. sterilis* cytoplasm displayed heterosis for seed production. A gain in female fertility also has been reported in some alloplasmic lines of maize (Khedra and Bhalla 1976), wheat (Jones et al. 1998), sorghum (Senthil et al. 1998), and pearl millet (Yadav 1999).

It is noteworthy that in alloplasmic lines nucleocytoplasmic heterosis is not degraded from one generation to another, because it is not based on gene segregation. Whatever advantage an alien cytoplasm has in conjunction with the resident nucleus, it will persist over generations. This is in contrast to heterosis based on nuclear gene interaction, which is most pronounced in first generation hybrids.

The cytoplasm in speciation via hybridization

Whereas the nucleocytoplasmic interactions can cause reduced hybrid fitness, these interactions also may alter character expressions without destroying their functionality, as noted below. Therefore, the cytoplasmic variable may impact speciation through its influence on the success or failure of hybrid lineages (diploid or polyploid), because some of their attributes may depend on their cytoplasmic background, much as they may in first and advanced generations, and backcross hybrids.

The effect of nucleocytoplasmic interactions on character expression is best exemplified in floral traits. For example, reciprocal crosses between *Epilobium luteum* and *E. hirsutum* yield hybrids with small petals (4.6 mm by 3.4 mm) when *E. hirsutum* is the cytoplasmic donor (Michaelis 1954). When *E. luteum* is the cytoplasmic donor, petals of hybrids are very much larger (10.7 mm by 9.2 mm). Reciprocal crosses between *Streptocarpus wendlandii* and *S. rexii* yield F_1 hybrids with corollas averaging 4.9 cm in length when the former is the maternal parent versus 3.1 cm when the latter is the maternal parent (Oehlkers 1964). Finally, the cross *Gilia tenuiflora* × *G. exilis* produces hybrids with corolla limbs 13–15 mm in diameter versus 8–9 mm in the reciprocal cross (Grant 1956).

Floral pigmentation also may be influenced by a hybrid's cytoplasmic constitution. For example, when *G. exilis* is the egg parent in the aforementioned cross the corolla throat of hybrids is tinged with red, whereas no red marking are present in hybrids when *G. tenuiflora* is the egg parent. Another reciprocal difference is seen in *Streptocarpus* (Oehlkers 1964). When *S. michelmorei* is crossed as the egg parent with *S. wendlandii*, flowers of hybrids have a brilliant yellow spot. The spot is weak or absent in reciprocal hybrids.

Vegetative traits of hybrids also may be influenced by their cytoplasmic constitutions. Hybrids from crosses between *Gilia tenuiflora* and *G. exilis* may be nearly twice as tall when the former is the maternal parent (Grant 1956). Also when the former is the maternal parent, stems of hybrids have glandular pubescence their entire length, whereas in the reciprocal hybrid glands are confined to upper branches of the inflorescence. Hybrids from the cross *Epilobium luteum* × *E. hirsutum* have a branched, decumbent growth form in contrast to hybrids from the reciprocal cross that have an erect, pyramidal form (Michaelis 1954).

Perhaps the most striking evidence for the cytoplasmic influence on the performance of wild hybrids comes from the work at the Carnegie Institution of Washington, where plants were grown in field plots at three stations along an elevational gradient. Reciprocal differences in leaf width, date of flowering, width of basal rosettes, and vigor were documented in F_1 hybrids between subalpine and foothill races of *Potentilla glandulosa* (Clausen and Hiesey 1958). Reciprocal differences for biomass were found in hybrids between some populations of *Mimulus cardinalis*, and reciprocal differences in light-saturated photosynthetic rates were found in hybrids between *M. cardinalis* and *M. lewisii* (Hiesey et al. 1971).

Additional evidence of the effect of the cytoplasmic genome on hybrid fitness in nature comes from *Ipomopsis*. Campbell and Waser (2001) crossed *I. tenuituba* and *I. aggregata* in both directions, and planted their seeds into a range of habitats across a zone of intergradation between the species. In parental sites the proportion of F_1 hybrids that survived to age five or flowering was much higher when *I. aggregata* was the egg parent than when *I. tenuituba* was the egg parent. Notably, in hybrid sites there was no reciprocal difference.

Reciprocal differences in hybrid fitness also have been observed in greenhouse studies on *Iris* (Burke et al. 1998*a*). The growth and clonal reproduction of F_1 hybrids from the cross *I. fulva* × *I. hexagona* was roughly 40% less than hybrids from the reciprocal cross. Distorted segregation ratios in F_2 populations with the cytoplasmic backgrounds of each species also speak to viability differentials based on nucleocytoplasmic interactions (Burke et al. 1998*b*).

The cytoplasmic constitution of hybrids also may affect their fertility. For example, first-generation hybrids between *Phlox drummondii* and *P. cuspidata* are almost completely sterile when the former is the egg parent (Levin unpublished.). However, hybrids are about 25% pollen fertile and set some seeds when *P. cuspidata* is the egg parent.

If reciprocal hybrids have different properties, these differences are likely to be retained in allopolyploids sharing the same nuclear genomes but differing in cytoplasmic constitution. Experimental demonstrations that reciprocal differences indeed are retained would be most welcome. We know that traits governed by nuclear genes may be affected by polyploidy, but what happens to nucleocytoplasmic interactions is unclear.

The best information on this matter comes from the tetraploid *Tragopogon miscellus*, whose diploid progenitors are *T. dubius* and *T. pratensis* (Soltis et al. 1995). Ownbey and McCollum (1953) obtained reciprocal hybrids from these diploids. When *T. dubius* was the egg parent, ligules of the hybrids were long. When *T. pratensis* served as the egg parent, ligules of hybrids were short. Both cytoplasms are present in *T. miscellus*, which has had multiple origins. Some plants have the *T. dubius* cytoplasm and have long ligules; others have the *T. pratensis* cytoplasm and have short ligules (Soltis and Soltis 1989).

The cytoplasmic constitution of allopolyploids may affect their fertility as well as their phenotype. Magoon et al. (1958) made reciprocal crosses between the diploids *Solanum pinnatisectum* and *S. jamesii*, and observed that the pollen fertility was ca. 93% regardless of the direction of the cross. However, when both hybrids had their chromosome numbers doubled with colchicine, those with the *S. jamesii* cytoplasm had a mean pollen fertility of 31% versus 73% in polyploids with *S. pinnatisectum* cytoplasm.

The cytoplasmic genome of an allopolyploid also may affect the level of pairing between partially homologous chromosomes. When the cytoplasm of *Triticum aestivum* (breadwheat) was replaced with cytoplasms of *T. timopheevi*, *Aegilops sharonensis*, *Ae. variabilis*, and *A. juvenalis*, homoeologous pairing in breadwheat × triticale hybrids increased (Wang et al. 1999). Conversely, when the cytoplasm of *T. aestivum* was replaced with that of *Ae. bicornis* homoeologous pairing declined. The cytoplasm of *Ae. kotschyi* inhibited both homoeologous and homologous chromosome pairing, and in turn greatly diminished plant fertility.

The recombination potential of allopolyploids may be influenced by the chiasma frequency during gametogenesis, as well as by the level of homoeologous pairing; and cytoplasmic constitution of the polyploid may affect the former. For example, hybrids from the cross *Elymus canadensis* × *E. trachycaulis* have a chiasma frequency of ca. 26 per pollen mother cell (PMC), while hybrids from the reciprocal cross average ca. 22 chiasmata per PMC (Aung and Walton 1990).

Differences in the attributes of allopolyploids that are related to their cytoplasmic genomes need not be restricted to the phenotype, fertility, and chromosome pairing. Alternate haplotypes may differ in ecological tolerances and niche breadth, phenology, relationships with mutualists and pests, and more. Only future studies will tell.

Divergent fitnesses and ecological attributes due to the cytoplasmic variable may explain in part why a given allopolyploid has the cytoplasmic genome of one parent and not the other. The presence of one genome or another need not be an accidental by-product of the crossing pattern of parental species.

If one type of cytoplasm were superior to others, we would expect it to prevail in a range of nuclear backgrounds. With this in mind, it is interesting to consider the cytoplasmic background of 28 allopolyploids in the Triticeae (Poaceae) that contain the nuclear genome **St**. This genome is found in diploid *Pseudoroegneria*. It occurs in tetraploids, hexaploids, and octaploids in combination with the nuclear genomes of *Agropyron*, *Australopyrum*, *Hordeum*, and *Elymus*. Some of the polyploids have three different nuclear genomes. Based on the sequencing of the chloroplast *ndhF* gene, all of the polyploids with the **St** nuclear genome have the **St** cytoplasm of *Pseudoroegneria* (Redinbaugh et al. 2000). It is surprising that none of the other cytoplasms associated with the several other nuclear genomes are evident in the polyploids. Whether superiority or serendipity is the cause remains to be determined.

Epilogue

We are just beginning to appreciate that the coupling of nuclear and organelle genomes from different species may have a mild to profoundly negative effect on the viability, growth, and fertility of hybrids. Unfortunately, most of the information at hand is on domesticated plants. Our understanding of plant speciation would profit substantially if we performed reciprocal crosses (and backcrosses when possible) when investigating the character and strength of post-zygotic barriers in wild plants. Not only would we have some indication of whether nucleocytoplasmic incompatibility was involved, but we would also see whether character expression was transformed from one functional state to another. New functional states may be important in the success of hybrids or allopolyploids. Finally, we should not ignore the possible roles of cytoplasmic genes in adaptation, as demonstrated in *Chamaecrista fasciculata* (Galloway and Fenster 1999; 2001).

We would like to determine the relative strength of nuclear-nuclear incompatibilities and nucleocytoplasmic incompatibilities in the genesis of post-zygotic barriers. Comparing reciprocal first-generation hybrids is not sufficient, because the latter incompatibility may occur in the presence of the former. The two incompatibilities are not mutually exclusive. The formation of alloplasmic lines provides much more insight, because only one nuclear genome is present. Thus any fitness reduction could not be attributed to nuclear-gene conflicts. This approach is better than the production of cybrids, which contain only one nuclear genome, but a resident mitochondrial genome coupled with an alien chloroplast genome.

Nuclear-cytoplasmic interactions are poorly explored in polyploids. We could learn much from the genesis of reciprocal hybrids followed by chromosome doubling. Then we could see the effect of the cytoplasmic background on seed germination rates, survivorship to flowering, and the number of seeds per plant in natural sites. It would not be surprising to find that the net reproductive rates of populations (calculated from these variables) differed between cytotypes. We also would benefit from growing these synthetics along environmental gradients and establishing to what extent their ecological preferences and niche breadth are divergent.

Finally, there is the matter of cytoplasmic introgression. Cytoplasmic introgression has occurred or is occurring locally in many species combinations (Rieseberg et al. 1996). How is this possible given the obstacles which nucleocytoplasmic incompatibility may afford? Moreover, the rate of cytoplasmic replacement may be rapid. This has been shown in *Helianthus*, where the *H. petiolaris* cytoplasm has almost completely been replaced by *H. annuus* cytoplasm in some populations (Rieseberg and Soltis 1991). It would be most interesting to determine whether plants of *H. annuus* carrying *H. petiolaris* organelles have a fitness advantage over those with *H. annuus* organelles. These studies could be conducted in the field or under controlled culture conditions. It also would be interesting to see what obstacles there were to this introgression in a synthetic crossing program.

About half of the populations of *H. annuus* ssp. *texanus* examined have some plants with the chloroplast genome of *H. debilis* ssp. *cucumerifolius* (Rieseberg et al. 1990). As the former is thought to be a product of introgression involving *H. debilis*, it would be interesting to determine whether carriers of the alien cytoplasm were more or less fit than “pure” *H. annuus*. This would provide clues as to whether the alien cytoplasm is apt to increase in frequency or is in decline.

References

Adams, K.L., Qiu, Y-L, Stoutemyer, M., and Palmer, J.D. 2002. Punctuated evolution of mitochondrial gene content: high rate and variable rates of mitochondrial gene loss and transfer to the nucleus during angiosperm evolution. Proc. Natl. Acad. Sci. U.S.A. **99**: 9905–9912.

Asahi, T., Kumashiro, T., and Kubo, K. 1988. Constitution of mitochondrial and chloroplast genomes in male sterile tobacco obtained by protoplast fusion of *Nicotiana tabacum* and *Nicotiana debneyi*. Plant Cell. Physiol. **29**: 43–50.

Aung, T. and Walton, P.D. 1990. Morphology and cytology of the reciprocal hybrids between *Elymus trachycaulis* and *E. canadensis*. Genome **33**: 123–130.

Babiychuk, E., Schantz, R., Cherep, N., Weil, J.-H., Gleba, Y. and Kushmir, S. 1995. Alterations in chlorophyll a/b binding proteins in Solanaceae cybrids. Mol. Gen. Genet. **249**: 648–654.

Barraclough, T.G., and Savolainen, V. 2001. Evolutionary rates and species diversity in flowering plants. Evolution **55**: 677–683.

Beavis, W.D., and Frey, K.J. 1987. Expression of nuclear-cytoplasmic interactions and heterosis in quantitative traits of oats (*Avena* spp.). Euphytica **36**: 877–886.

Bousquet, J., Strauss, S.H., Doerkesen, A.H., and Price, R.A. 1992. Extensive variation in evolutionary rate of *rbcL* gene sequences among seed plants. Proc. Natl. Acad. Sci. U.S.A. **89**: 7844–7848.

Burke, J.M., Carney, S.E., and Arnold, M.L. 1998*a*. Hybrid fitness in the Louisiana irises:analysis of parental and F_1 performance. Evolution **52**: 37–43.

Burke, J.M., Voss, T.J., and Arnold, M.L. 1998*b*. Genetic interactions and natural selection in Louisiana irises. Evolution **52**: 1304–1310.

Campbell, D.R. and Waser, N.M. 2001. Genotype-by-environment interaction and the fitness of plant hybrids in the wild. Evolution **55**: 669–676.

Cheptou, P.-O., Berger, A., Blanchard, A., Collin, C., and Escarre, J. 2000. The effect of drought stress on inbreeding depression in four populations of the Mediterranean outcrossing plant *Crepis sancta* (Asteraceae). Heredity **85**: 294–302.

Clausen, J., and Hiesey, W.M. 1958. Experimental studies on the nature of species. IV. Genetic structure of ecological races. Publ. No. 615. Carnegie Institution of Washington.

Desplanque, B., Viard, F., Bernard, J., Forcioli, D., Saumitou-Laprade, P.J., Cugen, J., and Van Dijk, H. 2000. The linkage disequilibrium between chloroplast DNA and mitochondrial DNA haplotypes in *Beta vulgaris* ssp. *maritima* (L): the usefulness of both genomes for population genetic studies. Mol. Ecol. **9**: 41–154.

Dhitaphichit, P., Jones, P., and Keane, E.M. 1989. Nuclear and cytoplasmic gene control of resistance to loose smut (*Ustilago tritici* (Pers.) Rostr.) in wheat. Theor. Appl. Genet. **78**: 897–903.

Doebley, J. 1990*a*. Molecular evidence and the evolution of maize. Econ. Bot. **44** (supplement): 6–27.

Doebley, J. 1990*b*. Molecular systematics of *Zea* (Gramineae). Maydica **35**: 143–150.

Dumolin-Lapègue, S., Pemonge, M.-H., and Petit, R.J. 1998. Association between chloroplast and mitochondrial lineages in oaks. Mol. Biol. Evol. **15**: 1321–1331.

Edwards, J.W., and Coors, J. G. 1996. Teosinte cytoplasmic genomes: II. Performance of maize hybrids with teosinte cytoplasms. Crop Sci. **36**: 1092–1098.

Fishman, L., Kelly, A.J., Morgan, E., and Willis, J.H. 2001. A genetic map in the *Mimulus guttatus* species complex reveals transmission ratio distortion due to heterospecific interactions. Genetics **159**: 1701–1716.

Galloway, L.F., and Fenster, C.B. 1999. The effect of nuclear and cytoplasmic genes on fitness and local adaptation in an annual legume, *Chamaecrista fasciculata*. Evolution **53**: 1734–1743.

Galloway, L.F., and Fenster, C.B. 2001. Nuclear and cytoplasmic contributions to intraspecific divergence in an annual legume. Evolution **55**: 488–497.

Gaut, B.S., Clark, L.G., Wendel, J.F., and Muse, S.V. 1997. Comparisons of the molecular evolutionary process at *rbcL* and *ndhF* in the grass family (Poaceae). Mol. Biol. Evol. **14**: 769–777.

Grant, V. 1956. The genetic structure of races and species in *Gilia*. Adv. Genet **8**: 55–87.

Gray, J.C., Sullivan, J.A., Wang, J.-H., Jerome, C.A., and Maclean, D. 2003. Coordination of plastid and nuclear gene expression. Phil. Trans. Roy. Soc. Lond. B Biol. Sci. **358**: 135–145.

Hanson, M.R. 1991. Plant mitochondrial mutations and male sterility. Annu. Rev. Genet. **25**: 461–486.

Hiesey, W.M., Nobs, M., and Björkman, O. 1971. Experimental studies on the nature of species. V. Biosystematics, genetics, and physiological ecology of the *Erythranthe* section of *Mimulus*. Publ. No. 628. Carnegie Institution of Washington.

Ikeda, T.M., Shibasaka, M., and Tsunewaki, K. 1994. Respiratory activities, cytochrome composition, and cytochrome c oxidase subunit II levels of mitochondria from alloplasmic wheats having *Aegilops* cytoplasms. Plant Cell Physiol. **35**: 779–784.

Jan, C.C. (1992). Cytoplasmic-nuclear gene interaction for plant vigor in *Helianthus* species. Crop Sci. **32**: 320–323.

Jones, P, Keane, E.M., and Osborne, B.A. 1998. Effects of alien cytoplasmic variation on carbon assimilation and productivity in wheat. J. Exp. Bot. **49**: 1519–1528.

Khehra, A., and Bhalla, S.K. 1976. Cytoplasmic effects on quantitative characters in maize (*Zea mays* L.). Theor. Appl. Genet. **47**: 271–274.

Komaki, K. 1982. Inbreeding depression and concealed deleterious mutations in buckwheat populations, *Fagopyrum esculentum*. Jpn. J. Genet. **57**: 361–370.

Kushnir, S.G., Babichuck, E., Bannikova, M., Momot, V., Komarnitsky, I., Cherep, N., and Gleba, Y. 1991. Nucleo-cytoplasmic incompatibility in cybrid plants possessing an *Atropa* genome and a *Nicotiana* plastome. Mol. Gen. Genet. **225**: 225–230.

Leon, P., Arroyo, A., and Mackenzie, S. 1998. Nuclear control of plastid and mitochondrial development in higher plants. Annu. Rev. Plant Physiol. Plant Mol. Biol. **49**: 453–480.

Levin, D.A. 2000. The origin, expansion, and demise of plant species. Oxford University Press, New York.

Levin, D.A. 2002. The role of chromosomal change in plant evolution. Oxford University Press, New York.

Libby, W.J., McCutchan, B.G., and Millar, C.I. 1981. Inbreeding depression and selfs of redwood. Silvae Genet. **30**: 15–25.

Magoon, M. L., Cooper, D.C., and Hougas, R.W. 1958. Induced polyploids in solanums and their crossability with *S. tuberosum*. Bot. Gaz. **119**: 224–233.

Medgyesy, P. 1994. Cybrids — transfer of chloroplast traits through protoplast fusion between sexually incompatible Solanaceae species. *In* Somatic hybridization in crop improvement. *Edited by* Y.P.S Bajaj. Springer-Verlag, Berlin. pp. 72–85.

Michaelis, P. 1954. Cytoplasmic inheritance in *Epilobium* and its theoretical significance. Adv. Genet. **6**: 287–401.

Nakamura, C, Kasai, K., Kubota, Y., Yamagami, C., Suzuki, T., and Mori, N. 1991. Cytoplasmic diversity in alloplasmic common wheats with cytoplasms of *Triticum* and *Aegilops* revealed by photosynthetic and respiratory characteristics. Jpn. J. Genet. **66**: 471–483.

Oehlkers, F. 1964. Cytoplasmic inheritance in the genus *Streptocarpus* Lindley. Adv. Genet **12**: 329–370.

Olsen, M. S., and McCauley, D. E. 2000. Linkage disequilibrium and phylogenetic congruence between chloroplast and mitochondrial haplotypes in *Silene vulgaris*. Proc. R. Soc. Lond. B Biol. Sci. **267**: 1810–1808.

Ownbey, M., and McCollum, G.D. 1953. Cytoplasmic inheritance and reciprocal amphiploidy in *Tragopogon*. Am. J. Bot. **40**: 788–796.

Palmer, J.D., Adams, K.L., Cho, Y., Parkinson, C.L., Qiu, Y.-L., and Song, K. 2000. Dynamic evolution of plant mitochondrial genomes: mobile genes and introns and highly variable mutation rates. Proc. Natl. Acad. Sci. U.S.A. **97**: 6960–6966.

Perl, A., Aviv, D., and Galun, E. 1991. Nuclear-organelle interaction in *Solanum*: interspecific cybridizations and their correlation with a plastome dendrogram. Mol. Gen. Genet. **228**: 193–200.

Pollak, P.E 1991. Cytoplasmic effects on components of fitness in tobacco cybrids. Evolution **45**: 785–790.

Redinbaugh, M.G., Jones, T.A., and Zhang, Y. 2000. Ubiquity of the **St** chloroplast genome in **St**-containing Triticeae polyploids. Genome **43**: 846–852.

Reboud, X., and Zeyl, C. 1994. Organelle inheritance in plants. Heredity **72**: 132–140.

Rieseberg, L.H., Beckstrom-Sternberg, S., and Doan. K. 1990. *Helianthus annuus* ssp. *texanus* has chloroplast DNA and nuclear ribosomal RNA genes of *Helianthus debilis* ssp. *cucumerifolius*. Proc. Natl. Acad. Sci. U.S.A. **87**: 593–597.

Rieseberg, L.H.. and Soltis, D.E. 1991. Phylogenetic consequences of cytoplasmic gene flow in plants. Evol. Trends Plants **5**: 65–84.

Rieseberg, L.H., Whitton, J., and Linder, C.R. 1996. Molecular marker incongruence in plant hybrid zones and phylogenetic trees. Acta Bot. Neerl. **45**: 243–262.

Saumitou-Laprade, P., Cuguen, J., and Vernet, P. 1994. Cytoplasmic male sterility in plants: molecular evidence and nucleocytoplasmic conflict. Trends Ecol. Evol. **9**: 431–435.

Schmitz, U.K. 1988. Dwarfism and male sterility in interspecific hybrids of *Epilobium*. 1. Expression of plastid genes and the structure of the plastome. Theor. Appl. Genet. **75**: 350–356.

Schmitz, U.K., and Michaelis, G. 1988. Dwarfism and male sterility in interspecific hybrids of *Epilobium*. 2. Expression of mitochondrial genes and structure of the mitochondrial DNA. Theor. Appl. Genet. **76**: 565–569.

Schmitz-Linneweber, C., Regel, R., Du, T.G., Hupfer, H., Herrmann, R.G., and Maier, R.M. 2002. The plastid chromosome of *Atropa belladonna* and its comparison with that of *Nicotiana tabacum*: the role of plastid RNA editing in generating divergence in the process of plant speciation. Mol. Biol. Evol. **19**: 1602–1612.

Schuster, W., and Brennicke, A. 1994. The plant mitochondrial genome: physical structure, information content, RNA editing, and gene migration to nucleus. Annu. Rev. Plant Physiol. Mol. Biol. **45**: 61–78.

Senthil, N., Ramasamy, P., and Falzlulluh Khan, A.K. 1998. Fertility restoration and heterosis involving different cytoplasms in sorghum (*Sorghum bicolor* L.) Moench hybrids. J. Genet. Plant Breed. **52**: 339–342.

Soltis, D.E., and Soltis, P.S. 1989. Allopolyploid speciation in *Tragopogon*: insight from chloroplast DNA. Am. J. Bot. **76**: 1119–1124.

Soltis, P.S., Plunkett, G.M., Novak, S.J., and Soltis, D.E. 1995. Genetic variation in *Tragopogon* species: additional origins of the allotetraploids *T. mirus* and *T. miscellus* (Compositae). Am. J. Bot. **82**: 1329–1341.

Suguira, M. 1992. The chloroplast genome. Plant Mol. Biol. **19**: 149–169.

Tsitrone, A., Kirkpatrick, M., and Levin, D.A. 2003. A model of chloroplast capture. Evolution **57**: 1776–1782.

Wang, G.-Z., Miyashita, N.T. and Tsunewaki, K. 1997. Plasmon analyses of *Triticum* (wheat) and *Aegilops*: PCR-single strand conformational polymorphism (PCR SSCP) analyses of organellar DNAs. Proc. Natl. Acad. Sci. U.S.A. **94**: 14570–14577.

Wang, T., Liu, W., He, C., Zhang, G., Shi, R., Xu, X., and Li, J. 1999. Effect of alien cytoplasms on meiosis in the F_1 hybrids between alloplasmic lines of *Triticum aestivum* cv. *Chinese spring* and rye or triticale. Cytologia **64**: 285–291.

Washington, W.J., and Maan, S.S. 1974. Disease reaction with alien cytoplasms. Crop Sci. **14**: 903–905.

Wendel. J.F., and Doyle, J.J. 1998. Phylogenetic incongruence: window into genome history and molecular evolution. *In* Molecular Systematics of Plants. II. DNA Sequencing. *Edited by* D.E. Soltis, P.S. Soltis, and J.J. Doyle. Kluwer Academic Publishers, Boston, MA. pp. 265–296.

Wilson, P., and Driscoll, C.J. 1983. Hybrid wheat. *In Heterosis*. *Edited by* R. Frankel. Springer-Verlag, New York. pp. 94–123.

Yadav, O.P. 1999. Heterosis and combining ability in relation to cytoplasmic diversity in pearl millet. Ind. J. Genet. Plant Breed. **59**: 445–450.

Zubko, M.K., Zubko, E.I., Ruban, A.V., Adler, K., Mock, H-P., Misera, S., Gleba, Y.Y., and Grimm, B. 2001. Extensive developmental and metabolic alterations in cybrids *Nicotiana tabacum* (+*Hyoscyamus niger*) are caused by complex nucleo cytoplasmic incompatibility. Plant J. **25**: 627–639.

6
Discussion report: answered and unanswered questions in plant adaptation

Brian C. Husband

"The problems faced by pre- and post-genomic genetics are therefore much the same — they all involve bridging the chasm between genotype and phenotype."

Brenner 2000

Introduction

Biologists have been studying plant adaptation within an evolutionary context for over 140 years. The evolutionary synthesis of the last century was an important turning point in the study of adaptation in that it marked the integration of genetics with the fields of systematics and ecology, and fueled much debate regarding the approaches to studying adaptation (Gould and Lewontin 1976). Now, in the age of molecular biology and genomics, it is useful again to take stock and reflect on the successes and shortcomings in our understanding of adaptive mechanisms and their products. Our comprehension might reasonably be measured by our ability to predict the pressures imposed by the selective environment, the agents of selection and their relative strengths, and the adaptive response to selection in plant populations. We would further hope to make predictions about the genetic and genomic architecture of adaptive traits, including the roles of standing genetic variation versus novel mutation, the contributions of structural and regulatory elements in genetic changes, as well as the role of epigenetic phenomena. As summarized below, it is clear that we have made strides in these endeavours. At the same time, the state of our understanding is humbling and increasingly so as new techniques and avenues for research continue to unfold before us.

What follows is a synthesis and summary of a discussion on "Answered and Unanswered Questions in Plant Adaptation" that took place among participants of the Molecular Genetics and Ecology of Plant Adaptation conference held in Vancouver, Canada, December 11–13, 2003. For ease of reading, the ideas discussed are organized into three conceptual categories, concerned with (1) phenotypic variation and the chasm between genotype and phenotype, (2) the adaptive process, and (3) the products of adaptive evolution. For each section, we begin with a summary of progress towards an understanding of the topic. We then point to some areas in which our knowledge is incomplete or lacking. We have attempted to integrate the contributions of participants by providing some general context and, in some cases, we extend the arguments beyond what occurred in the actual discussion.

Brian C. Husband. Department of Botany, University of Guelph, Guelph, Ontario N1G 2W1, Canada.
Correct citation: Husband, B.C. 2004. Discussion report: answered and unanswered questions in plant adaptation. *In* Plant Adaptation: Molecular Genetics and Ecology. Proceedings of an International Workshop held December 11–13, 2002, in Vancouver, British Columbia, Canada. *Edited by* Q.C.B. Cronk, J. Whitton, R.H. Ree, and I.E.P. Taylor. NRC Research Press, Ottawa, Ontario. pp. 38–41.

Phenotypic variation and the "chasm" between genotype and phenotype

What we know: long tradition of trait analysis in botany

Biologists have long been studying the sources and determinants of phenotypic variation, as it is ultimately the fuel for adaptive divergence. This tradition is rooted in the rich history of morphological study in plants, including the study of intra- and inter-specific variation. Darwin (1862) and Haberlandt (1884) initiated the study of evolutionary morphology. For instance, Haberlandt noted the widespread occurrence of a particular sort of leaf anatomy, characterized by a wreath-like arrangement of mesophyll and bundle sheath cells around the major veins, which he called "Kranz". He noted that this morphology occurred in certain fast-growing grasses and other plants that we now know to have evolved C4 photosynthesis. While the phenotypic and ecological correlates of C3 versus C4 photosynthesis are well studied (e.g., Ehleringer et al. 1999), recent studies are now elucidating the genes and developmental pathways that distinguish these mechanisms (e.g., Rossini et al. 2001). We have also become reasonably adept at characterizing variation, although the problem of atomizing complex traits remains a recognized difficulty. At the same time, we are more aware of the interconnectedness and multivariate nature of morphological, physiological and developmental characters in plants, and have succeeded in incorporating aspects of this into our research methodologies. For example, rarely do we consider selection acting on a single character without considering indirect pressures operating via correlated characters. Furthermore, concepts such as 'phenotypic integration' and 'reaction norms' are being used to reflect variation in multivariate phenotype and patterns of selection on complex characters. A much stronger theoretical framework is now available for studying phenotypic plasticity (Schlichting and Murren, this volume, Chapter 4), an area in which plants and animals differ markedly. In animals, the developmental trajectory to the adult phenotype is unitary, whereas in modular plants, the developmental trajectory involves a continuous feedback loop between form and environment via signal transduction pathways.

There is also a long tradition of classical genetics (and later quantitative genetics), which seeks to link phenotype with genotype. Indeed, many of the benchmark studies of classical and quantitative genetics involve plants. More recently, a variety of approaches have been developed to test the role of selection in shaping a particular character, from the functional analyses of evolutionary ecology to the comparisons of population divergence among neutral and quantitative loci to the evaluation of synonymous and nonsynonymous base substitutions in gene sequences. In addition, microarray technology will likely have a similar if not larger impact, as it enables one to evaluate the patterns of gene expression associated with particular traits and environments. This technique is becoming increasingly accessible for a wide variety of model species.

What we don't know: chasm between genotype and phenotype

However, despite all this, there is still a chasm between phenotype and genotype. For instance, some ecologically important phenotypes are complex and polygenic (e.g., Lin and Ritland 1997) — although interestingly, very often there is a major QTL explaining some 30% of the variation, accompanied by very many QTLs of minor effect, as in tomato (Lipman and Tanksley 2001). Although the genetic bases of such traits are amenable to quantitative genetic description, most are 'black boxes' with respect to their genetic and molecular basis – that is, the number, location, and mode of action of genes underlying their expression. A more precise understanding of the molecular basis of phenotypic variation will be necessary for testing hypotheses regarding the pathway and tempo of adaptive evolution, and for predicting the constraints that may regulate phenotypic diversity. A better understanding of the molecular causes of phenotypes will also provide, for the first time, a means of forecasting how phenotypes will be influenced by the environment and genetic background.

Techniques such as quantitative trait mapping, as exemplified by speakers at this conference (Olsen and Purugganan, this volume, Chapter 7; Borevitz, this volume, Chapter 8; Reboud et al., this volume, Chapter 17), are already providing significant insights into the number, location and action of loci that are associated with adaptive phenotypes. As powerful as these techniques appear to be, concern was expressed about the resolution of such methods and their ability to detect loci of minor effect. Microarray technology, although formidable, presents considerable problems. Many genes may be found to have altered expression between adaptive genotypes or phenotypes, but which ones are truly important, and which are downstream consequences of changes at selected loci is impossible to tell without detailed knowledge of interacting gene networks.

Even if we can adequately describe the phenotype and its genic and molecular basis, it is a separate challenge to determine whether traits have undergone selection in the past or currently bear an association with fitness (Elle, this volume, Chapter 14; Matthews, this volume, Chapter 18). Beyond the association of traits and fitness, we hope to understand the functional relationships between variation in phenotype and fitness, that is, what is it about a particular character state that confers a fitness advantage or disadvantage? To understand this requires some understanding of the agents of selection – those aspects of the selective environment that create variation in fitness (Elle, this volume, Chapter 14). Clearly ecological studies involving experimental manipulations of genotypes and of the environment will continue to be integral beyond the comparative approaches commonly used to quantify selection in populations today.

The adaptive process: ecomolecular interaction

What we know: a good theoretical framework

Arguably, our greatest achievements in understanding adaptation thus far have come from our understanding of the process: natural selection. Specifically, we understand the significance of phenotypic variation, variation in heritability, and the phenotype-fitness correlation for the action of natural selection, and we can reasonably measure these components in natural populations. More importantly, population genetic and quantitative genetic models enable us to quantify the relationships between fitness and a suite of characters and, hence, estimate the magnitude and direction of selection (phenotype-fitness surface) acting either directly or indirectly on plant traits. Numerous studies have been conducted, some descriptive and others in an experimental context, which not only confirm that selection occurs, but also provide insights into its intensity, complexity and heterogeneity. Additional quantitative genetic models (e.g., breeders equation (Reale, McAdam, Boutin and Berteaux 2003), G matrix (Jones, Arnold and Burger 2003; Agrawal, Brodie and Rieseberg 2001; Phillips, Whitlock and Fowler 2001)) have allowed us to forecast the rate and extent of character change under a given selective regime, and depending on the genetic variances and covariances among associated traits. Under some circumstances, these models have performed remarkably well, at least over the short term.

Wright, in his shifting balance theory, provided a compelling conceptual picture of populations evolving on an adaptive landscape with multiple fitness peaks. He envisioned drift acting as a facilitator to move populations across low fitness valleys (maladapted gene combinations), such that selection could then push them up a new peak. Although this model has received recent attention from mathematical biologists, empirical evidence for this idea is weak or incomplete.

What we don't know: tempo and ecological reality of adaptation

Despite having documented the existence of natural selection, we have yet to adequately understand the role of this process in producing the phenotypic radiations observed within many plant groups. Key questions include: at what rate does adaptation occur? What determines the tempo of adaptation? That is, to what extent is adaptive divergence limited by mutation and the content of the genome rather than ecology and the selective pressures imposed on organisms? If the latter, what ecological circumstances favour adaptive radiations? Adaptive radiations on islands versus continental habitats suggest that ecological opportunity, afforded by young islands, may be extremely important in promoting diversification. Additional investigations into the effect of genomic changes such as polyploidy may help to further understand the relative importance of genes versus ecology in adaptive radiation (Husband, this volume, Chapter 15).

A related question concerns the relative importance of selection versus stochastic forces in adaptation. Moreover, there is some indication that biologists are beginning to explore these processes beyond the strict criteria of the shifting balance theory. For example, circumstances in which frequency- or density-dependent thresholds modulate adaptation to different equilibria may be important targets of study for evaluating the interactions between selection and stochastic processes. In addition, others are re-examining the nature of the adaptive landscape, and considering, for example, the implications of a landscape with fitness ridges connecting adaptive peaks, on which drift and mutation operate (Belotte, Curien, Maclean and Bell 2003; Cruzan and Rhode, this volume, Chapter 9), or fitness landscapes that themselves evolve as other species evolve (Otto and Nuismer 2004). Approaches such as these may not lead to a resolution of the role of shifting balance but rather identify new mechanisms by which drift may contribute to significant evolutionary change. Despite the fertile effect that the concept of fitness landscapes have had on the subject we still don't know what any individual fitness landscape looks like in detail!

In essence, our knowledge of the specific pathways by which certain adaptations arise is far from clear. More attention is required to the steps involved in evolutionary transitions from one state (maladaptation) to another (adaptation) and the ecological and genetic forces that regulate the transition. Studies of invading species, or species colonizing new environments may be particularly useful for studying the process by which plant populations, initially in a maladaptive state, respond demographically and genetically to novel environments.

In addition, much of what we know about adaptation is cobbled together from individual studies, each on a different system, with its own idiosyncrasies and unique historical, genetic and environmental circumstances. This is to a large extent inevitable, and indeed may be no bad thing as an understanding of the natural history of a system is an essential part of understanding the adaptive processes of an individual system. However, this does create a problem of "seeing the wood for the trees" and generating a general framework for the understanding of adaptation. Two approaches can help this. What is needed are opportunities to study adaptation under controlled conditions in a replicated design within single biological systems. Experimental evolution using organisms with rapid generation times may be most valuable in revealing cues to the general patterns of the adaptive process. Secondly, the pooling of multiple studies, to look for overall generalizations will also be increasingly relevant as more individual studies enter the literature. An example of a pooled study is that of Rieseberg et al. (2002) who used the quantitative trait locus (QTL) sign test for 572 traits from 86 studies to reveal significantly fewer antagonistic QTLs than expected under neutrality. This is good evidence for the importance and ubiquity of natural selection in divergent evolution. If the direction of selection at QTLs for a trait are all in the same direction then selection is implied, otherwise there would be a random mixture of co-directional and antagonistic effects.

Curiously, while models showing how neutral variation can exist are well worked out (Rand 1994), recent molecular studies modeling gene sequence evolution between different species have not found many good, unequivocal examples of neutrality. Instead, a mixture of purifying and positive selection appears to be common (Ree et al. 2004). In many cases genes have been shown to be under selection yet we don't know for sure what the agent of selection is, although speculation is easy. Neutrality may be more of a reality at the intraspecific population level. Certainly more data are needed to test null hypotheses of neutrality in a variety of systems and evolutionary scales.

At the other end of the evolutionary scale, that of deep evolution, studies of comparative gene sequences may also be illuminating, if the relevant genes can be identified. Major transitions of evolution, such as the evolution of the leaf, are commonly believed to be adaptively driven, but without any certain knowledge, as the ancient environments and organisms are long gone. However molecular evidence for selection on developmental pathways responsible for major transitions would be illuminating. In terms of deep evolution then, more data are also needed, preferably in the form of completely sequenced genomes. When there are 30–50 completely sequenced plant genomes spread out through the plant kingdom, from algae to mosses, lycopods, ferns and primitive seed plants, then it will be relatively easy to study deep evolution (Cronk 2002).

The products of adaptive evolution

What we know: syndromes, patterns, generalities

Following on the previous points, a common goal for many is to determine whether there are generalities in adaptation, not just in its mode of action but also in the responses to selection. At the phenotypic level, the existence of good phylogenies and methods for comparative analysis of traits in a phylogenetic context (Harvey and Pagel 1991) have given us a firm basis for generalization. Broadly speaking homology is seen as a unique evolutionary event, useful in that it allows us to reconstruct evolutionary history, but about which generalizations are impossible. Homoplasy, on the other hand, is the result of running the evolutionary experiment several times with similar results — exposing generalities of evolution. It is now possible to map pollination syndromes, and other examples of parallel evolution, onto phylogenies to make quantitative statements about evolutionary forces. Another aspect of phylogenetics, which has revolutionized our understanding of evolution, is the realization from island studies that morphological evolution and radiations can occur very rapidly (Baldwin and Sanderson 1988). However, the adaptive nature of such radiations, although, widely assumed, is purely speculative.

What we don't know: genetic architecture and cryptic adaptation

Although there has been much progress with the analysis of traits, and the pattern of trait evolution is now evident, the ecological, genetic and developmental determinants of character evolution are often still obscure and wanting experimental study. To what extent is adaptation repeatable versus conditional upon the genetic background? And, are there such things as syndromes of characters associated with particular selective agents? Did the phenotype arise for the purpose it currently performs or has its function been modified?

At the genetic level, the characteristics of the "adapted genome" are still woefully obscure. We still have too little information on the molecular basis of phenotypic variation. And we know little about how this significant variation is disposed within the genome, and how varying genes interact: i.e., the physical, epistatic and epigenetic components of "genetic architecture". The significance of locations of genes and the influence of genome duplication on the creation and maintenance of variation will be important topics for future study.

Conclusion

Adaptation refers to the process by which populations increase their mean fitness. The adaptive phenotype is one that confers a high relative fitness in a particular environment. Through the evolutionary synthesis and beyond, we have made great strides in our understanding. We understand the conditions for the operation of selection and can use quantitative models to predict over the short term how populations may respond to selection. In addition we have developed methodologies for understanding adaptive diversity through the use of phylogenies, which can assist in identifying evolutionarily independent events. However, it will take some time before this discipline is transformed from being a retrospective science to a truly predictive one. For that we must develop a clear understanding of the genetic basis of traits important in adaptation and the relationship between phenotype, genotype and environment. Furthermore, we need to place this molecular program into an ecological context, so as to identify the genes that influence fitness and how selection acts on them. These directions are currently being explored, primarily through the application of molecular and genomic tools to a limited number of model organisms. The search for generality in different ecological circumstances may depend on the availability of such techniques to a broader range of organisms.

References

Agrawal, A.F., Brodie, E.D. III, and Rieseberg, L.H. 2001 Possible consequences of genes of major effect: transient changes in the G-matrix. Genetica **112–113**: 33–43.

Baldwin, B.G. and Sanderson, M.J. 1998. Age and rate of diversification of the Hawaiian silversword alliance (Compositae). Proc. Natl. Acad. Sci. U.S.A. **95**: 9402–9406.

Cronk, Q.C.B. 2001. Plant evolution and development in a post-genomic context. Nat. Rev. Genet. **2**: 607–619.

Darwin, C. 1862. The various contrivances by which orchids are fertilized by insects. 2nd ed. J. Murray, London.

Haberlandt, G. 1884. Physiologische Pflanzenanatomie (Physiological Plant Anatomy). Engelmann, Leipzig.

Harvey, P.H., and Pagel, M.D. 1991. The Comparative Method in Evolutionary Biology. Oxford Series in Ecology and Evolution, vol. 1. Oxford University Press, London.

Jones, A.G, Arnold, S.J., and Burger R. 2003. Stability of the G-matrix in a population experiencing pleiotropic mutation, stabilizing selection, and genetic drift. Evolution **57**: 1747–1760.

Lin, J.-Z., and Ritland, K. 1997. Quantitative trait loci differentiating the outbreeding *Mimulus guttatus* from the inbreeding *M. platycalyx*. Genetics **146**: 1115–1121.

Lippman, Z., and Tanksley, S.D. 2001. Dissecting the genetic pathway to extreme fruit size in tomato using a cross between the small-fruited wild species *L. pimpinellifolium* and *L. esculentum*, var. Giant Heirloom. Genetics **158**: 413–422.

Nesbitt, T.C., and Tanksley, S.D. 2001. fw2.2 directly affects the size of developing tomato fruit with secondary effects on fruit number and photosynthate distribution. Plant Physiol. **127**: 575–583.

Otto, S.P., and Nuismer, S.L. 2004. Species interactions and the evolution of sex. Science (Washington, D.C.) **304**: 1018–1820.

Phillips, P.C., Whitlock, M.C., and Fowler, K. 2001. Inbreeding changes the shape of the genetic covariance matrix in *Drosophila melanogaster*. Genetics **158**: 1137–1145.

Rand, D.M., Dorfsman, M.L., and Kann, L.M. 1994. Neutral and non-neutral evolution of *Drosophila* mitochondrial DNA. Genetics **138**: 741–756.

Reale, D., McAdam, A.G., Boutin, S., and Berteaux, D. 2003. Genetic and plastic responses of a northern mammal to climate change. Proc. R. Soc. Lond. B Biol. Sci. **270**: 591–596.

Ree, R.H., Citerne, H.L., Lavin, M., and Cronk, Q.C. 2004. Heterogeneous selection on LEGCYC paralogs in relation to flower morphology and the phylogeny of *Lupinus* (Leguminosae). Mol. Biol. Evol. **21**: 321–331.

Rieseberg, L.H., Widmer, A., Arntz, A.M., and Burke, J.M. 2002. Directional selection is the primary cause of phenotypic diversification. Proc. Natl. Acad. Sci. U.S.A. **99**: 12242–12245.

Rossini, L., Cribb, L., Martin, D. J., and Langdale, J.A. 2001. The maize Golden2 gene defines a novel class of transcriptional regulators in plants. Plant Cell **13**: 1231–1244.

APPROACHES TO THE STUDY OF PLANT ADAPTATION

7
Plant population genomics, linkage disequilibrium mapping, and the genetics of adaptation

Kenneth M. Olsen and Michael D. Purugganan

Abstract: Understanding the genetic architecture of adaptive variation in species remains a major challenge in plant evolutionary genetics. Linkage disequilibrium (LD) mapping techniques, including candidate gene association analysis, provide new strategies for the identification and isolation of genes that underlie phenotypic variation in plant species. These LD mapping technologies require detailed information on the levels and patterns of single nucleotide polymorphisms (SNPs) in genomes, the evolutionary forces that structure this SNP variation within species, and genome-wide patterns of linkage disequilibrium. Studies in two plant systems — maize and *Arabidopsis thaliana* — have provided insights into the evolutionary forces that shape plant SNP variation. Wider studies of SNP variation in plant systems may permit the development of coherent LD mapping strategies that can help bridge the genotype-phenotype gap in studies of adaptive variation.

Introduction

Understanding the genetic basis of adaptation remains a central objective and a significant challenge for evolutionary genetics at the dawn of the 21st century. Over the last decade, intensive efforts have been made in characterizing intraspecific levels and patterns of molecular diversity in genes, and it is these studies that have provided us with the greatest insights into the nature of the evolutionary forces that shape gene structure. The study of nucleotide variation at specific genes, including those that underlie physiological and developmental phenotypes, has given us unprecedented glimpses into the nature of plant genetic variation and the histories of plant populations.

At the organismal level, phenotypic variation has been amply documented by over a century of evolutionary and ecological research, and some of this variation has also been shown to be adaptive in nature. The development of effective approaches for the isolation and functional characterization of genes underlying this phenotypic variation remains a major objective of plant genome research (Tanksley 1993; Remington et al. 2001; Paterson 2002). Much of this variation is quantitative in nature, and identifying the genes or quantitative trait loci (QTLs) underlying phenotypic variation lies at the heart of not only evolutionary genetics, but also studies of functional genomics and plant breeding (Tanksley 1993; Mauricio 2001; Remington et al. 2001; Paterson 2002). In this paper, we describe an emerging approach for identifying the genetic basis of phenotypic variation, linkage disequilibrium (LD) mapping, and the relationship of this approach to the more commonly used approach of QTL linkage mapping.

Identifying the genetic basis of phenotypic variation: Linkage-based QTL mapping

Linkage-based QTL mapping has become a standard method for examining the genetic architecture underlying phenotypic variation. This approach is now routinely employed in species for which genetic maps are available, including both crop species and wild species. QTL mapping is essentially a statistical method for localizing chromosomal regions where genetic variation can be associated with measurable phenotypic variation. Two individuals showing phenotypic differences for a trait are crossed. Their progeny are then either selfed, backcrossed to one parent, or crossed with each other to generate a large mapping population, in which the parental alleles are shuffled into different genetic backgrounds. Associations between phenotypes and genetic marker variation are identified, and the specific genomic locations of the genetic markers can then be inferred by comparison with the genetic map.

For addressing questions on the genetics of adaptive divergence, one of the most successful applications of QTL mapping has involved crop species and traits selected upon during domestication. Foundational studies using the cereal crop maize and its wild relatives elucidated the genetic bases of a variety of domestication traits, including plant architecture and inflorescence developmental traits (Dorweiler et al. 1993; Doebley et al. 1997). Many crops share a suite of 'domestication traits'. Among cereal crops, for example, there has been parallel selection for increased seed size and loss of seed disarticulation (shattering) upon ripening. As QTL studies have spread beyond maize to include other cereal crop species, it has become possible to assess the degree to which parallel selection pressures have acted on orthologous QTLs in related species. A substantial degree of QTL shar-

Kenneth M. Olsen and Michael D. Purugganan[1]. Department of Genetics, Box 7614, North Carolina State University, Raleigh, NC 27695, U.S.A.
Correct citation: Olsen, K.M., and Purugganan, M.D. 2004. Plant population genomics, linkage disequilibrium mapping, and the genetics of adaptation. *In* Plant Adaptation: Molecular Genetics and Ecology. Proceedings of an International Workshop held December 11–13, 2002, in Vancouver, British Columbia, Canada. *Edited by* Q.C.B. Cronk, J. Whitton, R.H. Ree, and I.E.P. Taylor. NRC Research Press, Ottawa, Ontario. pp. 45–52.

[1]Corresponding author e-mail address: michaelp@unity.ncsu.edu

ing has been observed in cereal crops such as maize, rice and sorghum (Paterson et al. 1995). A similar pattern of QTL sharing is also seen in crops of the Solanaceae, including tomato, eggplant, pepper, and potato (Doganlar et al. 2002). Thus, it appears that independent domestication events may have involved artificial selection on many of the same QTLs.

Among wild species, QTL mapping in the model organism *Arabidopsis thaliana* has provided important inferences into the genetics of adaptation. QTLs have been identified for a variety of life history traits that, if underlying natural phenotypic variation, would be expected to have key ecological and evolutionary relevance. Examples include chemical defense (Kliebenstein et al. 2002), timing of reproduction (e.g., El-Assal et al. 2002; Ungerer et al. 2002), inflorescence and floral morphology (Juenger et al. 2000); and fruit and seed characteristics (Alonso-Blanco et al. 1999). Unlike crop species, where a relatively limited number of QTLs are involved in domestication traits, multiple QTLs (often 10 or more) are typically found for these traits in *A. thaliana*, with complex epistatic interactions among them. One recent study (Weinig et al. 2002) specifically sought to test the degree to which the flowering time QTLs that were identified under controlled laboratory conditions corresponded to those detected in a natural ecological setting. Interestingly, a substantial proportion of the QTLs with major effects in the lab was undetectable in the field, and vice versa; in addition, QTLs in the field differed between seasonal environments. These results suggest a level of complexity in the genetic interactions underlying natural adaptive divergence that would remain undetected with standard QTL mapping approaches under controlled growth conditions.

QTL mapping in wild plant species other than *A. thaliana* have also provided significant insights into the evolutionary genetics of adaptation and adaptive divergence. One of the most fruitful areas of research has focused on the evolution of traits underlying reproductive isolation, and the connection between intraspecific adaptive divergence and the process of speciation. Studies involving two species in the genus *Mimulus* have proved particularly insightful (Bradshaw et al. 1995, 1998; Schemske et al. 1999). *Mimulus lewisii* and *M. cardinalis* are interfertile and sympatric, but are reproductively isolated in nature due to differences in floral traits and resulting pollinator preferences (the former species is pollinated by bumblebees, the latter by hummingbirds). A relatively few QTLs of major effect can account for major phenotypic differences in floral traits between these species. This finding suggests that the barrier to reproduction between these species may be attributable to a small number of 'speciation' QTLs. Studies in other species, including Louisiana irises (reviewed by Arnold 2000) and *Helianthus* species (e.g., Kim and Rieseberg 1999) have focused on natural hybridization and the genes underlying ecological differentiation between hybridizing species. This approach can be used in directly assessing the relationship between interspecific introgression and the movement of QTLs for environment-specific adaptations.

While an important tool for examining the genetics of adaptation, QTL approaches also suffer from a number of drawbacks. First, QTL analysis requires that a genetic map be available for a species, which limits its application outside of crops and model taxa. In addition, multiple generations of crossing are required to generate mapping populations after the initial cross, and large populations of progeny must be maintained. These factors hinder the use of this approach in physically large species and in perennial plants. QTL mapping also suffers from limited resolution and power for identifying genes underlying phenotypic variation. The chromosomal region identified as a QTL is typically on the order of 5–10 cM, even for species with well resolved genetic maps. Even in the small genome of *A. thaliana*, this often translates into a physical region of 1–2 Mb. Thus, it is no small task to pinpoint a specific candidate locus (or possibly loci) within a QTL, although fine-mapping studies with large recombinant populations are possible (see for example Yano et al. 2000) and may permit the isolation of adaptive genes.

A final drawback of QTL mapping is particularly problematic for studies aimed at examining the genetics of adaptation in natural species. Only those QTLs that segregate between the two individuals used the initial cross may be identified. This leaves the vast majority of naturally occurring genetic variation in a species undetected. Moreover, since the QTLs identified from a particular cross may vary in their expression in different genetic backgrounds and different environmental conditions (see *Arabidopsis* example above), the actual ecological relevance of QTLs identified from a particular cross remains tenuous at best.

Linkage disequilibrium mapping: a genomics strategy for exploiting natural variation

The development of genomic approaches in evolutionary biology, most notably the availability of high-throughput DNA sequencing at multiple loci, offers new strategies for studying the genetic basis of adaptive evolution. One such approach is linkage disequilibrium (LD) mapping (Terwilliger and Weiss 1998; Kruglyak 1999; Buckler and Thornsberry 2002; Rafalski 2000*a, b*; Weiss and Clark 2002). Like standard QTL mapping approaches, linkage disequilibrium (LD) mapping infers associations between genotypes and phenotypic variation by examining genetic polymorphisms that have been shuffled into different genetic backgrounds. With standard QTL mapping, the polymorphisms under consideration are limited to those present in the two parents, and the shuffling of these polymorphisms is limited to the recombination events occurring during the establishment of a mapping population. In contrast, LD mapping exploits the naturally occurring genetic variation present within a large sampling of individuals in a species, and the history of recombination represented among those samples.

Two aspects of molecular variation in eukaryotic genomes provide the foundation for the LD approach to QTL mapping. The first is the potentially large amounts of single nucleotide polymorphisms (SNPs) present in the genomes of individuals within a population or species (Schaefer and Hawkins 1998; Cargill et al. 1999; Kwok 2001; Syvanen 2001). In humans, SNP variation is ~0.5% per nucleotide site (Cargill et al. 1999), while in maize the variation is closer to 1–2% per nucleotide site (Tenaillon et al. 2001). On an applied level, the very high densities of SNPs in a genome have made them a favorite molecular marker for fine-mapping studies (Rafalski 2000*a, b*). But more fundamentally, SNPs are the basic units of genomic diversity, and understanding the evolutionary dynamics of plant genomes involves assessing the levels, patterning and distribution of these single nucleotide polymorphisms (Aquadro 1992; Nordborg and Innan 2002; Rosenberg and Nordborg 2002).

Evolutionary studies of SNPs can reveal how population history, breeding system and selection affect variation at genetic loci, and they can delineate the mechanisms leading to the evolutionary diversification of genomes (Aquadro 1992; Nordborg and Innan 2002; Rosenberg and Nordborg 2002). Such studies rest on the theoretical foundation provided by the coalescent (Kingman 1982), which provides an analytical framework for examining genetic variation in an explicit genealogical context. This framework makes it possible to statistically test whether specific genes are evolving in a manner consistent with neutral evolution, or whether they have been targets of selection (Kreitman and Akashi 1995; Hudson and Kaplan 1988; Hudson et al. 1997).

The second aspect of genomic diversity pertinent to modern mapping approaches is linkage disequilibrium (LD), which is defined as nonrandom associations between alleles of different loci in a population (Weir 1996; Nordborg and Tavaré 2002). When a mutation initially arises in a population, it is automatically associated ('in disequilibrium') with all of the alleles present in the genome of the individual which gave rise to the mutation. If this mutation persists during evolution through genetic drift or selection, associations with other alleles are gradually eroded by segregation and recombination, so that over time the mutation is in linkage disequilibrium only with alleles that are in close physical proximity (see Fig. 1). There are several quantitative measures of LD (Weir 1996), and it has been known for some time that LD is affected by various evolutionary and demographic forces, including selection, recombination, population admixtures, inbreeding, and bottlenecks (Weir 1996; Nordborg and Tavaré 2002; Weiss and Clark 2002). Recent work in humans suggests that LD is strong within blocks of allelic sites or "haplotype blocks", which may be ~60–100 kb in length (Daly et al. 2001; Gabriel et al. 2002).

LD mapping exploits the abundance of natural SNP variation and the fact that any mutation that causes a phenotypic change and that persists in a population should be in linkage disequilibrium with a suite of polymorphisms that are in close physical proximity to it (Terwilliger and Weiss 1998; Kruglyak 1999; Jorde 2000). In LD mapping, a population is genotyped for markers (most often SNP markers) that span a genomic region of interest. The SNP markers are then tested against a specific phenotype to determine whether a statistical correlation exists between marker genotypes and a particular trait. A significant association between a specific marker(s) and a trait phenotype may arise either because the nucleotide polymorphism causes the phenotypic difference, or because the marker is in linkage disequilibrium with the causal polymorphism (Terwilliger and Weiss 1998; Kruglyak 1999; Jorde 2000).

LD mapping has several clear advantages over traditional QTL mapping approaches. First, it can survey the variation in a large population, and not simply the two progenitors of a mapping population. Second, by making use of historical recombination, one may be able to localize QTLs in a genome to a higher degree of resolution than is possible with the same number of individuals using traditional QTL linkage analysis. Third, the technique can be used without developing new mapping populations. LD mapping can thus potentially achieve higher resolutions with greater efficiency than linkage-based QTL mapping techniques, by taking advantage of both the array of genomic diversity within a species, and the large amounts of historical recombination that has occurred within and between populations during evolution. It is these features that have excited the interest of geneticists, and have led to concerted efforts to develop and exploit LD mapping in identifying genes (Terwilliger and Weiss 1998; Kruglyak 1999; Long et al. 1998, 2000; Jorde 2000; Martin et al. 2000; Puca et al. 2001; Thornsberry et al. 2001; Geiger-Thornsberry et al. 2002; Tabor et al. 2002) and create genome-wide haplotype maps (Daly et al. 2001).

In practice, LD mapping can be used either in genome scans or in candidate gene association studies (Terwilliger and Weiss 1998; Kruglyak 1999; Tabor et al. 2002). In genome scans, either the entire genome or a specific genomic region can be analyzed with molecular markers of sufficient density that they help localize the QTL. In a candidate gene association study, a candidate gene for a given trait may have been previously identified, and the association is carried out in the context of polymorphic markers localized within this specific functional candidate gene (Tabor et al. 2002). The former approach obviously requires much more extensive knowledge of an organism's genome structure and organization than the latter approach.

Fig. 1. Graphical illustration of linkage disequilibrium. Assume two types of chromosomal haplotypes in the population (light and dark), with a mutation (triangle) arising in one type (dark). Over evolutionary time, recombination between light and dark chromosomes results in a population in which the mutation is in strongest LD to markers closest to it (between the dashed lines).

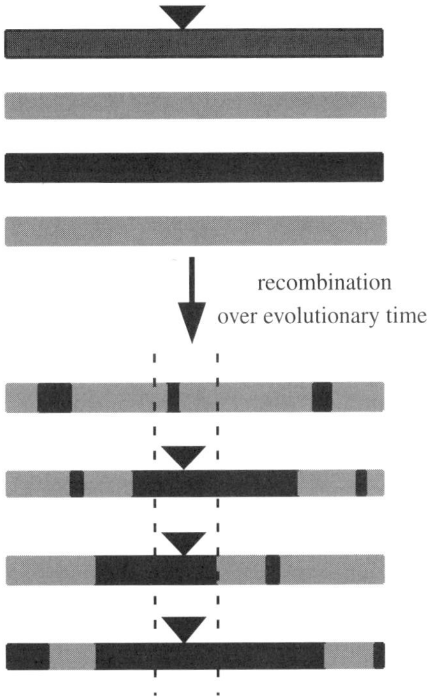

Population genomics of plants

Successful application of LD mapping requires not only the identification of genetic variation in a genome, but also knowledge of the evolutionary forces that have shaped these patterns of variation (Terwilliger and Weiss 1998; Kruglyak 1999; Jorde 2000). Before coherent LD mapping strategies can be developed, investigators must address three fundamental population genomic questions for their study system: (i) What is the evolutionary history and population structure of the target species? (ii) How are polymorphisms distributed in the genome? (iii) What is the extent of linkage disequilibrium in the genome?

The answers to these questions are known in some detail in only two plant species: the model genetic organism *Arabidopsis thaliana* and the crop species maize (*Zea mays)*. As noted above, genome-wide silent site SNP diversity for maize is about 1–2% on average. However, the nucleotide diversity of North American inbred maize lines appears to be reduced in comparison to exotic races; this suggests a genetic bottleneck associated with the establishment of these lines and a narrower range of diversity for most commercial maize cultivars (Tenaillon et al. 2001, 2002). Gene-specific reductions in SNP diversity have also been observed, and in several cases they may be attributed to directional selection during the course of domestication. Examples include the promoter of *tb1* (Wang et al. 1999), the anthocyanin regulatory gene *C1* (Hanson et al. 1996) and several members of the starch biosynthetic pathway (Whitt et al. 2002). Taken together, these studies in maize serve as indicators as to how variation in patterns of SNP diversity can be used to draw inferences about the evolutionary forces shaping gene structure (see also Hudson et al. 1997; Barton 2000; Nielsen 2001).

Linkage disequilibrium in maize decays, on average, at ~1 kb (Tenaillon et al. 2001; Remington et al. 2001), a narrow scale that suggests that LD mapping by genome scans may not be feasible in maize due to the high density of markers that would be required. These levels of LD, however, are sufficient to establish whether polymorphisms within candidate genes are associated with specific phenotypic variants. The success of a candidate gene association strategy in maize is illustrated by work on the *D8* gene, where several polymorphisms were found to be correlated with flowering time variation in a set of ~100 inbred lines (Thornsberry et al. 2001). This study was also unique in demonstrating in practical terms the importance of knowing population structure when undertaking LD mapping studies (Thornsberry et al. 2001; Pritchard et al. 2000; Pritchard and Donnelly 2001).

Arabidopsis thaliana differs from maize in two important respects: it is not domesticated, and it is a largely self-fertilizing species (selfing rate ~99%). The difference in breeding system has had particularly noticeable effects on the genomic structure of this species. Levels of species-wide silent site SNP diversity are lower than that of maize (~0.7%) (Purugganan and Suddith 1999; Aguade 2001), and most of the genetic variation is partitioned between, rather than within, populations (Bergelson et al. 1998). These patterns conform to the expectations of classical population genetic theory for inbreeding species. There is also an excess of rare polymorphisms within the species compared to expectations under neutrality (Purugganan and Suddith 1999). These rare polymorphisms include many nonsynonymous SNPs that appear to be kept at low frequency due to negative selection (Bustamante et al. 2002). Data on nucleotide polymorphisms, however, do provide some evidence for the role of positive or balancing selection in some genes in *A. thaliana*, including genes in the floral developmental pathway (Olsen et al. 2002) and disease resistance loci (Tian et al. 2002; Stahl et al. 1999).

Self-fertilization generates very high levels of homozygosity and thus very low levels of effective recombination in *A. thaliana*. This lack of effective recombination has resulted in LD that is dramatically more extensive than that of maize, with LD decay occurring at a range about ~150–200 kb (Hagenblad and Nordborg 2002; Nordborg et al. 2002). At least one study in a 170-kb *A. thaliana* region, however, shows no consistent negative correlation of LD with distance, suggesting that gene conversion may play a part in SNP patterning (Haubold et al. 2002). The difference in LD patterns between maize and *A. thaliana* indicate that different strategies for association mapping must be considered for these two species. In maize, large scale genome scanning is probably impractical, but the LD patterns indicate that candidate gene association studies can be readily accomplished (Thornsberry et al. 2001). In contrast, such candidate gene associations may be more difficult in *A. thaliana*, given that any association between a gene marker and a trait can, in principle, simply reflect LD that extends through several tens of kilobases of sequence (Hagenblad and Nordborg 2002). However, the large extent of LD in this selfing species may provide the basis for genome-wide scans for genes underlying natural phenotypic variation (Nordborg et al. 2002; Hagenblad and Nordborg 2002).

Outside of maize and *A. thaliana*, the levels and patterning of SNP variation are known for a few genes in a handful of other plant species, including *Arabidopsis lyrata* (Schierup et al. 2001; Savolainen et al. 2000), rice (Sato et al. 2001; Olsen and Purugganan 2002), pearl millet (Gaut and Clegg 1993) wild barley (Lin et al. 2001), and *Lycopersicon* (Nesbitt and Tanksley 2002; Baudry et al. 2001). Intensive studies of population genomics in these and other plant species may provide the basis for LD mapping in other taxa, as well as insights into the evolution of plant genes and genomes.

SNPs and association mapping: an illustrative example from *Oryza sativa*

The study of SNPs and the challenges of their application in LD mapping of adaptive traits is illustrated by recent work on domesticated Asian rice, *Oryza sativa.* Rice has features that are common to both maize and *Arabidopsis*. Like the former, it is a domesticated species with a history of strong selection and human-induced gene flow. Like the latter, it is a largely selfing species. It is unclear how these two facets of rice's evolutionary history have combined to shape genomic diversity in this crop plant.

Like most plant species of evolutionary and ecological interest, very little is known about the levels, patterns and genomic distribution of SNP diversity in rice (Buckler et al. 2001). There are active attempts to identify SNPs in rice

Table 1. SNP variation in rice genes.

		π^a		
Gene	Length (kb)	*japonica*	*indica*	Total[b]
Waxy	2.7	0.0044	0.0037	0.0045
RGRC2	1.1	0.0006	0.0022	0.0024
OSH3(A)	1.1	0.0002	0.0054	0.0056
OSH3(B)	0.4	0.0066	0.0044	0.0073

[a]Silent site nucleotide diversity, which is the mean number of SNP differences per site for two randomly chosen alleles.

[b]Species-wide SNP diversity.

Fig. 2. Haplotype networks of the rice *Waxy* and *RGRC2* gene. Haplotypes are indicated in letters, with the number of individuals of each haplotype in ssp. *indica* and ssp. *japonica* indicated. Each line between dots indicates a mutational change corresponding to a SNP.

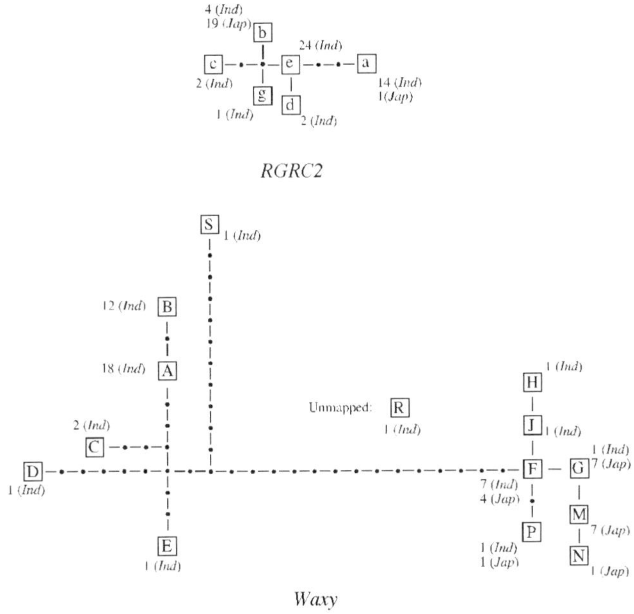

(Nasu et al. 2002; Blair et al. 2003), but only two studies have been published that characterize the levels and patterning of SNPs in rice genes within a defined evolutionary genomic context (Sato et al. 2001; Olsen and Purugganan 2002). The results of these studies are summarized in Table 1. These levels are two- to ten-fold lower than levels observed for maize nuclear genes and comparable to levels seen in *A. thaliana*. It is unclear whether there is a general difference in SNP levels between the two cultivated subspecies of rice, ssp. *japonica* and ssp. *indica*, although at least two of the four genes (*RGRC2* and *OsH3* region A) show a dramatically lower nucleotide diversity level in ssp. *japonica*.

Two rice genes, *Waxy* and *RGRC2*, have been examined in an explicit genealogical framework (see Fig. 2), and their haplotype trees show very different patterns of genetic divergence (Olsen and Purugganan 2002). The *Waxy* gene reveals two distinct haplotype groups within *O. sativa*, with the haplotypes found in ssp. *japonica* confined to one group (see Fig. 2). In contrast, there appears to be little differentiation among *RGRC2* haplotypes. Moreover, the frequency spectrum for SNPs differs between genes and between subspecies (data not shown). The small number of published studies on SNP variation in rice makes it difficult to determine which pattern is prevalent among rice genes as a result of their shared evolutionary history, and which features arise from gene-specific evolutionary forces such as selection (Nielsen 2001; Nordborg and Innan 2002).

Our understanding of linkage disequilibrium in rice is even more tenuous than our knowledge of SNP diversity. There are no published studies on the strength and extent of LD in rice. As a selfing species similar to *A. thaliana*, one might expect LD to extend to long ranges in the genome, and the data from *Waxy* confirm that LD extends further than 2.5 kb within this gene.

As a small test case for qualitative (rather than quantitative) association at the *Waxy* locus, we conducted a simple association study on the presence or absence of amylose in endosperm starch. The absence of amylose is responsible for the glutinous rice phenotype. Phenotypic associations with SNPs in the *Waxy* gene and in the unlinked gene *RGRC2* are shown in Fig. 3. The causal polymorphism for the absence of amylose is known to be a splice donor site mutation in *Waxy* intron 1 (position 1600 in Fig. 3) (Isshiki et al. 1998; Hirano et al. 1998; Olsen and Purugganan 2002). A successful association analysis should identify this polymorphism and those physically near it, but not polymorphisms in an unlinked gene. Our results confirm that SNPs in *Waxy* (including the causal polymorphism) show significant associations with the phenotype in a collection of >100 rice landraces, while those in *RGRC2* do not (Fig 3; see also Olsen and Purugganan 2002). As expected, only some SNPs within 1 kb of the causal polymorphism show significant associations with the phenotype. This reflects that fact that statistical significance of associations will vary depending on the frequency of flanking polymorphisms; rare polymorphisms are in weaker LD and are less likely to show a significant association. Thus, the choice of markers and marker densities is critical for successful LD mapping. Sufficient sampling is required to identify adequate numbers of SNPs, to establish the extent of their disequilibria, and to determine their distributions — not only across genes but also among the different populations within a species.

Conclusions: bridging the genomic gap

A fundamental goal of evolutionary genetics is to understand the molecular genetic basis of functional trait evolution. However, between the molecular genetic level and the organismal level, there currently exists a gap in our power of genetic resolution (Fig. 4). At the organismal level, ecological genetics provides insights into the nature of selection on phenotypic traits, but typically with no genetic resolution. Since these traits usually differ between individuals in nature, the genetic scale of these studies therefore encompasses the whole genome (~10^8–10^9 bps). Linkage-based QTL mapping allows resolution to subchromosomal regions, but current experimental designs invariably limit the resolution of these studies to genetic scales of ~10^6–10^7 for most wild plant species. At the finest level, molecular population ge-

Fig. 3. LD mapping of the glutinous phenotype in rice. Association of SNPs in *Waxy* and the unlinked *RGRC2* gene with the glutinous phenotype. Position and type of SNPs are indicated. The negative logarithm of the probability of an association is also given. Boxes are exons. The significance threshold is calculated with Bonferroni correction; all associations above the dashed line are significant under this criterion.

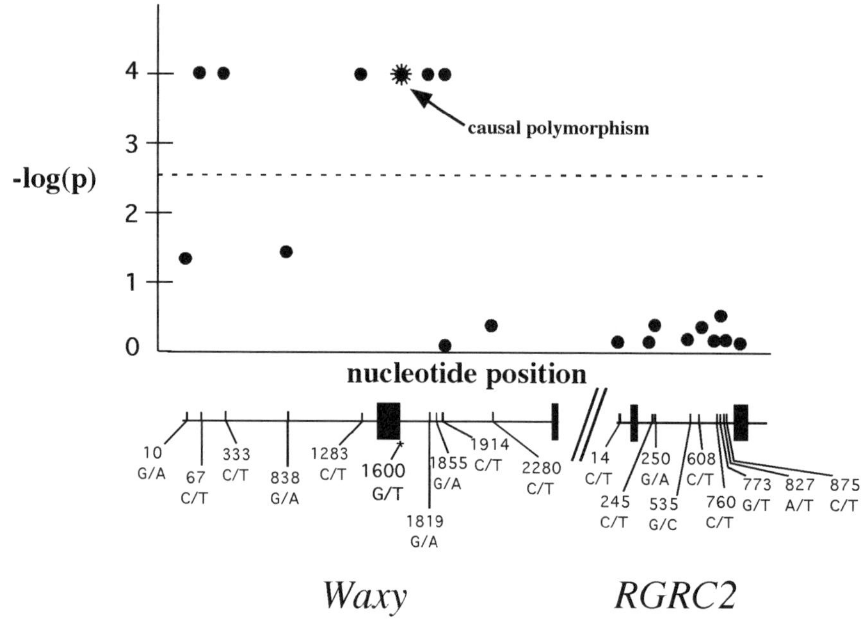

netics can examine the evolutionary forces shaping levels and patterns of nucleotide variation within specific genes, providing a glimpse of the nature of evolution at a genic scale (~10^0 to 10^4 basepairs). Thus, there is a 'genomic gap' in resolution between QTLs and the specific causal genes within them (Fig. 4).

Bridging this 'genomic gap' will permit investigators to routinely link variation at specific genes with variation at the phenotypic level, and to routinely assemble genotype-phenotype maps. Genomic technologies today provide a method to bridge this gap, not only in model plant species but wild taxa as well. The study of SNP variation and its application to LD mapping provides the best strategy currently available for mapping phenotypes onto molecular genotypes, and has already shown some success in flies (Long et al. 1998, 2000; Geiger-Thornsberry and Mackay 2002) and humans (Fullerton et al. 2000), as well as in maize (Thornsberry et al. 2001). For plants, existing data from maize and *A. thaliana* suggest that population genomic historics and life history characteristics favoring moderate levels of LD will be most amenable for association mapping. LD must be sufficient to allow the identification of statistically significant associations with phenotypic variation, yet not so extensive that the causal polymorphisms within an LD block cannot be localized. For species with suitable population genomic structure, this genomic mapping strategy might finally make it possible to understand the molecular genetic basis of adaptive diversification.

Fig. 4. Bridging the genotype-phenotype gap. The genomic scale of various methods and approaches is indicated. The gap between molecular population genetic variation and mapped QTL regions can potentially be bridged by LD-based scanning and candidate gene association studies, as well as fine-scale mapping.

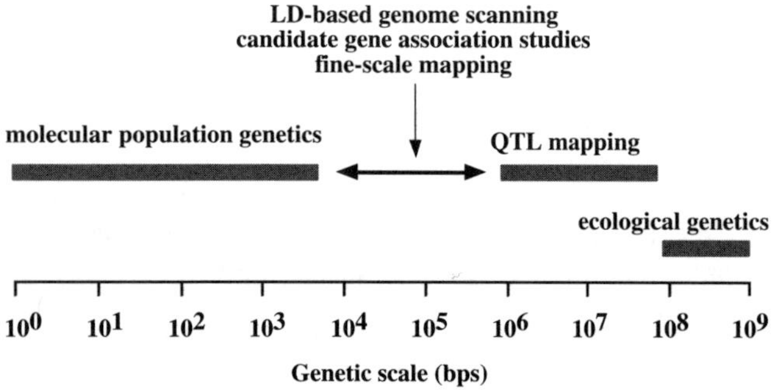

Acknowledgements

The authors would like to acknowledge funding from the US National Science Foundation Plant Genome Research and Integrated Challenges in Environmental Biology Programs for partial support of this work.

References

Aguade, M. 2001. Nucleotide sequence variation at two genes of the phenylpropanoid pathway, the *FAH1* and *F3H* genes, in *Arabidopsis thaliana*. Mol. Biol. Evol. **18**: 1–9.

Alonso-Blanco, C., Blankestijn-de Vries, H., Hanhart, C.J., and Koornneef, M. 1999. Natural allelic variation at seed size loci in relation to other life history traits of *Arabidopsis thaliana*. Proc. Natl. Acad. Sci. U.S.A. **96**: 4710–4717.

Aquadro, C.F. 1992. Why is the genome variable-insights from *Drosophila*. Trends Genet. **8**: 355–362.

Arnold, M.L. 2000. Anderson's paradigm: Louisiana irises and the study of evolutionary phenomena. Mol. Ecol. **9**: 1687–1698.

Barton, N.H. 2000. Genetic hitchhiking. Phil. Trans. Roy. Soc. Lond. B Biol. Sci. **355**: 1553–1562.

Baudry, E., Kerdelhue, C., Innan, H., and Stephan, W. 2001. Species and recombination effects on DNA variability in the tomato genus. Genetics **158**: 1725–1735.

Bergelson, J., Stahl, E., Dudek, S., and Kreitman, M. 1998. Genetic variation within and among populations of *Arabidopsis thaliana*. Genetics **148**: 1311–1323.

Bradshaw, H.D. Jr., Wilbert, S.M., Otto, K.G., and Schemske, D.W. 1995. Genetic mapping of floral traits associated with reproductive isolation in monkeyflowers (*Mimulus*). Nature **376**: 762–765.

Bradshaw, H.D., Otto, K.G., Frewen, B.E., McKay, J.K., and Schemske, D.W. 1998. Quantitative trait loci affecting differences in floral morphology between two species of monkeyflower (*Mimulus*). Genetics **149**: 367–382.

Buckler, E., Thornsberry, J., and Kresovich, S. 2001. Molecular diversity, structure and domestication of grasses. Genet. Res. **77**: 213–218.

Buckler, E.S., and Thornsberry, J.M. 2002. Plant molecular diversity and applications to genomics. Curr. Opin. Plant Biol. **5**: 107–111.

Bustamante, C.D., Nielsen, R., Sawyer, S.A., Olsen, K.M., Purugganan, M.D., and Hartl, D.L. 2000. The cost of inbreeding in *Arabidopsis*. Nature **416**: 531–534.

Cargill, M., Altshuler, D., Ireland, J., Sklar, P., Ardlie, K., Patil, N., Lane, C.R., Lim, E.P., Kalyanaraman, N., Nemesh, J., Ziaugra, L., Friedland, L., Rolfe, A., Warrington, J., Lipshutz, R., Daley, G.Q., and Lander, E.S. 1999. Characterization of single-nucleotide polymorphisms in coding regions of human genes. Nature Genet. **22**: 231–238.

Daly, M., Rioux, J., Schaffner, S., Hudson, T.. and Lander, E.S. 2001. High-resolution haplotype structure in the human genome. Nature Genet. **29**: 229–232.

Doebley, J., Stec, A., and Hubbard, L. 1997. The evolution of apical dominance in maize. Nature **386**: 485–488.

Doganlar, S., Frary, A., Daunay, M.-C., Lester, R.N., and Tanksley, S.D. 2002. Conservation of gene function in the Solanaceae as revealed by comparitive mapping of domestication traits in eggplant. Genetics **161**: 1713–1726.

Dorweiler, J., Stec, A., Kermicle, J., and Doebley, J. 1993. *Teosinte glume architecture 1*: a genetic locus controlling a key step in maize evolution. Science (Washington, D.C.) **262**: 233–235.

El-Assal, S., Alonso-Blanco, C., Peeters, A.J.M., Raz, V., and Koornneef, M. 2001. AQTL for flowering time in *Arabidopsis* reveals a novel allele of *CRY2*. Nature Genet. **29**: 435–440.

Fullerton, S.M., Clark, A.G., Weiss, K.M., Nickerson, D.A., Taylor, S.L. et al., 2000. Apolipoprotein E variation at the sequence haplotype level: implications for the origin and maintenance of a major human polymorphism. Am. J. Hum. Genet. **67**: 881–900.

Gabriel, S.B., Schaffner, S.F., Nguyen, H., Moore, J.M., Roy, J., Blumenstiel, B., Higgins, J., DeFelice, M., Lochner, A., Faggart, M., Liu-Cordero, S.N., Rotimi, C., Adeyemo, A., Cooper, R., Ward, R., Lander, E.S., Daly, M.J., Altshuler, D. 2002. The structure of haplotype blocks in the human genome. Science (Washington, D.C.) **296**: 2225–2229.

Gaut, B.S. and Clegg, M.T. 1993. Nucleotide polymorphism in the Adh1 locus of pearl millet. Genetics **135**: 1091–1097.

Geiger-Thornsberry, G.L., and Mackay, T.F.C. 2002. Association of single-nucleotide polymorphisms at the *Delta* locus with genotype by environment interaction for sensory bristle number in *Drosophila melanogaster*. Genet. Res. **79**: 211–218.

Hanson, M.A., Gaut, B.S., Stec, A., Fuerstenberg, S.I., Goodman, M.M., Coe, E. and Doebley, J.F. 1996. Evolution of anthocyanin biosynthesis in maize kernels: The role of regulatory and enzymatic loci. Genetics **143**: 1398–1407.

Hagenblad, J., and Nordborg, M. 2002. Sequence variation and haplotype structure surrounding the flowering time locus *FRI* in *Arabidopsis thaliana*. Genetics **161**: 289–298.

Haubold, B., Kroymann, J., Ratzka, A., Mitchell-Olds, T., and Wiehe, T. 2002. Recombination and gene conversion in a 170-kb genomic region of *Arabidopsis thaliana*. Genetics **161**: 1269–1278.

Hirano, H.-Y., Eiguchi, M., and Sano, Y. 1998 A single base change altered the regulation of the waxy gene at the post-transcriptional level during the domestication of rice. Mol. Biol. Evol. **15**: 978–987.

Hudson, R.R. and Kaplan, N.L. 1988. The coalescent process in models with selection and recombination. Genetics **120**: 831–840.

Hudson, R.R., Saez, E.G., and Ayala, F.J. 1997. DNA variation at the *Sod* locus of *Drosophila melanogaster*: An unfolding story of natural selection. Proc. Natl. Acad. Sci. U.S.A. **94**: 7725–7729.

Isshiki, M., Morino, K. , Nakajima, M. , Okagaki, R. J., Wessler, S.R., Izawa, T., and Shimamoto, K.1998. A naturally occurring functional allele of the rice *waxy* locus has a GT to TT mutation at the 5′ splice site of the first intron. Plant J. **15**: 133–138.

Jorde,L.B. 2000. Linkage disequilbrium and the search for complex disease genes. Genome Res. **10**: 1435–1444.

Juenger, T., Purugganan, M.D., and Mackay, T.F.C. 2000. Quantitative trait loci for floral morphology in *Arabidopsis thaliana*. Genetics **156**: 1379–1392.

Kim, S.-C., and Rieseberg, L.H. 1999. Genetic architecture of species differences in annual sunflowers: implications for adaptive trait introgression. Genetics **153**: 965–977.

Kingman, J.F.C. 1982 On the genealogy of large populations. J. Appl. Probab. **19A**: 27–43.

Kliebenstein, D.J., Figuth, A., and Mitchell-Olds, T. 2002. Genetic architecture of plastic methyl jasmonate responses in *Arabidopsis thaliana*. Genetics **161**: 1685–1696.

Kreitman, M., and Akashi, H. 1995. Molecular evidence for natural selection. Ann. Rev. Ecol. Syst. **26**: 403–422.

Kruglyak, L. 1999. Prospects for whole genome linkage disequilibrium mapping of common disease genes. Nature Genet. **22**: 139–144.

Kwok, P. 2001. Methods for genotyping single nucleotide polymorphisms. Ann. Rev. Genomics Human Genet. **2**: 235–258.

Lin, J.Z., Borwn, A.H.D. and Clegg, M.T. 2001. Heterogeneous geographic patterns of nucleotide sequence diversity between two alcohol dehydrogenase genes in wild barley (Hordeum vulgare subspecies spontaneum). Proc. Natl. Acad. Sci. U.S.A **98**: 531–536.

Long, A.D., Lyman, R.F., Langley, C.H., and Mackay, T.F.C. 1998. Two sites in the *Delta* gene region contribute to naturally occurring variation in bristle number in *Drosophila melanogaster*. Genetics **149**: 999–1017.

Long, A.D., Lyman, R.F., Morgan, A.H., Langley, C.H., and Mackay, T.F.C. 2000. Both naturally occurring insertions of transposable elements and intermediate frequency polymorphisms at the achaete-scute complex are associated with variation in bristle number in *Drosophila melanogaster*. Genetics **154**: 1255–1269.

Martin, E.R., Gilbert, J.R., Lai, E.H., Riley, J., Rogala, A.R., Slotterbeck, B.D., Sipe, C.A., Grubber, J.M., Warren, L.L., Conneally, P.M., Saunders, A.M., Schmechel, D.E., Purvis, I., Pericak-Vance, M.A., Roses, A.D., Vance, J.M. 2000. Analysis of association at SNPs in the *APOE* region. Genomics **63**: 7–12.

Mauricio, R. 2001. Mapping quantitative trait loci in plants: Uses and caveats for evolutionary biology. Nat. Rev. Genet. **2**: 370–381.

Nasu, S., Suzuki, J., Ohta, R., Hasegawa, K., Yui, R., Kitazawa, N., Monna, L., and Minobe, Y. 2002. Search for and analysis of single nucleotide polymorphisms SNPs in rice *Oryza sativa, Oryza rufipogon* and establishment of SNP markers. DNA Res. **9**: 163–171.

Nesbitt, T.C., and Tanksley, S.D. 2002. Comparative sequencing in the genus Lycopersicon: Implications for the evolution of fruit size in the domestication of cultivated tomatoes. Genetics **162**: 365–379.

Nielsen, R. 2001. Statistical tests of selective neutrality in the age of genomics. Heredity **86**: 641–647.

Nordborg, M., and Innan, H. 2002. Molecular population genetics. Curr. Opin. Plant Biol. **5**: 69–73.

Nordborg, M., and Tavare, S. 2002. Linkage disequilibrium: what history has to tell us. Trend. Genet. **18**: 83–90.

Nordborg, M., Borevitz, J.O., Bergelson, J., Berry, C.C., Chory, J., Hagenblad, J., Kreitman, M., Maloof, J.N., Noyes, T., Oefner, P.J., Stahl, E.A., and Weigel, D. 2002. The extent of linkage disequilibrium in *Arabidopsis thaliana*. Nature Genet. **30**: 190–193.

Olsen, K.M., and Purugganan, M.D. 2002. Molecular evidence on the origin and evolution of glutinous rice. Genetics **162**: 941–950.

Paterson, A.H., Lin, Y.-R., Li, Z., Schertz, K.F., Doebley, J.F., Pinson, S.R.M., Liu, S.C., Stansel, J.W., and Irvine, J.E.. 1995. Convergent domestication of cereal crops by independent mutations at corresponding genetic loci. Science (Washington, D.C.) **269**: 1714–1718.

Paterson, A.H. 2002. What has QTL mapping taught us about plant domestication? New Phytol. **154**: 591–608.

Puca, A.A, Daly, M.J., Brewster, S.J., Matise, T.C., Barrett, J., Shea-Drinkwater, M., Kang, S., Joyce, E., Nicoli, J., Benson, E., Kunkel, L.M., and Perls, T. 2001. A genome-wide scan for linkage to human exceptional longevity identifies a locus on chromosome 4. Proc. Natl. Acad. Sci. U.S.A. **98**: 10505–10508.

Pritchard, J.K., Stephens, M., and Donnelly, P. J. 2000. Inference of population structure using multilocus genotype data. Genetics **155**: 945–959.

Pritchard, J.K., and Donnelly, P.J. 2001. Case-control studies of association in structured or admixed populations. Theor. Popul. Biol. **60**: 227–237.

Purugganan, M.D., and Suddith, J.I. 1999. Molecular population genetics of floral homeotic loci: Departures from the equilibrium-neutral model at the *APETALA3* and *PISTILLATA* genes of *Arabidopsis thaliana*. Genetics **151**: 839–848.

Purugganan, M.D. 2000. The molecular population genetics of regulatory genes. Mol. Ecol. **9**: 1451–1461.

Rafalski, A. 2002*a*. Applications of single nucleotide polymorphisms in crop genetics. Curr. Opin. Plant Biol. **5**: 94–100.

Rafalski, J.A. 2002*b*. Novel genetic mapping tools in plants: SNPs and LD-based approaches. Plant Sci. **162**: 329–333.

Remington, D.L., Ungerer, M.C., and Purugganan, M.D. 2001. Map-based cloning of quantitativ e trait loci: Progress and prospects. Genet. Res. **78**: 213–218.

Remington, D,L., Thornsberry, J.M., Matsuoka, Y., Wilson, L.M., Whitt, S.R., Doebley, J., Kresovich, S., Goodman, M.M., and Buckler, E.S. 2001. Structure of linkage disequilibrium and phenotypic associations in the maize genome. Proc. Natl. Acad. Sci. U.S.A. **98**: 11479–11484.

Rosenberg, N.A., and Nordborg, M. 2002. Genealogical trees, coalescent theory and the analysis of genetic polymorphisms. Nat. Rev. Genet. **3**: 380–390.

Sato, Y., Fukuda, Y. and Hirano, H. 2001. Mutations that cause amino acid substitutions at the invariant positions in homeodomain of *OSH3/KNOX* protein suggest artificial selection during rice domestication. Genes Genet. Syst. **76**: 381–392.

Savolainen, O., Langley, C.H., Lazzaro, B.P., and Freville, H. 2000. Contrasting patterns of nucleotide polymorphism at the alcohol dehydrogenase locus in the outcrossing *Arabidopsis lyrata* and the selfing *Arabidopsis thaliana*. Mol. Biol. Evol. **17**: 645–655.

Schafer, A.J., and Hawkins, J.R. 1998. DNA variation and the future of human genetics. Nature Biotechnol. **16**: 33–39.

Schemske, D.W., and Bradshaw, H.D., Jr. 1999. Pollinator preference and the evolution of floral traits in monkeyflowers (*Mimumus*) Proc. Natl. Acad. Sci. U.S.A. **96**: 11910–11915.

Schierup, M.H., Mable, B.K., Awadalla, P., and Charlesworth, D. 2001. Identification and characterization of a polymorphic receptor kinase gene linked to the self-incompatibility locus of *Arabidopsis lyrata*. Genetics **158**: 387–399.

Stahl, E.A,, Dwyer, G., Mauricio, R., Kreitman, M., and Bergelson, J. 1999. Dynamics of disease resistance polymorphism at the *Rpm1* locus of *Arabidopsis*. Nature **400**: 667–671.

Syvanen, A. 2001. Accessing genetic variation: Genotyping single nucleotide polymorphisms. Nat. Rev. Genet. **2**: 930–942.

Tabor, H., Risch, N., and Myers, R. 2002. Candidate-gene approaches for studying complex genetic traits: practical considerations. Nat. Rev. Genet. **3**: 391–396.

Tanksley, R. 1993. Mapping polygenes. Ann. Rev. Genet. **27**: 205–233.

Tenaillon, M., Sawkins, M., Long, A., Gaut, R., Doebley, J., and Gaut, B.S. 2001. Patterns of DNA sequence polymorphism along chromosome 1 of maize *Zea mays* ssp *mays* L.. Proc. Natl. Acad. Sci. U.S.A. **98**: 9161–9166.

Tenaillon, M. Sawkins, M.C., Anderson, L., Stack, S., Doebley, J., and Gaut, B.S. 2002. Patterns of diversity and recombination along chromosome 1 of maize *Zea mays* ssp. *mays* L.. Genetics **162**: 1401–1413.

Terwilliger, J.D., and Weiss, K.M. 1998. Linkage disequilibrium mapping of complex disease: fantasy or reality? Curr. Opin. Biotechnol. **9**: 578–594.

Thornsberry, J.M., Goodman, M.M., Doebley, J., Kresovich, S., Nielsen, D., and Buckler E. 2001. *Dwarf8* polymorphisms associate with variation in flowering time. Nature Genet. **28**: 286–289.

Tian, D.C., Araki, H., Stahl, E., Bergelson, J., Kreitman, M. 2002. Signature of balancing selection in *Arabidopsis*. Proc. Natl. Acad. Sci. U.S.A. **99**: 11525–11530.

Ungerer, M.C., Halldorsdottir, S.S., Modliszewski, J.L., Mackay, T.F.C., and Purugganan, M.D. 2002. Quantitative trait loci for inflorescence development in *Arabidopsis thaliana*. Genetics **160**: 1133–1151.

Wang, R.L., Stec, A., Hey, J., Lukens, L., and Doebley, J. 1999. The limits of selection during maize domestication. Nature **398**: 236–239.

Weinig, C., Ungerer, M.C., Dorn, L.A., Kane, N.C., Toyonaga, Y., Halldorsdottir, S.S., Mackay, T.F.C., Purugganan, M.D., and Schmitt, J. 2002. Novel loci control variation in reproductive timing in *Arabidopsis thaliana* in natural environments. Genetics **162**: 1875–1884.

Weir, B.S. 1996. Genetic Data Analysis II. Sinauer, Sunderland, MA.

Weiss, K., and Clark, A.G. 2002. Linkage disequilibrium and the mapping of complex human traits. Trends Genet. **18**: 19–24.

Whitt, S.R., Wilson, L.M., Tenaillon, M.I., Gaut, B.S., and Buckler, E.S. 2002. Genetic diversity and selection in the maize starch pathway. Proc. Natl. Acad. Sci. U.S.A. **99**: 12959–12962.

Yano, M., Katayose, Y., Ashikari, M., Yamanouchi, U., Monna, L., Fuse, T., Baba, T., Yamamoto, K,, Umehara, Y., Nagamura, Y., and Sasaki, T. 2000. Hd1, a major photoperiod sensitivity quantitative trait locus in rice, is closely related to the *Arabidopsis* flowering time gene CONSTANS. Plant Cell **12**: 2473–2483.

8 Genomic approaches to identifying quantitative trait loci: lessons from *Arabidopsis thaliana*

Justin Borevitz

Introduction

Genomics tools such as genome and cDNA/EST sequencing, large mutant collections, and microarrays are powerful means by which the molecular genetic basis of phenotypic traits may be discovered. Their potential benefit to our understanding of evolutionary biology is great. To date, applications of these tools have been limited mainly to model organisms such as *Arabidopsis thaliana*. There are, of course, good reasons why certain species are model systems: *A. thaliana* has a relatively small, completely sequenced genome (The Arabidopsis Genome Initiative 2000), a rapid life cycle, and other resources such as knock-out mutant collections (Alonso et al. 2003; Sessions et al. 2002), and tens of thousands of markers accessible through convenient web-based search engines (e.g., see http://www.arabidopsis.org). But what other "non-model" species lack in laboratory convenience they make up in possessing traits of evolutionary interest. In time, as techniques are refined, as research facilities become better-equipped, and as costs decrease, genomics tools will come within reach of a more diverse range of study species, thus opening the door to studies of the molecular basis of adaptations, key innovations, and other traits that may be ecologically well understood but are currently intractable from the standpoint of their genetic underpinnings.

In this review, some recent Quantitative Trait Locus (QTL) studies in *Arabidopsis thaliana* and other species are described, focusing on QTLs that have been cloned and identified. I then discuss other genomics tools that may be used to discover the underlying basis of natural variation, progressing from coarse- to fine-grained approaches. These include: (*i*) genome-wide association mapping, in which the unit of inheritance is the ancestral haplotype; (*ii*) genotyping using oligonucleotide microarrays; (*iii*) bulk segregant analysis, an approach to rapidly map quantitative variation using pools of extreme lines; (*iv*) fine mapping and transcriptional profiling as a means of delimiting a short list of candidate genes; and finally, (*v*) functional tests of the candidates to confirm their role in the QTL.

Justin Borevitz[1]. Plant Biology, Salk Institute, 10010 N. Torrey Pines Road, La Jolla, CA 92037, U.S.A.
e-mail address: borevitz@salk.edu
Correct citation: Borevitz, J. 2004. Genomic approaches to identifying quantitative trait loci: lessons from *Arabidopsis thaliana*. *In* Plant Adaptation: Molecular Genetics and Ecology. Proceedings of an International Workshop held December 11–13, 2002, in Vancouver, British Columbia, Canada. *Edited by* Q.C.B. Cronk, J. Whitton, R.H. Ree, and I.E.P. Taylor. NRC Research Press, Ottawa, Ontario. pp. 53–60.

[1]Current address: Evolution and Ecology, University of Chicago, 920 East 58th St., Chicago, IL 60637-5415, U.S.A. e-mail address: borevitz@uchicago.edu

QTL mapping in *Arabidopsis*

Quantitative trait locus analysis provides a means to simultaneously map several loci and their interactions in a segregating cross (Doerge 2002; Mauricio 2001; Paterson 1998). QTL mapping can be performed in multiple environments to determine if QTL have environment-specific effects. Several crosses between natural *Arabidopsis thaliana* accessions have been developed into sets of recombinant inbred lines (RILs) (Alonso-Blanco et al. 1998*b*; Deslandes et al. 1998; Lister and Dean 1993; Loudet et al. 2002; Wilson et al. 2001) and more are under way (e.g., see http://www.naturalvariation.org and http://www.natural-eu.org). The RILs that are available at stock centers have gone through at least 6 generations of selfing, and contain publicly available genotype information (e.g., see http:/ /www.natural-eu.org).

The Cvi × Ler RIL (Alonso-Blanco et al. 1998*b*) set has been particularly useful for mapping QTLs for a variety of traits, including flowering time, seed size, circadian rhythms, seed storability, and inflorescence development. Four major QTLs that have environment-specific effects on flowering time were found in this population (Alonso-Blanco et al. 1998*a*). In that study, it was shown that the *Early Day length Insensitive* (*EDI*) locus has a much larger flower promoting effect in short days; in addition, two other flowering time QTLs (*FLF* and *FLG*) that are linked on chromosome 5 show a major epistatic interaction that is suppressed by vernalization. QTL analyses of seed size identified eleven important loci, six of which seem to affect only seed size while the others are pleiotropic and co-localized with QTLs for other life history traits (Alonso-Blanco et al. 1999). QTLs for leaf circadian rhythms have been found, two at novel loci and two that overlap candidate genes (Swarup et al. 1999). Analysis of seed storability traits identified four QTLs, none of which mapped to loci important for variation in seed oligosaccharides, ruling out the hypothesis that variation in seed oligosaccharides was responsible for differences in seed storability in the Cvi/Ler population (Bentsink et al. 2000). Analyses of 13 inflorescence development traits identified 63 QTLs; again, some were pleiotropic while others affected only specific traits (Ungerer et al. 2002).

QTLs have also been mapped for hypocotyl length light response in the Cvi/Ler RIL population (Borevitz et al. 2002). In that study, 12 QTLs were found by mapping responses to several light environments; some were specific to certain photoreceptor pathways, while others acted downstream of multiple photoreceptors. The QTL map positions suggest new light response loci such as *LIGHT1* (chromosome 1 27cM) as well as probable new alleles of known photomorphogenic loci such as *PHYB* (*LIGHT2*). These results contrast with light response QTLs that have been mapped in the Col/Ler RIL population (Yanovsky et al. 1997), in which only minor differences in light environment response were found. Two very low fluence response loci (*VLF1/2*) were identified that cause variation in cotyledon opening under pulses of far-red light. Neither *VLF1* nor *VLF2* was known from laboratory mutagenesis studies.

Plant QTLs that have been identified

There has been recent substantive progress in identifying the genes corresponding to QTLs in *Arabidopsis*, exemplified by studies of light response and photoreceptors. The *EDI* locus, identified in a cross between Cvi and L*er,* is a gain of function change in the CRYPTOCHROME2 (CRY2) photoreceptor (El-Din El-Assal et al. 2001). Loss-of-function alleles of *cry2* cause late flowering (Guo et al. 1998), while a conservative amino acid substitution in the Cvi allele of CRY2 causes early flowering and day length insensitivity by rendering the CRY2 protein more stable in short day conditions when it is normally degraded. Screening *Arabidopsis* accessions for quantitative differences in light response (Maloof et al. 2001) identified one accession with dramatically altered far-red light phenotype. Like *cry2*, this is due to a change in a single (conserved) amino acid, in this case in the PHYTOCHROME A photoreceptor, rendering the protein more stable in light. It is also known that a 14 bp deletion in a natural allele of the PHYTOCHROME D photoreceptor quantitatively alters light response (Aukerman et al. 1997). Finally, QTL analyes suggest that natural alleles at the PHYTOCHROME B locus also play a role in quantitative variation (Borevitz et al. 2002). It is striking to note that natural variation in light response is prevalent at the level of the photoreceptors, as laboratory-induced mutations in the photoreceptors have severe pleiotropic defects. Natural alleles thus appear to modify a subset of downstream responses, e.g. hypocotyl length light response but not flowering time or vice versa.

The *FRIDIGA* (*FRI*) (Johanson et al. 2000) and *FLOWERING LOCUS C* (*FLC*) (Michaels and Amasino 1999) are QTLs that interact to cause late flowering. These genes have been identified in *Arabidopsis* and may also be important in *Brassica* (Osborn et al. 1997; Schranz et al. 2002). Their combined effect is suppressed by vernalization. *FRI* and *FLC* were identified as QTLs in a cross between H51 and L*er* accessions (Clarke et al. 1995); they do, however, segregate as simple Mendelian loci in appropriate backgrounds. Two independent loss-of-function alleles (caused by deletions) at *FRI* are common (Johanson et al. 2000), however many others exist (Le Corre et al. 2002). Studies to determine functional allelic variation at *FLC* are in progress.

Two QTLs important for the amount and type of glucosinolates, secondary metabolites that are involved in herbivore deterrence, have been identified in *Arabidopsis*. The *APO2/3* locus contains expression polymorphisms at two tandemly duplicated genes; L*er* expresses *APO3* while Col and Cvi express *APO2* (Kliebenstein et al. 2001) which cause glucosinolates to be modified differently. A deletion polymorphism resulting in a loss of *APO2* function is present in ecotypes that only make a glucosinolate precursor. Another locus, *MAM1,* encodes a methylthioalkylmalate synthase and is responsible for variation in glucosinolate side chain elongation (Kroymann et al. 2001). The loss-of-function L*er* allele contains several single nucleotide polymorphisms.

A few QTLs have been cloned from other plant species (reviewed in Maloof 2003). The first QTL cloned from maize, *Teosinte branched1* (*Tb1*)*,* is involved in morphological changes accompanying the early domestication of maize from its wild progenitor, teosinte (Doebley et al. 1997). The *Tb1* gene product encodes a *CYCLOIDEA* homolog (Luo et al. 1996), which acts as a repressor of lateral organ outgrowth and determines the sex of the inflorescence. The maize allele of *Tb1* is expressed at higher levels than the teosinte allele and causes lateral inflorescences to be tipped by ears. The promoter of *Tb1* has been shown to have dramatically reduced levels of variation in maize compared to wild teosintes, suggesting the molecular location of selection by ancient agriculturists (Wang et al. 1999). The specific nucleotide changes in the promoter of *Tb1* responsible for the difference in expression have not yet been identified.

In tomato (*Lycopersicon esculentum*), a major QTL for fruit weight, *fw2.2*, has been cloned (Frary et al. 2000). *fw2.2* encodes a protein with predicted structural homology to the human oncogene c-H-ras p21. *fw2.2* is semi-dominant, and the large-fruit allele is expressed earlier and at higher levels in young fruit (Cong et al. 2002), perhaps due to differences at the promoter. No evidence for selection at *fw2.2* in *Lycopersicon* was found (Nesbitt and Tanksley 2002). Another tomato QTL, brix, controlling soluble solids in tomato fruits has also been cloned (Fridman et al. 2000). The location of the *Brix9-2-5* QTL was narrowed down to a 484 bp interval by high-resolution mapping. Across this narrow interval are several polymorphisms in the intron of a fruit-specific apoplastic invertase gene. When the wild allele is introgressed into domesticated tomato it leads to increased fructose and sucrose levels.

Several QTLs that control heading date (flowering time) in rice have recently been cloned by fine mapping and candidate gene approaches. *Hd1* is important for short day length perception, which promotes heading in rice (Yano et al. 2000). *Hd1* encodes a homolog of *CONSTANTS* (CO) which when mutated in *Arabidopsis* causes late flowering in long days (Putterill et al. 1995). In contrast, natural alleles of *Hd1*, which contain frame shift deletions, are early heading in long day field conditions. *Hd6* was detected as a minor QTL in advanced backcross progeny of the cross where *Hd1* was identified. Positional cloning identified an early stop codon in *Hd6* which encodes the α subunit of protein kinase CK2 (Takahashi et al. 2001). Hd6 is involved in day length sensitivity as functional alleles increase the heading time in long days but not in short days. In *Arabidopsis*, the α-subunit of protein kinase CK2 phosphorylates *CIRCADIAN CLOCK ASSOCIATED1*, an important regulator of circadian rhythms and flowering time in *Arabidopsis* (Sugano et al. 1998; Wang and Tobin 1998). An additional rice heading date QTL, *Hd3a* (Kojima et al. 2002*)* has been reported to encode a homolog of the *Arabidopsis FLOWERING LOCUS T* gene (Kardailsky et al. 1999; Kobayashi et al. 1999). The less functional allele of *Hd3a* contains deletions in the 3′ untranslated region and a single amino acid substitution. All heading date QTLs from rice that have been identified so far are homologs of genes known to be important for flowering time in *Arabidopsis*. These examples illustrate how studies of *Arabidopsis* can provide candidate genes for cloning QTLs from other species.

Genomic approaches to identifying QTLs

Haplotype mapping

With the completion of genome sequences from disparate organisms, attention is now focusing on variation in the sequence and how this contributes to phenotypic diversity. It is now common to determine the specific changes responsible for simple Mendelian traits in most organisms. As the era of single gene analysis nears completion, the next stage will be to determine the genetic basis of complex, multigenic traits. Unlike Mendelian traits, determining the genetic nature of complex traits is difficult since the overall phenotype is the result of the action and interaction of many genes with environmental effects. To detect multiple genes with small effects that may be dependant on the genetic background and/or the environment, a statistical linkage is sought between the quantitative measure of the complex trait and a region of the chromosome (Mackay 2001). This is straightforward in model organisms, where large experimental populations exist and environments can be held constant; however, this is much more difficult in natural populations. In outbred populations association mapping is used, whereby many unrelated individuals can be selected at similar ages and from similar environments (Goldstein and Weale 2001). Association studies would be more powerful than linkage analysis in determining the location of complex disease-causing genes if all variants could be identified and typed in a large population (Risch and Merikangas 1996). At the moment this is not feasible, but hope lies in exploiting the nonrandom nature of sequence variation resulting from linkage disequilibrium (LD). In the absence of LD, each polymorphism is independent and would need to be typed individually; with LD, the genotype at one marker is correlated with the genotype at another. This relaxes the requirement that each marker be typed individually but depends on the extent of LD.

Haplotype analysis thus represents a transition away from a single nucleotide-based approach. Rather than testing single changes in isolation, haplotypes are now used as the unit of inheritance. This has greater biological significance because no variant is an island, as it were; the local genetic context needs to be considered. However, the marker resolution required for the successful creation of a haplotype map depends heavily on the extent of LD, which has been studied in humans, *Drosophila*, and *Arabidopsis* (Borevitz and Nordborg 2003). In humans, the genome consists of blocks, 10–100 kb in size, with limited haplotype diversity that are separated by recombination hotspots (Dawson et al. 2002; Gabriel et al. 2002; Patil et al. 2001; Reich et al. 2001). In contrast, blocks of extensive LD have not been seen in *Drosophila*; instead, recombination is much more rampant, reducing LD to a scale of hundreds of base pairs (Long et al. 1998). In *Arabidopsis*, LD is on the scale of that in humans, 10–100 kb (Nordborg et al. 2002). Haplotypes in *Arabidopsis* are clearly defined due to a major reduction in effective recombination caused by selfing, because most recombination events happen between homozygous loci and thus chromosome segments are not broken up. However, the effective mutation rate is only partially affected by selfing (Nordborg 2000). The resulting high ratio of diversity to recombination rate makes LD mapping in *Arabidopsis* particularly powerful, because haplotype blocks are more extensive and well-defined.

Empirical studies highlight the unexpectedly large variance in the extent of LD (Nordborg et al. 2002). Not all markers at close distances show high levels of LD; as a result, placing markers at a uniform density, as determined by the mean extent of LD, will not be that useful in determining all haplotypes (Goldstein and Weale 2001). Determining the marker resolution required to identify all common haplotypes in an organism is difficult; certainly a high density will initially be required, e.g. 3–10 kb (300,000–1,000,000 markers) for humans (Gabriel et al. 2002).

Genotyping using microarrays

Currently, the rate limiting step for QTL analysis and haplotype mapping is genotyping, as hundreds of lines need to be typed at hundreds to thousands of markers. The use of arrays for genotyping is attractive due to its highly parallel nature: each line can be assayed for all markers simultaneously, making the cost dependent on the number of lines rather than the total number of markers. Several array methods have been used for genotyping. Initially, previously identified polymorphisms, usually Single Nucleotide Polymorphisms (SNPs), could be arrayed in a such a way that alternative hybridization to each allele could be distinguished. Cho et al (1999) used a high density Affymetrix SNP array that was designed against 412 *Arabidopsis* SNPs. Subsequent use of this array against many *Arabidopsis* accessions revealed 163 SNPs robust enough to be typed from all lines (Nordborg et al. 2002). Spotting oligonucleotides to glass arrays can also be effective, often at reduced cost (Stickney et al. 2002), but these may suffer from reduced quality and reproducibility. In general, array-based methods require a collection of markers to be identified by other means, usually sequencing. One approach for discovery and genotyping uses resequencing arrays or variation detector arrays (VDA) where 25 bp oligonucleotides are designed to every base and three oligonucleotides are included to detect possible mismatches at a central base (Cutler et al. 2001). These arrays were used to survey variation on human chromosome 21 (Patil et al. 2001).

A more attractive approach to both gene discovery and genotyping makes use of expression arrays, which were originally intended for monitoring the expression of thousands of genes. Total genomic DNA hybridized to these arrays in independent replicates identified thousands of markers in yeast (Winzeler et al. 1998). This approach initially was not pursued further because it was thought that complex genomes would not be amenable to such a simple method due to high levels of cross hybridization, but recent work has extended this technique to *Arabidopsis*, which has a genome 10 times larger and more complex than that of yeast (Borevitz et al. 2003). I see no reason why this technique could not be extended to organisms with higher genome complexity, a simple caveat being that increased numbers of replicates may be required to improve the overall signal-to-noise ratio. Markers identified by total genomic DNA hybridization are termed Single Feature Polymorphisms (SFPs), which differ from SNPs in that the specific nature of the DNA variation is not known. Just as with RFLPs (Restriction Fragment Length Polymorphism) and AFLPs (Amplified Fragment Length Polymorphisms), knowledge of the exact substitution is not necessary for their function as markers. However since the 25bp sequence of the SFPs is known for the reference strain these alleles can be directly placed on the physical map.

Detection of population polymorphisms

Array genotyping is a useful method for assessing genealogical relationships among a collection of individuals spanning geographic and ecological regions, an important component of many evolutionary questions. Among populations within a single species, analysis of a single gene may not be sufficient, as individual gene histories may differ from the actual genealogy (Rosenberg and Nordborg 2002). For example, in *Arabidopsis,* different trees are inferred from different loci (Aguade 2001; Innan et al. 1996; Kawabe and Miyashita 1999; Kawabe et al. 2000; Olsen et al. 2002) and genome-wide anonymous markers show no discernible phylogeny (Breyne et al. 1999; Erschadi et al. 2000; Miyashita et al. 1999; Nordborg et al. 2002). Array genotyping is beneficial in that it is not biased toward any single locus, and can thus reveal population histories based on genome-wide data. Indeed, if a locus differs significantly from the overall genealogy, it is suggestive of selection at that locus.

Array genotyping may be effective for comparisons of related species, even when the array has not been designed to a particular species. In this case SFPs cannot be ordered on a physical chromosome map, but are treated simply as genetic markers in much the same way as AFLPs and RFLPs. Array hybridization with two subspecies of *Brassica oleracea*, ssp. *botrytis* (cauliflower) and *italica* (broccoli), to *Arabidopsis* arrays reveal thousands of SFPs (unpublished data), and comparisons between more phylogenetically distant taxa may potentially work on a subset of conserved oligonucleotide features. A large mapping population from the related species typed with SFPs will order the anonymous markers into linkage groups, resulting in a high density genetic map. Given a known sequence of the reference SFP, the physical map can thus be anchored to the genetic map.

Bulk segregant mapping

Mapping in specific crosses usually takes the form of detecting linkage of a phenotype with genotype. In the simplest case, qualitative Mendelian variants segregate as anonymous molecular markers (e.g., SNPs). Perfect correlation between the phenotypic class (e.g., tall or short) and marker class indicates very tight linkage and nearby markers decrease in correlation with distance. For quantitative traits, this commonly takes the form of log likelihood odds scores, with a maximum score at the most likely position for a single gene, and decreasing scores at more distant locations. A drawback to these methods is their reliance on measuring the trait and many genome wide markers from many individual plants.

Bulk segregant mapping is a rapid alternative to individual genotyping that involves selective pooling, in which clearly distinguishable "mutant" and "wild-type" pools delimit the bulking strategy (Michelmore et al. 1991). The basic principle is that loci unlinked to the trait of interest will have a roughly equal parental chromosomal contribution in each of the pools. If markers across the genome could be accurately typed, then most loci would show an equal parental chromosome contribution in the pools but near the mutation would be enriched for "mutant" parental alleles in the mutant pool and enriched for "wild-type" parental alleles in the wild-type pool thus identifying the location of the mutation. A related approach for quantitative traits, so-called extreme mapping, involves pooling the extreme tails of the distribution for the trait of interest (Tanksley 1993).

Chip genotyping is ideally suited for the bulk segregant approach as it allows many markers distributed across the genome to be simultaneously estimated. Here each SFP is a quantitative hybridization signal that can estimate the contribution of each parental allele to the pool. The accuracy of any single SFP is low, but this is compensated by the vast number of markers surveyed (usually >10 per cM). Extreme Array Mapping (XAM) (D. Wolyn and J. Borevitz, in preparation) is a method in development for mapping quantitative traits. Simulations show that a single additive QTL that explains 20% of the variance can also be mapped to a 6cM region when extreme lines are chosen from 1000 F2 individuals.

At the time of writing, bulk segregant mapping with chip genotyping has been used to rapidly locate the positions of three novel qualitative mutations and one QTL in addition to the *ERECTA* locus (Borevitz et al. 2003), and I expect many more to be revealed by this method in the future. To further recommend this approach, I note that analysis tools are available in the form of R scripts (see http://naturalvariation.org/sfp); furthermore, when the parental genotypes are known (from 3 hybridizations each to ensure confidence), only two further hybridizations are required to map a mutation making this method very cost effective.

Fine mapping using chip genotyping

Once mutations or QTLs have been localized to a specific region of the genome, e.g. within 6 cM (rough mapping), the next step involves zooming in on the gene itself, or fine mapping. At this stage, many progeny (usually 1,000 or more) must be screened for recombination events, so that close breakpoints can be obtained. This step is straightforward and involves genotyping at only two flanking markers spanning the region. Once recombinants are identified (120 or so), the rate-limiting step is to determine where each particular recombination event took place on the physical chromosome map. Chip genotyping can also be used at this stage. Again, pools of mutant and wild-type recombinant plants can be made and hybridized to arrays. Since each chromosome in the pool contains a recombination event, there is much greater localizing power at this second stage of chip genotyping than at the initial stage of bulk segregant mapping; one can expect to be able to identify an interval <50kb with this approach. Both simulation and empirical studies are in progress to determine the effectiveness of fine mapping with chip genotyping.

Expression profiling

Many mutations that result from mutagenesis also alter the expression of the gene. Natural alleles at a QTL may also be differentially expressed; examples of this include *Tb1* (Doebley et al. 1997), *fw2.2* (Cong et al. 2002; Frary et al. 2000), *FLC* (Michaels and Amasino 1999), *FRI* (Johanson et al. 2000), and *APO2/3* (Kliebenstein et al. 2001). But differential expression patterns are not always the cause of the observed phenotypic differences, as shown by studies of *EDI* (El-Din El-Assal et al. 2001), *PHYA*, (Maloof et al. 2001), and *MAM1* (Kroymann et al. 2001), in which the protein is found to be altered between alleles. Global profiles of gene expression are easily performed on microarrays and thus are a useful tool in QTL studies. Transcriptional differences that map to QTL regions suggest candidate genes; however, when they map to other loci they may be part of the downstream response, and are observed as the molecular phenotype of the QTL.

Several experimental designs using transcriptional profiling can be used to identify candidate genes. A direct approach is to try and predict a time point, tissue, and environment relevant to the phenotype of interest where differential expression could be detected. With sufficient replication, statistical tests can be performed on each gene to determine those that are differentially expressed (Baldi and Long 2001; Tusher et al. 2001). These differentially expressed genes may, however, simply be general expression level polymorphisms and not be at all related to the phenotype. An extension of this approach is to look for expression level differences between genotypes that depend on environment, tissue, or time point. In this case, the two genotypes are profiled from two or more conditions with replicates. Gene expression analysis can now identify those genes regulated by the condition, and/or by genotype, or by an interaction between the two. Genes that show a significant interaction may be the most likely candidates, as the expression level polymorphism is related to the trait, and especially when the gene maps to the QTL region. For example, *fw2.2* and *Tb1* show expression differences only in the right tissue and time point (Cong et al. 2002; Hubbard et al. 2002).

The choice of genotypes for QTL profiling experiments are several-fold and consist of direct profiling of parental lines, extreme pools of lines segregating the phenotype, and specific QTL introgression lines. Using parental lines should identify the broadest collection of genotype-dependent differences in gene expression, but these may not be specific to the trait of interest. Pools of lines showing extreme phenotypes are advantageous because they are pre-selected for a specific trait. Expression differences between the parental lines in genes not linked to the trait will be averaged away in the pools of extreme lines. The most specific genotypes are near isogenic lines (NILs) that differ only at a single QTL region. Here gene expression polymorphisms that map to the NIL region are candidates, while others act downstream of the QTL in the introgression line. One drawback to this approach is that NILs often take several generations to construct. Some inbred mapping lines contain specific regions of residual heterozygosity (so-called heterogeneous inbred families, or HIFs) (Alonso-Blanco and Koornneef 2000); these can also be useful for expression profiling.

An important issue regarding expression analysis between different accessions is that polymorphisms in the DNA/RNA sequence will lead to mis-hybridization and will be interpreted erroneously as differences in expression level. Although true differences in expression level and polymorphisms within genes can both point out good candidates, it is desirable to know the nature of the observed difference in hybridization. Oligonucleotide arrays are more sensitive to small hybridization polymorphisms than cDNA arrays, but insertion/deletion polymorphisms will affect both array types. When several oligos are designed to a given gene or a given exon of a gene, mis-hybridization of a single oligo may be less of a factor. However new methods for analysis of gene expression take account of individual oligo hybridization intensity in the model (Chu et al. 2002). SFPs identified by DNA hybridization can be excluded from analysis of RNA or perhaps even identified in the RNA data itself. In addition, alternative splicing may be identified if oligos are directed across exons.

Candidate genes and confirmation of the QTL

Candidate genes can be determined once a rough map position has been determined or after fine mapping has narrowed the interval, and having a complete genome sequence is advantageous at this point. Literature searches, expression profiling, investigating the knock out phenotype, and direct sequencing of potential genes are other means by which candidates may be identified. A wealth of information about signaling mechanisms is known for several well-studied traits in *Arabidopsis* and this can also be used to help identify candidate genes by considering the phenotype and map position. One example that illustrates this point is the determination of *PHYTOCHROME B* as a candidate gene for the *LIGHT2* QTL (Borevitz et al. 2002). Traditional mutagenesis studies in *Arabidopsis* have been useful in identifying the right candidate genes for rice QTL (Kojima et al. 2002; Takahashi et al. 2001; Yano et al. 2000). Polymorphisms, predicted by DNA hybridization to arrays, is another means to refine a list of candidate genes: SFPs in coding regions may also represent functional changes, and the clustering of several SFPs within a gene is suggestive of a potential deletion. For example, hundreds of genes were found to be potentially deleted between two strains of *Arabidopsis* at a stringent threshold (Borevitz et al. 2003).

Reverse genetics tools, such as a near saturating collection of knockout insertions, now exist for *Arabidopsis* (Alonso et al. 2003; Sessions et al. 2002). Loss of function phenotypes for most genes within an already narrowed QTL interval can be screened with the use of these collections. The final confirmation of a cloned QTL comes from transgenics, in which alleles of a single gene from one parent can be introduced into the reciprocal background and/or into a null mutant background. This begins the next step in determining the

Fig 1. Scheme for identifying QTL genes. First, haplotype mapping or QTL mapping is used to identify genomic regions responsible for the trait. This can be done by QTL mapping in recombinant inbred lines or by bulk segregant mapping of extreme pools with chip genotyping. Heterogeneous inbred families (HIFs) or near-isogenic lines (NILs) are used to confirm QTLs and are a source for fine mapping and/or expression profiling. Array hybridization can identify polymorphisms in genes and serve as a background for gene expression studies. Expression profiling can also be done directly on pools of extreme lines, HIFs, or NILs. Gene expression differences, single feature polymorphisms (SFPs) in coding regions, and predicted functions of genes in the QTL regions suggest candidates. These are tested by direct sequencing, quantitative complementation tests, knockout and overexpression studies, and finally transformation with alternative alleles.

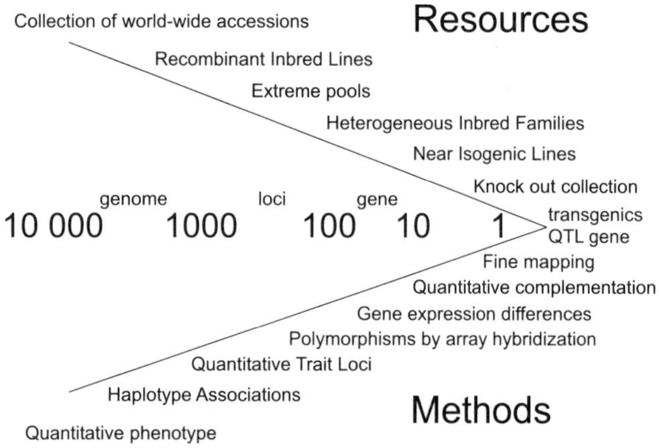

specific change(s) responsible for the QTL, ranging from the obvious (e.g., a deletion or stop codon) to the more difficult, e.g. involving regulatory or amino acid changes that require further testing.

Prospects for non-model organisms

It is still a lengthy process, even in model organisms, to identify the genes corresponding to a QTL. Is there any hope, then, for other "non-model" species? I foresee a light at the end of the tunnel: first, because many of the genomics techniques discussed here are becoming more routine and less expensive; second, resources for non-model species are becoming increasingly available; and third, novel techniques will be developed that will overcome what are currently rate-limiting steps. Many universities now have microarray core facilities and there is competition in the Biotechnology sector (Borevitz et al. 2003) driving costs down. Bacterial artificial chromosome (BAC) and cDNA libraries have been made from several species. Single-pass sequencing for thousands of cDNA clones (expressed sequence tags, or ESTs) and BAC ends can be generated at a reasonable cost; oligos designed to these can be put on arrays. Array options include spotted arrays, which have high cost initially due to oligo synthesis, and high density arrays, where oligos are synthesized *in situ*. New techniques to synthesize oligos at high density on arrays that do not require the large initial investment of a mask are very attractive (Nuwaysir et al. 2002), as these can be used for the dual purpose of genotyping and expression analysis. High density genetic maps can thus be constructed from any organism, and the identification and typing of markers are no longer limiting steps. By arraying oligos designed to BAC ends they can be integrated onto the high density genetic map, thus greatly assisting in the creation of a physical contig once the QTL interval is defined. It may be possible to sequence these BACs directly to look for candidate genes, such as those, which are differentially expressed on arrays. Expression profiling can then be done on pools of extreme lines from multiple environments. The approaches reviewed here illustrate that the techniques and knowledge of the developmental pathways drawn from *Arabidopsis* will be of great help in identifying QTL candidate genes from other plant species. The first example of the extension of these techniques from *Arabidopsis* to other species has been rice; the next species will likely be maize, tomato, and the model legumes (*Medicago* and *Lotus*). My prediction is that in the future, more phylogenetically diverse species will be amenable to rapid QTL gene identification as research communities come together and create the genomics tools, as exemplified by the efforts of those studying *Mimulus*.

The approaches reviewed here illustrate that the techniques and knowledge of the developmental pathways drawn from *Arabidopsis* will be of great help in identifying QTL candidate genes from other plant species. The first example of the extension of these techniques from *Arabidopsis* to other species has been rice; the next species will likely be maize, tomato, and the model legumes (*Medicago* and *Lotus*). My prediction is that in the future, more phylogenetically diverse species will be amenable to rapid QTL gene identification as research communities come together and create the genomics tools, as exemplified by the efforts of those studying *Mimulus*.

References

Aguade, M., 2001. Nucleotide sequence variation at two genes of the phenylpropanoid pathway, the FAH1 and F3H genes, in *Arabidopsis thaliana*. Mol Biol Evol **18:** 1–9.

Alonso-Blanco, C., and Koornneef, M. 2000. Naturally occurring variation in *Arabidopsis*: an underexploited resource for plant genetics. Trends Plant Sci. **5:** 22–29.

Alonso, J. M., Stepanova, A.N., Leisse, T.J., Kim, C.J., Chen, H., Shinn, P., Stevenson, D.K., Zimmerman, J., Barajas, P., Cheuk, R., and 29 others. 2003. Genome-wide insertional mutagenesis of *Arabidopsis thaliana*. Science (Washington, D.C.) **301:** 653–657.

Alonso-Blanco, C., Blankestijn-De Vries, H., Hanhart, C.J. and Koornneef, M. 1999. Natural allelic variation at seed size loci in relation to other life history traits of *Arabidopsis thaliana*. Proc. Natl. Acad. Sci. U.S.A. **96:** 4710–4717.

Alonso-Blanco, C., El-Assal, S.E.-D., Coupland, G., and Koornneef, M. 1998*a*. Analysis of natural allelic variation at flowering time loci in the Landsberg erecta and Cape Verde Islands ecotypes of *Arabidopsis thaliana*. Genetics **149:** 749–764.

Alonso-Blanco, C., Peeters, A.J.M., Koornneef, M., Lister, C., Dean, C., Van Den Bosch, N., Pot, J., and Kuiper, M.T.R. 1998*b*. Development of an AFLP based linkage map of Ler, Col and Cvi *Arabidopsis thaliana* ecotypes and construction of a Ler/Cvi recombinant inbred line population. Plant J. **14:** 259–271.

Aukerman, M.J., Hirschfeld, M., Wester, L., Weaver, M., Clack, T., Amasino, R.M., and Sharrock, R.A. 1997. A deletion in the PHYD gene of the *Arabidopsis* Wassilewskija ecotype defines a role for phytochrome D in red/far-red light sensing. Plant Cell **9:** 1317–1326.

Baldi, P., and Long, A. D. 2001. A Bayesian framework for the analysis of microarray expression data: regularized t-test and statistical inferences of gene changes. Bioinformatics **17:** 509–519.

Bentsink, L., Alonso-Blanco, C., Vreugdenhil, D., Tesnier, K., Groot, S.P., and Koornneef, M. 2000. Genetic analysis of seed-soluble oligosaccharides in relation to seed storability of *Arabidopsis*. Plant Physiol. **124:** 1595–1604.

Borevitz, J.O., Liang, D., Plouffe, D., Chang, H.S., Zhu, T., Weigel, D., Berry, C.C., Winzeler, E., and Chory, J. 2003. Large-scale identification of single-feature polymorphisms in complex genomes. Genome Res **13:** 513–523.

Borevitz, J.O., Maloof, J.N., Lutes, J., Dabi, T., Redfern, J.L., Trainer, G.T., Werner, J.D., Asami, T., Berry, C.C., Weigel, D., and Chory, J. 2002. Quantitative trait loci controlling light and hormone response in two accessions of *Arabidopsis thaliana*. Genetics **160:** 683–696.

Borevitz, J.O., and Nordborg, M. 2003. The impact of genomics on the study of natural variation in *Arabidopsis*. Plant Physiol. **132:** 718–725.

Breyne, P., Rombaut, D., Van Gysel, A., Van Montagu, M., and Gerats, T. 1999. AFLP analysis of genetic diversity within and between *Arabidopsis thaliana* ecotypes. Mol. Gen. Genet. **261:** 627–634.

Cho, R.J., Mindrinos, M., Richards, D.R., Sapolsky, R.J., Anderson, M., Drenkard, E., Dewdney, J., Reuber, T.L., Stammers, M., Federspiel, N., and 12 others. 1999. Genome-wide mapping with biallelic markers in *Arabidopsis thaliana*. Nature Genet. **23:** 203–207.

Chu, T.M., Weir, B., and Wolfinger, R. 2002. A systematic statistical linear modeling approach to oligonucleotide array experiments. Math. Biosci. **176:** 35–51.

Clarke, J.H., Mithen, R., Brown, J.K., and Dean, C. 1995. QTL analysis of flowering time in *Arabidopsis thaliana*. Mol. Gen. Genet. **248:** 278–286.

Cong, B., Liu, J., and Tanksley, S.D. 2002. Natural alleles at a tomato fruit size quantitative trait locus differ by heterochronic regulatory mutations. Proc. Natl. Acad. Sci. U.S.A. **99:** 13606–13611.

Cutler, D.J., Zwick, M.E., Carrasquillo, M.M., Yohn, C.T., Tobin, K.P., Kashuk, C., Mathews, D.J., Shah, N.A., Eichler, E.E., Warrington, J.A., and Chakravarti, A. 2001. High-throughput variation detection and genotyping using microarrays. Genome Res. **11:** 1913–1925.

Deslandes, L., Pileur, F., Liaubet, L., Camut, S., Can, C., Williams, K., Holub, E., Beynon, J., Arlat, M.. and Marco, Y. 1998. Genetic characterization of *RRS1*, a recessive locus in *Arabidopsis thaliana* that confers resistance to the bacterial soilborne pathogen *Ralstonia solanacearum*. Mol. Plant-Microbe Interact. **11:** 659–667.

Doebley, J., Stec, A., and Hubbard, L. 1997. The evolution of apical dominance in maize. Nature **386:** 485–488.

Doerge, R.W. 2002. Mapping and analysis of quantitative trait loci in experimental populations. Nat. Rev. Genet. **3:** 43–52.

El-Din El-Assal, S., Alonso-Blanco, C., Peeters, A.J., Raz, V., and Koornneef, M. 2001. A QTL for flowering time in *Arabidopsis* reveals a novel allele of CRY2. Nature Genet. **29:** 435–440.

Erschadi, S., Haberer, G., Schoeniger, M., and Torres-Ruiz, R. A. 2000. Estimating genetic diversity of *Arabidopsis thaliana* ecotypes with amplified fragment length polymorphisms (AFLP). Theor. Appl. Genet. **100:** 633–640.

Frary, A., Nesbitt, T.C., Grandillo, S., Knaap, E., Cong, B., Liu, J., Meller, J., Elber, R., Alpert K.B., and Tanksley, S.D. 2000. *fw2.2*: a quantitative trait locus key to the evolution of tomato fruit size. Science (Washington, D.C.) **289:** 85–88.

Fridman, E., Pleban T., and Zamir, D. 2000. A recombination hotspot delimits a wild-species quantitative trait locus for tomato sugar content to 484 bp within an invertase gene. Proc. Natl. Acad. Sci. U.S.A. **97:** 4718–4723.

Guo, H., Yang, H., Mockler T.C., and Lin, C. 1998. Regulation of flowering time by *Arabidopsis* photoreceptors. Science (Washington, D.C.) **279:** 1360–1363.

Hubbard, L., McSteen, P., Doebley, J. and Hake, S. 2002. Expression patterns and mutant phenotype of *teosinte branched1* correlate with growth suppression in maize and teosinte. Genetics **162:** 1927–1935.

Innan, H., Tajima, F., Terauchi R., and Miyashita, N.T. 1996. Intragenic recombination in the Adh locus of the wild plant *Arabidopsis thaliana*. Genetics **143:** 1761–1770.

Johanson, U., West, J., Lister, C., Michaels, S., Amasino R., and Dean, C. 2000. Molecular analysis of FRIGIDA, a major determinant of natural variation in *Arabidopsis* flowering time. Science (Washington, D.C.) **290:** 344–347.

Kardailsky, I., Shukla, V.K., Ahn, J.H., Dagenais, N., Christensen, S.K., Nguyen, J.T., Chory, J., Harrison M.J., and Weigel, D. 1999. Activation tagging of the floral inducer FT. Science (Washington, D.C.) **286:** 1962–1965.

Kawabe, A., and Miyashita, N.T. 1999. DNA variation in the basic chitinase locus (ChiB) region of the wild plant *Arabidopsis thaliana*. Genetics **153:** 1445–1453.

Kawabe, A., Yamane, K., and Miyashita, N.T. 2000. DNA polymorphism at the cytosolic phosphoglucose isomerase (PgiC) locus of the wild plant *Arabidopsis thaliana*. Genetics **156:** 1339–1347.

Kliebenstein, D.J., Lambrix, V.M., Reichelt, M., Gershenzon, J., and Mitchell-Olds, T. 2001. Gene duplication in the diversification of secondary metabolism: tandem 2-oxoglutarate-dependent dioxygenases control glucosinolate biosynthesis in *Arabidopsis*. Plant Cell **13:** 681–693.

Kobayashi, Y., Kaya, H., Goto, K., Iwabuchi M., and Araki, T. 1999. A pair of related genes with antagonistic roles in mediating flowering signals. Science (Washington, D.C.) **286:** 1960–1962.

Kojima, S., Takahashi, Y., Kobayashi, Y., Monna, L., Sasaki, T., Araki, T., and Yano, M. 2002. Hd3a, a rice ortholog of the *Arabidopsis* FT gene, promotes transition to flowering downstream of Hd1 under short-day conditions. Plant Cell Physiol. **43:** 1096–1105.

Kroymann, J., Textor, S., Tokuhisa, J.G., Falk, K.L., Bartram, S., Gershenzon, J., and Mitchell-Olds, T. 2001. A gene controlling variation in *Arabidopsis* glucosinolate composition is part of the methionine chain elongation pathway. Plant Physiol. **127:** 1077–1088.

Le Corre, V., Roux, F., and Reboud, X. 2002. DNA polymorphism at the FRIGIDA gene in *Arabidopsis thaliana*: extensive nonsynonymous variation is consistent with local selection for flowering time. Mol. Biol. Evol. **19:** 1261–1271.

Lister, C., and Dean, C. 1993. Recombinant inbred lines for mapping RFLP and phenotypic markers in *Arabidopsis thaliana*. Plant J. **4:** 745–750.

Loudet, O., Chaillou, S., Camilleri, C., Bouchez, D., and Daniel-Vedele, F. 2002. Bay-0 × Shahdara recombinant inbred line population: a powerful tool for the genetic dissection of complex traits. Theor. Appl. Genet. **104:** 1173–1184.

Luo, D., Carpenter, R., Vincent, C., Copsey L., and Coen, E. 1996. Origin of floral asymmetry in *Antirrhinum*. Nature **383:** 794–799.

Maloof, J. N., 2003. QTL for plant growth and morphology. Curr. Opin. Plant Biol. **6:** 85–90.

Maloof, J.N., Borevitz, J.O., Dabi, T., Lutes, J., Nehring, R.B., Redfern, J.L., Trainer, G.T., Wilson, J.M., Asami, T., Berry, C.C., Weigel, D. and Chory, J. 2001. Natural variation in light sensitivity of *Arabidopsis*. Nature Genet. **29:** 441–446.

Mauricio, R. 2001. Mapping quantitative trait loci in plants: uses and caveats for evolutionary biology. Nat. Rev. Genet **2:** 370–381.

Michaels, S.D., and Amasino, R.M. 1999. FLOWERING LOCUS C encodes a novel MADS domain protein that acts as a repressor of flowering. Plant Cell **11:** 949–956.

Michelmore, R.W., Paran, I., and Kesseli, R.V. 1991. Identification of markers linked to disease-resistance genes by bulked segregant analysis: a rapid method to detect markers in specific genomic regions by using segregating populations. Proc. Natl. Acad. Sci. U.S.A. **88:** 9828–9832.

Mitchell-Olds, T. 2001. *Arabidopsis thaliana* and its wild relatives: A model system for ecology and evolution. Trends Ecol. Evol. **16:** 693–700.

Miyashita, N. T., Kawabe, A., and Innan, H. 1999. DNA variation in the wild plant *Arabidopsis thaliana* revealed by amplified fragment length polymorphism analysis. Genetics **152:** 1723–1731.

Nesbitt, T.C., and Tanksley, S.D. 2002. Comparative sequencing in the genus *Lycopersicon*. Implications for the evolution of fruit size in the domestication of cultivated tomatoes. Genetics **162:** 365–379.

Nordborg, M., Borevitz, J.O., Bergelson, J., Berry, C.C., Chory, J., Hagenblad, J., Kreitman, M., Maloof, J.N., Noyes, T., Oefner, P.J., Stahl E.A., and Weigel, D. 2002. The extent of linkage disequilibrium in *Arabidopsis thaliana*. Nature Genet. **30:** 190–193.

Nuwaysir, E.F., Huang, W., Albert, T.J., Singh, J., Nuwaysir, K., Pitas, A., Richmond, T., Gorski, T., Berg, J. P., Ballin, J., and 13 others. 2002. Gene expression analysis using oligonucleotide arrays produced by maskless photolithography. Genome Res **12:** 1749–1755.

Olsen, K.M., Womack, A., Garrett, A.R., Suddith J.I., and Purugganan, M.D. 2002. Contrasting evolutionary forces in the *Arabidopsis thaliana* floral developmental pathway. Genetics **160:** 1641–1650.

Osborn, T.C., Kole, C., Parkin, I.A., Sharpe, A.G., Kuiper, M., Lydiate, D.J., and Trick, M. 1997. Comparison of flowering time genes in *Brassica rapa*, *B. napus* and *Arabidopsis thaliana*. Genetics **146:** 1123–1129.

Paterson, A. H. 1998. Molecular dissection of complex traits. CRC Press, Boca Raton, FL.

Patil, N., Berno, A.J. Hinds, D.A., Barrett, W.A., Doshi, J.M., Hacker, C.R., Kautzer, C.R., Lee, D.H., Marjoribanks, C., McDonough, D.P., and 12 others. 2001. Blocks of limited haplotype diversity revealed by high-resolution scanning of human chromosome 21. Science (Washington, D.C.) **294:** 1719–1723.

Putterill, J., Robson, F., Lee, K., Simon, R., and Coupland, G. 1995. The CONSTANS gene of *Arabidopsis* promotes flowering and encodes a protein showing similarities to zinc finger transcription factors. Cell **80:** 847–857.

Rosenberg, N.A., and Nordborg, M. 2002. Genealogical trees, coalescent theory and the analysis of genetic polymorphisms. Nat. Rev. Genet. **3:** 380–390.

Schranz, M.E., Quijada, P., Sung, S.B., Lukens, L., Amasino, R., and Osborn, T.C. 2002. Characterization and effects of the replicated flowering time gene FLC in *Brassica rapa*. Genetics **162:** 1457–1468.

Sessions, A., Burke, E., Presting, G., Aux, G., McElver, J., Patton, D., Dietrich, B., Ho, P., Bacwaden, J., Ko, C., and 12 others. 2002. A high-throughput *Arabidopsis* reverse genetics system. Plant Cell **14:** 2985–2994.

Stickney, H.L., Schmutz, J., Woods, I.G., Holtzer, C.C., Dickson, M.C., Kelly, P.D., Myers, R M., and Talbot, W.S. 2002. Rapid mapping of zebrafish mutations with SNPs and oligonucleotide microarrays. Genome Res **12:** 1929–1934.

Sugano, S., Andronis, C., Green, R.M., Wang Z.Y., and Tobin, E.M. 1998. Protein kinase CK2 interacts with and phosphorylates the *Arabidopsis* circadian clock-associated 1 protein. Proc. Natl. Acad. Sci. U.S.A. **95:** 11020–11025.

Swarup, K., Alonso-Blanco, C., Lynn, J.R., Michaels, S.D., Amasino, R M., Koornneef, M. and Millar, A. J. 1999. Natural allelic variation identifies new genes in the *Arabidopsis* circadian system. Plant J. **20:** 67–77.

Takahashi, Y., Shomura, A. Sasaki, T., and Yano, M. 2001. Hd6, a rice quantitative trait locus involved in photoperiod sensitivity, encodes the alpha subunit of protein kinase CK2. Proc. Natl. Acad. Sc.i U.S.A. **98:** 7922–7927.

Tanksley, S. D. 1993. Mapping polygenes. Annu. Rev. Genet. **27:** 205–233.

The Arabidopsis Genome Initiative. 2000. Analysis of the genome sequence of the flowering plant *Arabidopsis thaliana*. Nature **408:** 796–815.

Tusher, V.G., Tibshirani, R., and Chu, G. 2001. Significance analysis of microarrays applied to the ionizing radiation response. Proc. Natl. Acad. Sci. U.S.A. **98:** 5116–5121.

Ungerer, M.C., Halldorsdottir, S.S., Modliszewski, J.L., Mackay T.F., and Purugganan, M.D. 2002. Quantitative trait loci for inflorescence development in *Arabidopsis thaliana*. Genetics **160:** 1133–1151.

Wang, R.L., Stec, A., Hey, J., Lukens, L, and Doebley, J. 1999. The limits of selection during maize domestication. Nature **398:** 236–239.

Wang, Z.Y., and Tobin, E.M. 1998. Constitutive expression of the CIRCADIAN CLOCK ASSOCIATED 1 (CCA1) gene disrupts circadian rhythms and suppresses its own expression. Cell **93:** 1207–1217.

Wilson, I.W., Schiff, C.L., Hughes, D.E., and Somerville, S.C. 2001. Quantitative trait loci analysis of powdery mildew disease resistance in the *Arabidopsis thaliana* accession kashmir-1. Genetics **158:** 1301–1309.

Winzeler, E.A., Richards, D.R., Conway, A.R., Goldstein, A.L., Kalman, S., McCullough, M.J., McCusker, J.H., Stevens, D.A., Wodicka, L., Lockhart, D.J., and Davis, R.W. 1998. Direct allelic variation scanning of the yeast genome. Science (Washington, D.C.) **281:** 1194–1197.

Yano, M., Katayose, Y., Ashikari, M., Yamanouchi, U., Monna, L., Fuse, T., Baba, T., Yamamoto, K., Umehara, Y., Nagamura, Y,. and Sasaki, T., 2000. Hd1, a major photoperiod sensitivity quantitative trait locus in rice, is closely related to the *Arabidopsis* flowering time gene CONSTANS. Plant Cell **12:** 2473–2484.

Yanovsky, M.J., Casal, J.J., and Luppi, J.P. 1997. The VLF loci, polymorphic between ecotypes Landsberg erecta and Columbia, dissect two branches of phytochrome A signal transduction that correspond to very-low-fluence and high-irradiance responses. Plant J. **12:** 659–667.

9 Experimental analysis of adaptive landscape topographies

Mitchell B. Cruzan and Jennifer M. Rhode

Abstract: The metaphor of a fitness landscape of peaks and valleys has been a valuable heuristic tool for understanding the process of adaptive evolution. Wright envisioned landscapes of high relief, with fitness peaks separated by deep troughs and evolution occurring via peak shifts. Recently, Gavrilets extrapolated Wright's model to a multidimensional scale and demonstrated that high-fitness ridges connecting peaks may be a common feature of landscapes. In this view of a "holey adaptive landscape", populations can traverse ridges through mutation-drift processes alone and become reproductively isolated when they come to reside on opposite sides of a fitness "hole". In this paper, we discuss existing empirical evidence and experimental approaches to the assessment of adaptive landscape topography. We provide an example of landscape topography analysis using fitness distributions of parental and hybrid genotypes of herbaceous perennials in the *Piriqueta caroliniana* complex. In this study, the mean fitness of backcross hybrids under field conditions was relatively low, but some recombinant hybrids had fitnesses equal to or exceeding parental types. While these data are consistent with the presence of a low-fitness "hole" and a high-fitness "ridge" separating the parental types, details of the intervening topography are lacking. We propose that more rigorous analyses that integrate genomic approaches with experimental tests of fitness distributions will lead to a better understanding of fitness topographies and the process of adaptive evolution.

Introduction

While adaptive evolution is a central tenet of modern biology, studies of its associated processes remain difficult. The primary obstacle to such investigations is the genetic complexity that underlies individual traits; the "one gene, one product" model fails to describe most living systems. Research into adaptive evolution is further confounded by phenotypic plasticity, which allows complex morphological traits to be expressed differentially across space and time. In the absence of much experimental evidence, our view of adaptive evolution has been shaped by the heuristic tools of Wright (Wright 1931; 1932; 1988) and of Fisher (1930). Both authors used the metaphor of a fitness landscape with regions of low and high fitness to describe evolution in populations. These models, while not providing specific hypotheses for testing and analysis, summarize contrasting views of the adaptive evolution process.

Models of both Wright and Fisher allow fitness optima to shift in response to environmental changes and emphasize adaptive "hill climbing", but they differ in the relative importance of epistatic (unequal expression of a mutant allele in different backgrounds) and additive (equal effects of a mutant allele in a variety of genetic backgrounds) genetic variation. In Wright's model, interactions among alleles at different loci (epistasis) explain a large proportion of the genetic variation. Populations conforming to this model inhabit a rugged fitness surface with multiple high-fitness peaks, and their evolution is characterized by periods of stasis interrupted by rapid changes (peak shifts) (Simpson 1944; Arnold et al. 2001). Unlike Wright, Fisher envisioned evolution as a gradual process, with allelic substitutions enabling populations to move steadily across a relatively smooth incline towards a single fitness optimum. These contrasting views of genetic architecture of traits, and the varied topographies that result, continue to be a subject of active debate (e.g., Coyne et al. 1997; Wade and Goodnight 1998). Though opinions on topographic specifics differ, fitness surfaces are considered a useful metaphor for understanding adaptive evolution.

Wright's and Fisher's models of adaptive evolution predict that allele substitution will have different effects on genotypes occupying the same environment; Wright's rugged landscape is the consequence of fitness effects of specific combinations of alleles at separate loci, while Fisher's model predicts that alleles contributing to fitness differences will have similar effects regardless of genetic background. For example, alleles contributing to an increase in the size of floral displays may have positive fitness effects in a broad range of genotypes, but alleles affecting expression of characters that are part of a pollination syndrome (e.g., red petals) may only have positive fitness effects in the presence of alleles that produce other specific characters (e.g., tubular flowers). This simple example serves to illustrate the salient

Mitchell B. Cruzan[1] and Jennifer M. Rhode[2]. Department of Biology, Portland State University, Portland, OR 97207, U.S.A.
Correct citation: Cruzan, M.B., and Rhode, J.M. 2004. Experimental analysis of adaptive landscape topographies. *In* Plant Adaptation: Molecular Genetics and Ecology. Proceedings of an International Workshop held December 11–13, 2002, in Vancouver, British Columbia, Canada. *Edited by* Q.C.B. Cronk, J. Whitton, R.H. Ree, and I.E.P. Taylor. NRC Research Press, Ottawa, Ontario. pp. 61–69.
[1]Corresponding author e-mail address: Cruzan@pdx.edu
[2]Current address: Georgia College and State University, Department of Biology and Environmental Science, CBX 081, Herty Hall 202, Milledgeville, GA 31061, U.S.A.

differences between these views and highlights the fact that they need not be mutually exclusive.

Wright's vision of a rugged adaptive landscape has garnered support as it has been widely used as a contextual framework for theoretical and empirical investigations of adaptive evolution (Barton and Charlesworth 1984; Whitlock et al. 1995*b*; Wade and Goodnight 1998). This model, which is largely verbal with graphical depictions of the fitness surface, has been modified since its introduction (Wright 1931; Wright 1932). Wright's original depiction of a multilocus genetic surface (Wright 1932; Provine 1986) was later refined to an allele frequency surface for two loci (Wright 1988; Whitlock et al. 1995*a*), and this genetic landscape has been interpreted as a phenotypic surface for paired traits (Pearson 1903; Simpson 1944; Lande 1976; Arnold et al. 2001; Cruzan 2001). The key feature shared by these depictions of adaptive evolution is the critical role of epistasis; only interactions among alleles at different loci can produce a rugged fitness topography when the environment is homogeneous.

While theoretical analyses that assume a rugged fitness surface have become common (Barton and Charlesworth 1984; Wade and Goodnight 1998), empirical assessments of adaptive landscape topography are relatively scarce. Phenotypic landscapes are the most amenable to empirical analysis and have become popular as tools for making direct assessments of fitness topographies (Kingsolver 1988; Armbruster 1990; Cresswell and Galen 1991; Schluter 2000). Recent analyses of gene frequency landscapes have focused on the relative importance of epistasis (Whitlock et al. 1995*a*; Armbruster et al. 1997; Kim and Rieseberg 2001) and on evidence for the existence of specific gene combinations representing separate adaptive peaks (Wade and Goodnight 1991; Korona et al. 1994; Lenski and Travisano 1994; Cluster and Allard 1995; Burch and Chao 1999; Kaltz and Bell 2002). Most of these analyses indicate that adaptive landscapes are somewhat rugged, but the resolution of many of these studies are too low to assess the depth of valleys (i.e., the relative importance of epistasis) and the presence of ridges that may connect separate adaptive peaks.

Since many loci probably contribute to fitness, and numbers of potential interactions among them are exponentially higher, it has been suggested that Wright's vision of a two-locus, two-dimensional landscape may be inadequate to accurately represent fitness relationships among multilocus genotypes (Whitlock et al. 1995*a*; Gavrilets 1997*a*). Gavrilets and colleagues (Gavrilets 1997*b*; Gavrilets 1997*a*; Gavrilets 1999) approached this problem from a theoretical perspective and determined that epistasis among a large number of loci affecting fitness will result in a multidimensional genetic landscape that retains many of the characteristics of two-dimensional rugged landscapes. However, a critical difference develops at high levels of dimensionality. Here, clusters of adaptive genotypes become interconnected by "ridges" of high fitness, which are the direct consequence of epistasis among a large number of loci contributing to fitness. These ridges represent paths of high fitness that form a network of interconnected regions rather than the isolated peaks that are characteristic of systems with few dimensions. Based on these results, Gavrilets concluded that, with a sufficient number of interacting loci, populations could easily move between "peaks" by traversing "ridges" of high fitness through mutation-drift processes. Speciation would result when populations came to occupy opposite sides of a low fitness "hole" (i.e., such that crosses between them only produced inviable hybrids) without ever having to pass through an adaptive valley (Fig. 1).

Fig. 1. Graphical depiction of divergent evolution along the rim of a fitness "hole" conforming to Gavrilets' (1997*a*) metaphor of a holey adaptive landscape. The arrows indicate the expected relative success of crosses between the progenitor population (A) and populations at different stages of divergence (B–E). Hybridization between newly derived populations (A to B) results in a large proportion of second-generation hybrid genotypes falling along the fitness "ridge". For crosses to more divergent populations, an increasing number of second-generation hybrids display low fitness recombinant genotypes (i.e., they fall into the fitness "hole").

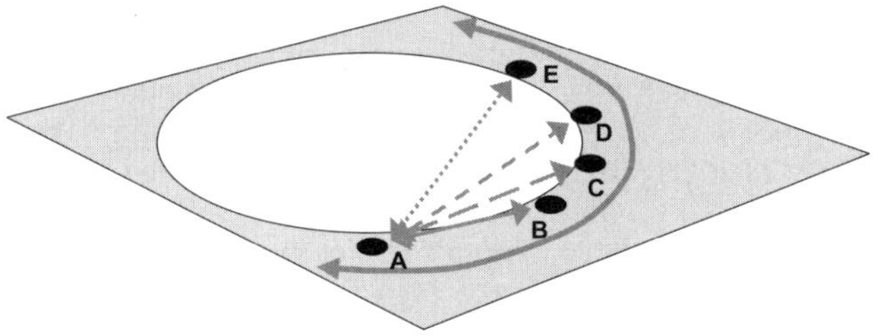

The metaphor of a holey adaptive landscape provides insights into speciation via accumulation of mutations in divergent lineages (Dobzhansky 1936; Muller 1942; Orr 1995; Coyne and Orr 1998; Orr 2001). In the Dobzhansky-Muller model for evolution of reproductive isolation, a lack of fitness effects of mutations that accumulate in separate lineages is contingent upon the presence of previously fixed alleles at different loci in the same genome (Orr 1995). However, when mutations from two different lineages are combined through hybridization, reduced fitness can result from interaction between mutant, divergent alleles (Orr 2001). Hence, speciation can proceed without loss of fitness via a chain of intermediate genotypes separated by single mutational steps. The Dobzhansky-Muller model is consistent with conditions expected under a holey adaptive landscape because it is dependent on the presence of ridges that allow population divergence and reproductive isolation to develop without severe fitness losses.

This view of a "holey" adaptive landscape is compelling because the resulting topology is a direct consequence of extrapolating from the simple two-dimensional case of Wright to a multidimensional level (Gavrilets 1997*a*). Moreover, there is empirical evidence that it is common to have ridges of high fitness that connect adaptive regions. For example, experiments have demonstrated that separate microbial populations derived from the same ancestral population will tend to diverge over time without losing fitness compared to progenitors (Korona et al. 1994; Lenski and Travisano 1994). Indeed, evolution of these populations was accompanied by fitness increases reminiscent of a mutation/selection process expected under the type of strictly additive genetic model considered by Fisher (1930). However, the microbial populations in these experiments can also undergo morphological divergence (Korona et al. 1994) as they occupy different regions of the adaptive landscape (rather than a single region as expected under Fisher's model). The resulting populations can also be characterized by different average fitnesses (adaptive peaks of differing heights) that are retained over many thousands of generations (Lenski and Travisano 1994), suggesting that they may have become trapped in regions of the adaptive landscape that are characterized by low connectivity (i.e., fewer ridges). While these experiments are intriguing, more analyses in a wider variety of organisms are needed before any general conclusions can be drawn concerning the ruggedness of adaptive landscapes and levels of connectivity among high fitness regions.

Empirical assessments of adaptive landscape topographies

As described above, most empirical studies of adaptive landscape topographies have focused on the relative importance of epistasis, which is indicated by the presence of multiple fitness peaks. Some authors have endeavored to "map" fitness surfaces by examining the character distributions in groups of closely related species (Kingsolver 1988; Armbruster 1990; Cresswell and Galen 1991; Vinogradov 1999; Schluter 2000). However, individual species are generally presumed to occupy single peaks, so this approach allows for assessment of the general topography without providing information on the specific features that separate pairs of taxa (i.e., whether they are separated by valleys or connected by ridges). Furthermore, Wright's model assumes that the environment is spatially uniform, so that adaptive peaks represent different solutions to the same set of environmental conditions. When environments differ, the apparent fitness peaks that represent taxa occupying different habitats are likely to be due to additive as well as epistatic genetic variation. The requirement of environmental homogeneity for a robust assessment of the contribution of epistasis to landscape topography renders comparison of more divergent lineages difficult, particularly if their habitat requirements are known to differ (e.g., Armbruster 1990; Vinogradov 1999). More detailed analyses of fitness landscapes of the intervening regions between recently diverged populations in similar environments are needed to more accurately discriminate among the proposed alternative topographies.

Strategies for landscape topography analysis

One promising approach for assessing adaptive landscape topography is the analysis of fitness distributions of hybrids derived from crosses between recently diverged lineages (Wright 1978; Geiger 1988; Whitlock et al. 1995*a*). The main goals of such analyses should be focused on establishing characteristics of the topographic features of regions separating pairs of divergent populations. For example, the mean fitness of early hybrid generations (i.e., F_1, F_2, and backcross hybrids) would indicate whether the average fitness elevations were lower than the presumed fitness peaks. The range of fitnesses among individual hybrid genotypes would help establish the depth of a valley or hole, occupied by the lowest fitness genotypes, or the elevation of a putative ridge, occupied by the highest fitness genotypes. Low fitnesses for all second-generation hybrid genotypes (F_2 and backcross) would suggest that a true valley exists between the divergent lineages and the local topography is rugged, or that divergence has proceeded to the point where all hybrid genotypes fall into regions of low fitness (Fig. 1). If divergence had not proceeded too far, the holey landscape model would predict both high and low fitness hybrids (Gavrilets 1997*b*), which would effectively indicate the presence of both a fitness ridge and a hole separating the two divergent lineages.

Hybrid fitness distributions could provide valuable information on local adaptive landscape topography, but there are several cautions that should be considered before proceeding with such experiments. First, this kind of analysis would be best conducted with recently diverged lineages. Focusing on closely related taxa would improve the chance that the information obtained pertains to a specific portion of the local topography and would avoid instances where divergence had proceeded too far to detect intervening ridges (Fig. 1). Secondly, the taxa being compared need to be from similar environments, and fitness assessments need to be conducted in their native habitats. Controlling for environmental variation as much as possible will help ensure that valid comparisons are being made across a single landscape and circumvent complications due to additive genetic variation for adaptation to different habitats. In this paper we provide an example of an analysis of fitness distributions of hybrids between two closely related and ecologically similar groups of populations. Our analyses endeavor to establish the basic characteristics of the topographical features separating two putative adaptive peaks and to identify the primary modes of adaptive evolution leading to diversification in this group.

Hybridization and evolution in *Piriqueta*

Our work focuses on two taxa within the *Piriqueta caroliniana* complex (Turneraceae). Plants in this group have been variously classified as separate taxa (Small 1933; Long and Lakela 1971) or have been lumped into a single subspecies (Arbo 1995). For purposes of our studies we have defined separate taxonomic units (morphotypes) that possess diagnostic morphological and ecological characters and have independent evolutionary histories (Maskas and Cruzan 2000). This study focuses on two of those morphotypes and their hybrid derivatives. The *caroliniana* morphotype, which is found in the northern range of the hybrid zone, has broad, hirsute leaves and a decumbent growth pattern. It is found in well-drained quartz sand soils and is associated with the turkey oak scrub habitats of northern Florida and southern Georgia. The *viridis* morphotype has narrow, smooth leaves and an erect growth pattern. It grows in poorly-drained limestone soils and is associated with slash pine and palmettos. Between the ranges of these parental types is a broad geographic region occupied by hybrid plants that display genotypes and phenotypes intermediate to those of *caroliniana* and *viridis* (Martin and Cruzan 1999). Morphological and genetic analyses have revealed that hybridization between *caroliniana* and *viridis* began around 5000 to 7000 ybp (Maskas and Cruzan 2000). The hybrid zone between these morphotypes is wide and appears to have expanded in recent history as *viridis* genes have introgressed northwards into the range of *caroliniana*. The dynamics of this hybrid zone and ultimate fate of the expansion process are still under investigation.

In this study we examine the fitness distributions of plants from crosses within and between the *caroliniana* morphotype (C) and a hybrid derivative of introgression between the *caroliniana* and *viridis* morphotypes (H). The hybrid derivative genotypes were collected from populations at the center of the hybrid zone that were morphologically uniform (Martin and Cruzan 1999) and had not been subjected to recent introgression from allopatric regions (Cruzan and Estill 2004; Rhode and Cruzan 2004). The C and H parental types were chosen as a good model system for studies of adaptive landscape topography as they meet all the criteria outlined above; (*i*) they are closely related, so hybrid crosses between them are highly successful and form viable hybrid offspring (Wang and Cruzan 1998; Rhode and Cruzan 2004), and (*ii*) they both occur in sandhill turkey oak scrub habitats, which are broadly similar in physical and vegetative characteristics between their respective locations in northern and central Florida (Myers 1990). *Viridis* (V) individuals were excluded from the analyses as their preferred habitat differed from that of C and H. For the purpose of this paper, all fitness comparisons were made between individuals planted in the same common garden at a field site in central Florida (the "red hill" area of Archbold Biological Station).

We compare the fitnesses of *caroliniana* (C), hybrid derivatives (H), F_1 (*caroliniana* × hybrid: CH, HC), and backcross hybrids (CH × C, CH × H) in an attempt to describe the adaptive landscape separating these distinct parental genotypes. Both the rugged and holey landscape models would predict that the average fitness of backcross generation hybrids would be lower than the parental genotypes. However, the distinction between these two models is indicated by the fitness distribution of the backcross hybrids; the presence of some individual genotypes with fitnesses that are equivalent to the parentals could be interpreted as evidence for a ridge that connects the putative adaptive peaks. Since a larger portion of the recombinant genotypes are likely to be located in fitness holes, we would expect the mean fitness of backcrosses to be lower even if ridges were present. A true adaptive valley, on the other hand, would be indicated if all backcross genotypes had lower fitnesses than the parentals. We analyze the fitness distribution of backcross hybrid genotypes between these divergent lineages in their natural environment to provide some preliminary evidence concerning the local fitness topography.

Materials and methods

Recently we began a series of field investigations to identify factors that contribute to hybrid success in the *Piriqueta caroliniana* species complex. In summer 2001, *caroliniana* (C) and hybrid derivative (H) individuals were collected from field sites in Florida and Georgia. Hand-pollinations were used to produce first (CH and HC, where the first letter indicates the maternal parent) and backcross generation hybrids (eight reciprocal combinations of C and H with CH and HC hybrids). At least 30 crosses per combination were attempted, yielding 3 to 24 fruits each. Seeds from these fruits were planted, and individuals were greenhouse-germinated in February 2002. Three to five cuttings per genotype were taken in May 2002, and established cuttings were planted in central Florida common gardens at the Archbold Biological Station in July. The site chosen for the transplant garden was typical of both the *caroliniana* and derivative hybrid habitat (sparsely vegetated regions in sandhills surrounded by turkey oak scrub vegetation). More than 350 cuttings representing 75 genotypes (mean = 4.73 cuttings per genotype) were planted in a randomized incomplete block design.

Plants were monitored every 6 weeks for vegetative growth and sexual reproduction over two field seasons. Monitoring occurred in July, August, September, and November of the first year and May, July, and August of the second year. Ratios of leaf width to length were calculated for each plant, averaged across clones, and used as a measure of phenotypic variation among individual genotypes. A vegetative growth component of fitness (aboveground biomass) was estimated as the product of the number of leaves and total stem length. This estimate of vegetative fitness did not vary among parental types in greenhouse plants (unpub. data), so we are confident that it represents biomass variation due to differences in vigor rather than differences in genetically constrained taxonomic characters. Total reproductive output was calculated as the sum of buds, flowers, and fruits present at each of the 7 monitoring periods over both seasons. Differences among parents, F_1 hybrids, and backcross hybrids for vegetative and reproductive components of fitness were assessed using SAS PROC GLM repeated measures analysis of variance (SAS 1999), which included contrast statements to test for differences between the backcross and F_1 hybrids and between the two parental genotypes (midparent value). In all tests, spatial location in the garden (plot) and genotype (nested within cross group) were treated as random factors.

Results

Repeated measures analysis of variance revealed significant differences in total above-ground biomass among the parental types and the two hybrid generations over the first season of growth (repeated measures $T = 4$, $F = 9.29$, $P < 0.0001$, 2/66 df for cross group tested over the variance among genotypes; Fig. 2). There were significant differences in above-ground biomass in the first season of growth for the contrasts between the backcross and parental genotypes (backcross < midparent: $F = 11.68$, $P < 0.0008$) and the backcross and F_1 genotypes (backcross < F_1: $F = 6.96$, $P < 0.0090$) but not the F_1 and parental genotypes (midparent = F_1: $F = 0.05$, $P = 0.8310$). Vegetative size at the end of the first season explained a large proportion of the variation in over-winter survival in this (logistic regression: chi square = 5.55, $P < 0.0185$) and a previous study (Rhode and Cruzan 2004). Mortality of the smallest individuals tended to equalize vegetative sizes among groups for the second season (repeated measures $T = 3$, $F = 1.00$, $P = 0.3704$, 2/213 df). Total reproductive output differed among hybrid and parental genotypes ($F = 4.79$, $P < 0.0091$, 2/229 df for cross group tested over the variance among genotypes; Fig. 3), with the

Fig. 2. Means and standard errors for the vegetative component of fitness (estimated aboveground biomass) and phenotypic differences (leaf width/length ratio) for the *caroliniana* morphotype (C), an advanced-generation hybrid derivative (H), and the F_1 (F1) and backcross (BC) hybrids from crosses between C and H of the *Piriqueta caroliniana* complex. For the C and H parental types and the F_1 hybrids, only means and standard errors are shown. For the backcross hybrids both the mean (triangle with the error bar) and individual observations of each genotype (means of field-grown cuttings: triangles without error bars) are shown.

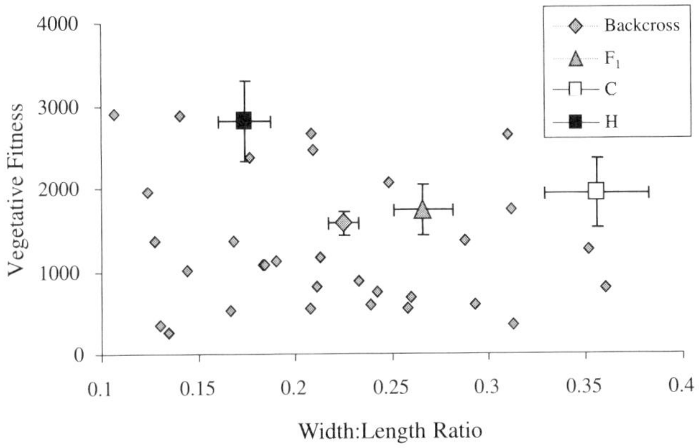

Fig. 3. The proportion of individual genotypes of the *caroliniana* morphotype (C), an advanced-generation hybrid derivative (H), and the F_1 (F1) and backcross (BC) hybrids from crosses between C and H of the *Piriqueta caroliniana* complex reproducing (i.e., producing flowers or fruits) during the first season.

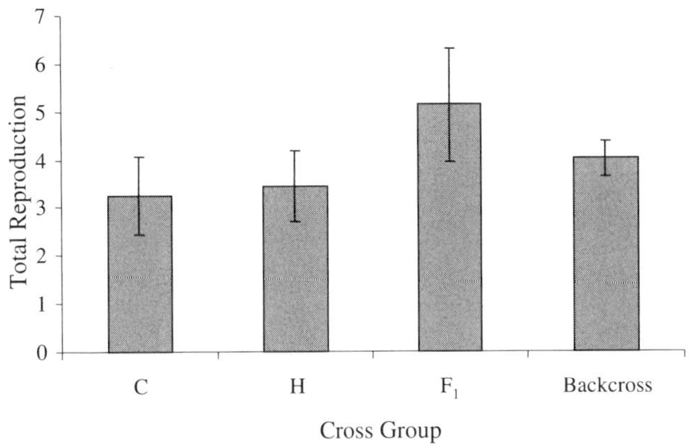

F_1 hybrids reproducing most ($F = 9.47$ with $P < 0.023$ and $F = 8.75$ with $P < 0.0043$ for F_1 vs. parental and backcross genotypes, receptively) and backcross and parental genotypes reproducing equally ($F = 0.20$, $P = 0.6550$). It is important to note that the parental genotypes used in this study were inbred (i.e., they were from within-population crosses); therefore, the observed high reproductive output for F_1 genotypes was probably partly due to heterosis, which was found to contribute significantly to F_1's fitness in a previous study (Rhode and Cruzan 2004). About half (47.88%) of the backcross genotypes never reproduced, but some backcrossed plants (16.5%) produced relatively large numbers of flowers and fruits (>10); only 10% of the parental genotypes produced >10 flowers and fruits. Though mean vegetative fitness of parentals was higher than that of hybrids, ten backcrossed genotypes (15% of all backcrosses) had vegetative fitness estimates exceeding that of parentals (Fig. 2).

Discussion

Our field studies revealed that the average vigor of early generation hybrids was relatively low, but some individuals were quite successful, with fitnesses equal to or exceeding those of parental genotypes. The presence of high-fitness hybrid genotypes across the phenotypic range that separates these divergent lineages is consistent with the hypothesis that adaptive ridges are present. Further, as expected for a holey fitness landscape, a large proportion of hybrids fell into regions of low fitness, indicating the presence of an adjoining hole. These patterns suggest that epistasis among divergent genomic elements is prevalent, but there are a sufficient number of viable genetic combinations to potentially form regions of high fitness to connect adaptive peaks.

In this study we examined fitness distributions of intermediate genotypes between divergent lineages that were derived from a past hybridization event. The model proposed by Gavrilets envisions gradual divergence along an adaptive surface through mutation-drift processes, but hybridization differs from this process since it simultaneously introduces a large number of alien genetic elements into a population. However, it is important to note that the adaptive landscape exists independently of the mode of derivation, so it is reasonable to test for the presence of a ridge even if it does not represent the path followed by the newly derived lineage. For example, it is possible that the introduction of a large number of foreign alleles via hybridization forced the population through a low-fitness hole. The analysis presented here is not meant to recover the actual evolutionary development of a new lineage but rather aims to evaluate the possibility that the new phenotype could have been derived without severe fitness losses.

Presumably the high-fitness hybrid genotypes detected in these crosses and in other studies (see reviews in Arnold and Hodges 1995; Arnold 1997) fall along an avenue that would facilitate divergence; however, details of the intervening fitness topography in this analysis are lacking. For example, if one or more ridges are present, it is not clear whether any of them connect to form a continuous path or whether these genotypes represent isolated fitness pinnacles (Fig. 4). Under ideal circumstances we would map genotypes on a cartographic space that was defined by the salient adaptive characters. Such an analysis would provide information on the arrangement of high fitness genotypes across the phenotypic landscape. Casual inspection could then indicate the location and height of ridges and the depth of holes. It may be possible to obtain an approximate realization of the local landscape by using pairs of composite variables that combine a number of phenotypic traits that appear to have adaptive relevance. However, it is equally likely that such an analysis could be misleading. For example, because of the complicated genetic determination of traits and phenotypic plasticity, similar phenotypes in recombinant hybrid populations

Fig. 4. High-fitness backcross genotypes and tests to discriminate between two adaptive landscape models. A. In the holey landscape model, intermediate hybrids fall along a ridge that separate the parental peaks (P_1 and P_2) so crosses between adjacent genotypes produce mostly high-fitness progeny (indicated by the oval with an X). B. In the rugged landscape model the recombinant hybrids are on pinnacles so crosses between them produce mostly low-fitness progeny.

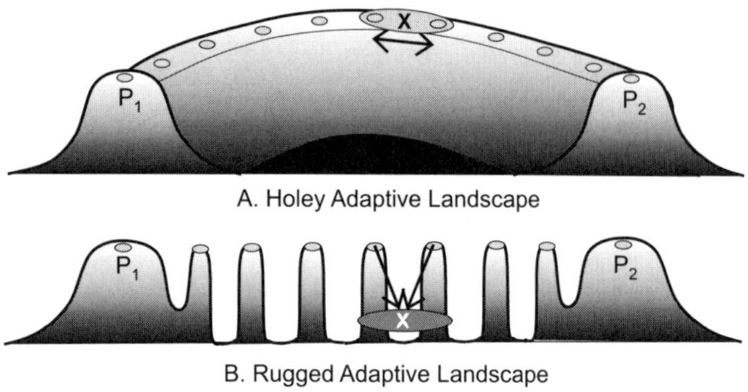

may be genetically disparate. The difficulty of choosing appropriate character combinations may render phenotypic approaches to the analysis of fitness topographies difficult and perhaps unproductive.

Genomic analysis of adaptive topographies

Accurate assessments of the topographic details separating pairs of divergent taxa may be possible using more direct analyses of genetic variation. Models of evolution on landscapes have assumed that population divergence occurs as a consequence of allelic substitutions at individual loci (Fisher 1930; Wright 1932; Wright 1988). For example, Gavrilets (Gavrilets 1997*a*; Gavrilets 1999) describes ridges connecting adaptive peaks as "chains" of genotypes separated by single allele substitutions. However, this does not refer to strictly neutral mutations; it is the relative importance of additive and epistatic effects on fitness and general characteristics of fitness topographies that are crucial for discriminating among models of adaptive evolution. Under the holey landscape model, ridges are manifest because new mutations are effectively neutral in their native genetic backgrounds but contribute to the delineation of holes if they have negative effects on fitness when in combination with novel alleles from divergent lineages. Testing the hypothesis that ridges connect putative adaptive peaks will require detailed information on the fitness consequences of different genetic combinations, which could be accomplished by the integration of genomic analyses with information on the growth and reproduction of hybrids.

An initial characterization of the landscape topography separating divergent taxa could be accomplished by analyzing the fitnesses and the genetic composition of their recombinant hybrids. While it may not be feasible to conduct genetic composition analyses for large numbers of transcribed regions, quantitative trait locus mapping techniques (Mackay 2001; Mauricio 2001) would allow a general characterization of recombinant genotypes. This information could be used to make an assessment of epistasis among genomic regions (e.g., Kim and Rieseberg 2001) and to define distances among individual genotypes, which will identify their relative position with respect to putative fitness peaks (i.e., parental genotypes) and to each other. Ideally one would endeavor to map individuals on a genotype space that would allow the identification of putative ridges or valleys. However, accurate multidimensional mapping is currently not feasible, and arraying genotypes along a linear scale (e.g., genetic hybrid index approaches: Cruzan and Arnold 1993; Arnold 1997; Carney et al. 2000) would not provide an accurate enough description of the genotypic space to allow an assessment of the fitness topography. Fortunately, the alternative models for adaptive landscapes can be differentiated by expected fitness distributions of progeny from crosses among recombinant hybrid genotypes.

Explicit tests for topographic features of adaptive landscapes can be made by integrating information from estimates of offspring fitness (preferably under a single field environment that is representative of the habitats of both taxa and with clonal replication of genotypes as in the *Piriqueta* example above), the genetic distances between their recombinant hybrid parents, and their position with respect to the putative adaptive peaks. One approach for describing fitness topographies would be to make assessments of progeny fitness from crosses among recombinant hybrids that lie at varying distances along the "rim" of a putative fitness hole (i.e., crosses between recombinant genotypes that display high fitness). An implicit assumption of the holey landscape model is that the fitness consequences of interactions among genomes arrayed along the circumference of an adaptive hole are directly related to the distance (i.e., number of genetic map differences) separating them (Gavrilets et al. 1998). If the holey landscape model is supported, then we would expect that crosses between more distant combinations would produce higher frequencies of low-fitness genotypes (Fig. 1). In accordance with this model, we would expect that increasing the genetic distance of a cross results in lower mean fitnesses and lower frequencies of high-fitness recombinant genotypes (i.e., fewer individuals landing on the ridge). Alternatively, if Wright's rugged landscape model holds, we would expect to see a low mean offspring fitness regardless of the genetic distance of the cross unless both recombinant hybrids are similar to one of the parental genotypes (i.e., they are both near one of the adaptive peaks).

A second test that discriminates between the rugged and holey adaptive landscape models could be made by crossing only genetically similar pairs of hybrids that are varying distances from the putative adaptive peaks (i.e., the parental genotypes). The holey landscape model predicts that the mean fitness of offspring from all crosses between high-fitness recombinant genotypes that are genetically similar should be comparable to the fitnesses of parentals (i.e., because the majority of offspring will fall on a ridge; Fig. 4A). The rugged landscape model, on the other hand, assumes that high-fitness recombinant genotypes would be indicative of a pinnacled topography (Fig. 4B), and the mean fitness of their offspring should be low unless they are in close proximity one of the parental's adaptive peaks. While it may be impossible to assess the fitness distributions of all possible pairs of genetically similar genotypes, judicious combinations of hybrids that are varying distances from the parentals would allow a general characterization of the intervening topography.

For example, analyses of a number of single-marker differences across the genome would give an indication of ridge "continuity" (i.e., whether the ridge was smooth or a "saw-toothed" arête). Such analyses could provide specific tests of hypotheses concerning the local fitness topography and would produce insights into modes of adaptive evolution and speciation.

Limitations of the holey adaptive landscape metaphor

Many aspects of the holey adaptive landscape model are appealing, as it is consistent with our understanding of the evolution of reproductive isolation and appears to be supported by the available empirical evidence. As with all theoretical constructs that achieve some level of generality, this model necessarily contains several simplifying assumptions and special conditions. For example, to allow the examination of a high level of dimensionality, Gavrilets (Gavrilets 1997*a*; Gavrilets and Gravner 1997) primarily considered two fitness levels (set at 0 or1). However, the general results of these analyses still hold when fitness is allowed to vary continuously (Gavrilets 1997*a*; Gavrilets 1999). The models that have been examined so far were based on fitness surfaces that were effectively flat (or nearly so) and pock-marked by holes, so the movement of populations along the surface was driven primarily by neutral processes. A more accurate depiction of a landscape might include peaks of different heights and ridges that were sloped or undulating to account for additive genetic effects on fitness, which would allow populations to explore the surface and ultimately move towards the highest peaks (as originally envisioned by Wright 1932). Such a modification could accommodate the observation of fitness increases in the first few thousand generations of long-term experiments on microbial populations (Korona et al. 1994; Lenski and Travisano 1994).

Another assumption used by Gavrilets to obtain an adaptive landscape with a high level of peak connectivity was that fitness effects of individual allelic combinations were randomly assigned. One possible effect of randomly assigning fitness states would be the creation of a fitness space with an approximately even distribution of high-fitness regions. Natural systems, on the other hand, are typically marked by aggregation at some spatial scale (Levin 1992). The consequence of the random distribution assumption in the holey landscape model may be that it would produce a relatively equal level of connectivity (i.e., density of ridges) across the multidimensional space. Clusters of high-fitness peaks, on the other hand, may produce a landscape for which the level of connectivity, and consequently the ease of peak shifts, varies across the landscape. Variation in the levels of connectivity, or unevenness in the general topography, may explain why some microbial populations apparently became "trapped" in regions of intermediate fitness and did not all converge on the same fitness optimum (Lenski and Travisano 1994).

The critical difference between the holey landscape model and Wright's original metaphor of adaptive evolution is the level of dimensionality. Under the assumptions delineated by Gavrilets, the multidimensional landscape did not attain a high level of connectivity until a large number of allelic differences among species were considered (i.e., more than 10,000 Gavrilets 1997*a*). One could argue that the assumption of such a large number of loci having epistatic effects on fitness is unreasonable, especially considering some recent studies that ascribe species differences to a smaller number of loci (Bradshaw et al. 1995; Kim and Rieseberg 1999). However, it is important to recognize that those studies only examined particular aspects of morphological differences among species e.g., floral traits (Bradshaw et al. 1995) and that such trait mapping experiments are biased towards detecting only regions possessing loci with large effects (Mackay 2001; Mauricio 2001). Furthermore, Gavrilets assumption of a random distribution of fitness effects may have biased his model from finding a "percolated" structure of high connectivity at lower dimensions. We expect that novel mutations appearing in lineages would conform to the existing genetic architecture (Orr 1995; Coyne and Orr 1998; Orr 2001), such that connections among adaptive regions may be a natural consequence of evolutionary divergence.

As with any theoretical construct, adaptive landscapes sacrifice accuracy to some degree to achieve generality. At present these heuristic tools effectively summarize our understanding of adaptive evolution, and any of the versions of this metaphor may apply to a particular case. It is also possible that some experimental results could be interpreted as supporting evidence for more than one of these models. For example, even though there were some high-fitness genotypes among the *Piriqueta* hybrids discussed above, the mean fitness was lower, so this result could be interpreted as supporting evidence for a rugged landscape. Additional analyses of the fitness distributions of hybrids in this and other systems will perhaps clarify this distinction and provide additional insights into the genetic architecture of species differences. The critical point in this case is that the motivation for particular experiments may not have existed had we not considered alternatives to traditional views of adaptive landscapes. The value of considering such models lies in their ability to direct our empirical investigations in novel directions, and hopefully towards a better understanding of the salient aspects of natural systems.

Acknowledgements

We wish express our thanks to S. Gavrilets and two anonymous reviewers for comments on earlier versions of this manuscript, and to K. McBreen, S. Handy, A. Fortner, M. Heard, B. Benz, J. Martin, K. McQuinn, and R. Roberts for technical assistance. Thanks also to personnel and colleagues at the Archbold Biological Field Station for valuable discussion and logistical assistance. This research was supported by NSF grant DEB-0080437 to MBC.

References

Arbo, M.M. 1995. Turneraceae — Parte I *Piriqueta*. The New York Botanical Garden, Bronx, New York.

Armbruster, P., Bradshaw, W.E., and Holzapfel, C.M. 1997. Evolution of the genetic architecture underlying fitness in the pitcher-plant mosquito, *Wyeomyia smithii*. Evolution **51**: 451–458.

Armbruster, W.S. 1990. Estimating and testing the shapes of adaptive surfaces: the morphology and pollination of *Dalechampia* blossoms. Am. Nat. **135**: 14–31.

Arnold, M.L. 1997. Natural hybridization and evolution. Oxford University Press, New York.

Arnold, M.L., and Hodges, S.A. 1995. Are natural hybrids fit or unfit relative to their parents? Trends Ecol. Evol. **10**: 67–71.

Arnold, S.J., Pfrender, M.E., and Jones, A.G. 2001. The adaptive landscape as a conceptual bridge between micro- and macroevolution. Genetica **112**: 9–32.

Barton, N.H., and Charlesworth, B. 1984. Genetic revolutions, founder events, and speciation. Annu. Rev. Ecol. Syst. **15**: 133–164.

Bradshaw, H.D., Wilbert, S.M., Otto, K.G. and Schemske, D.W. 1995. Genetic mapping of floral traits associated with reproductive isolation in monkeyflowers (*Mimulus*). Nature **376**: 762–765.

Burch, C.L., and Chao, L. 1999. Evolution by small steps and rugged landscapes in the RNA virus phi 6. Genetics **151**: 921–927.

Carney, S.E., Gardner, K.A., and Rieseberg, L.H. 2000. Evolutionary changes over the fifty-year history of a hybrid population of sunflowers (*Helianthus*). Evolution **54**: 462–474.

Cluster, P.D., and Allard, R.W. 1995. Evolution of ribosomal DNA (Rdna) genetic-structure in colonial Californian populations of *Avena-Barbata*. Genetics **139**: 941–954.

Coyne, J.A., Barton, N.H., and Turelli, M. 1997. A critique of Sewall Wright's shifting balance theory of evolution. Evolution **51**: 643–671.

Coyne, J.A., and Orr, H.A. 1998. The evolutionary genetics of speciation. Phil. Trans. Roy. Soc. Lond. B Biol. Sci. **353**: 287–305.

Cresswell, J.E., and Galen, C. 1991. Frequency-dependent selection and adaptive surfaces for floral character combinations: the pollination of *Polemonium viscosum*. Am. Nat. **138**: 1342–1353.

Cruzan, M.B. 2001. Adaptive landscapes. *In* Encyclopedia of Genetics. *Edited by* S. Brenner and J.H. Miller, Academic Press, New York. pp. 5–7.

Cruzan, M.B., and Arnold, M.L. 1993. Ecological and genetic associations in an *Iris* hybrid zone. Evolution **47**: 1432–1445.

Cruzan, M.B., and Estill, J.C. 2004. Hybridization in the *Piriqueta caroliniana* complex: patterns of gene flow across an expanding hybrid zone. (in review)

Dobzhansky, T. 1936. Studies on hybrid sterility. II. Localization of sterility factors in *Drosophila pseudoobscura* hybrids. Genetics **21**: 113–135.

Fisher, R.A. 1930. The genetical theory of natural selection. Clarendon Press, Oxford.

Gavrilets, S. 1997*a*. Evolution and speciation on holey adaptive landscapes. Trends Ecol. Evol. **12**: 307–312.

Gavrilets, S. 1997*b*. Hybrid zones with Dobzhansky-type epistatic selection. Evolution **51**: 1027–1035.

Gavrilets, S. 1999. A dynamical theory of speciation on holey adaptive landscapes. Am. Nat. **154**: 1–22.

Gavrilets, S., and Gravner, J. 1997. Percolation on the fitness hypercube and the evolution of reproductive isolation. J. Theor. Biol. **184**: 51–64.

Gavrilets, S., Li, H., and Vose, M.D. 1998. Rapid speciation on holey adaptive landscapes. Proc. R. Soc. Lond. B Biol. Sci. **266**: 817–824.

Geiger, H.H. 1988. Epistasis and heterosis. *In* Proceedings of the 2nd International Congress on Quantitative Genetics. *Edited by* B. Weir, E. Eisen, M. Goodman and G. Namkoong, Sinauer, New York. pp. 395–399.

Kaltz, O., and Bell, G. 2002. The ecology and genetics of fitness in *Chlamydomonas*. XII. Repeated sexual episodes increase rates of adaptation to novel environments. Evolution **56**: 1743–1753.

Kim, S.C., and Rieseberg, L.H. 1999. Genetic architecture of species differences in annual sunflowers: Implications for adaptive trait introgression. Genetics **153**: 965–977.

Kim, S.C., and Rieseberg, L.H. 2001. The contribution of epistasis to species differences in annual sunflowers. Mol. Ecol. **10**: 683–690.

Kingsolver, J.G. 1988. Thermoregulation, flight, and the evolution of wing pattern in Pierid butterflies: the topography of adaptive landscapes. Am. Zool. **28**: 899–912.

Korona, R., C. H. Nakatsu, L. J. Forney, and R. E. Lenski. 1994. Evidence for multiple adaptive peaks from populations of bacteria evolving in a structured habitat. Proc. Natl. Acad. Sci. U.S.A. **91**: 9037–9041.

Lande, R. 1976. Natural selection and random genetic drift in phenotypic evolution. Evolution **30**: 314–334.

Lenski, R.E., and Travisano, M. 1994. Dynamics of adaptation and diversification — a 10,000- generation experiment with bacterial populations. Proc. Natl. Acad. Sci. U.S.A. **91**: 6808–6814.

Levin, S.A. 1992. The problem of pattern and scale in ecology. Ecology **73**: 1943–1967.

Long, R.W., and Lakela, O. 1971. A Flora of Tropical Florida. University of Miami Press, Coral Gables, FL.

Mackay, T.F.C. 2001. The genetic architecture of quantitative traits. Annu. Rev. Genet. **35**: 303–339.

Martin, L., and Cruzan, M. 1999. Patterns of hybridization in the *Piriqueta caroliniana* complex in central Florida: evidence for an expanding hybrid zone. Evolution **53**: 1037–1049.

Maskas, S.D., and Cruzan, M.B. 2000. Patterns of intraspecific diversification in the *Piriqueta caroliniana* complex in eastern North America and the Bahamas. Evolution **54**: 815–827.

Mauricio, R. 2001. Mapping quantitative trait loci in plants: uses and caveats for evolutionary biology. Nat. Rev. Genet. **2**: 370–381.

Muller, H.J. 1942. Isolating mechanisms, evolution and temperature. Biol. Symp. **6**: 71–125.

Myers, R.L. 1990. Scrub and high pine. *In* Ecosystems of Florida. *Edited by* R.L. Myers and J.J. Ewel. University of Central Florida Press, Orlando, FL. pp. 150–193.

Orr, H.A. 1995. The population genetics of speciation: the evolution of hybrid incompatibilities. Genetics **139**: 1803–1813.

Orr, H.A. 2001. The genetics of species differences. Trends Ecol. Evol. **16**: 343–350.

Pearson, K. 1903. Mathematical contributions to the theory of evolution. XI. On the influence of selection on the variability and correlation of organs. Phil. Trans. Roy. Soc. Lond. A Phys. Sci. **200**: 1–66.

Provine, W.B. 1986. Sewall Wright and evolutionary biology. University of Chicago Press, Chicago, IL.

Rhode, J., and Cruzan, M. 2004. First generation hybrid breakdown in the *Piriqueta caroliniana* complex: the roles of heterosis, nuclear epistasis, and cytonuclear epistasis. (in review)

SAS. 1999. SAS/STAT user's guide, version 6, 4th ed. SAS Institute, Cary, NC.

Schluter, D. 2000. The ecology of adaptive radiation. Oxford University Press, New York.

Simpson, G.G. 1944. Tempo and mode in evolution. Columbia University Press, New York.

Small, J.K. 1933. Manual of the southeastern Flora. University of North Carolina Press, Chapel Hill, NC.

Vinogradov, A.E. 1999. Mountain ridges on adaptive landscapes and the predictability of evolution. Oikos **85**: 79–82.

Wade, M.J., and Goodnight, C.J. 1991. Wright's shifting balance theory: an experimental study. Science (Washington, D.C.) **253**: 1015–1018.

Wade, M.J., and Goodnight, C. J. 1998. Perspective: The theories of Fisher and Wright in the context of metapopulations: when nature does many small experiments. Evolution **52**: 1537–1553.

Wang, J., and Cruzan, M.B. 1998. Interspecific mating in the *Piriqueta caroliniana* (Turneraceae) complex: Effects of pollen load size and composition. Am. J. Bot. **85**: 1172–1179.

Whitlock, M.C., Phillips, P.C., Moore, B.G., and Tonsor, S.J. 1995*a*. Multiple fitness peaks and epistasis. Annu. Rev. Ecol. Syst. **26**: 601–629.

Whitlock, M.C., Phillips, P.C., Moore, F.B.G., and Tonsor, S.J. 1995*b*. Multiple fitness peaks and epistasis. Annu. Rev. Ecol. Syst. **26**: 601–629.

Wright, S. 1931. Evolution in mendelian populations. Genetics **16**: 97–159.

Wright, S. 1932. The roles of mutation, inbreeding, crossbreeding and selection in evolution. Proceedings of the Sixth International Congress of Genetics **1**: 356–366.

Wright, S. 1978. Evolution and genetics of populations. Volume 4. Variability within and among natural populations. University of Chicago Press, Chicago, IL.

Wright, S. 1988. Surfaces of selective value revisited. Am. Nat. **131**: 115–123.

10
Terpene synthases and the mediation of plant–insect ecological interactions by terpenoids: a mini-review

Dezene P.W. Huber and Jörg Bohlmann

Abstract: Many interactions between plants and insects are mediated in large part by an array of structurally diverse terpenoids, particularly monoterpenes and sesquiterpenes. These ubiquitous plant chemicals are formed in nature from a small number of precursors via the mevalonate pathway or the 2-C-methyl-D-erythritol 4-phosphate (MEP) pathway. An important committed step in terpenoid biosynthesis is completed by a class of enzymes, known as terpene synthases (TPS), that are associated with a large, apparently monophyletic TPS gene family. Terpenoid biosynthesis in plants may be constitutive or induced and may act to directly deter or harm herbivores or may be released by the plant to attract insect predators and parasitoids of the herbivores. The molecular and biochemical mechanisms governing the important role of floral scent terpenoids in pollination biology are beginning to be elucidated. Terpenoid volatiles are also exploited by herbivorous insects to aid them in their search for a suitable host. Because plants are able to produce a diverse mixture of terpenoids with varying function, and because they are able to regulate the constituents of the produced mixtures, they possess a considerable ability to adapt their defenses or signals in the face of potentially devastating herbivory or discriminating pollinators. Considering the importance of terpenoids to plant adaptation and survival, a better understanding of TPS and the mechanisms that regulate their activity and expression should provide new technologies for crop improvement and pest management. Recent work has revealed the enticing possibility of the existence of TPS in animals, and further work in this regard may provide interesting clues to the evolution of these enzymes in metazoans.

Dezene P.W. Huber and Jörg Bohlmann[1]. Biotechnology Laboratory, Department of Botany, and Department of Forest Sciences, University of British Columbia, #237 – 6174 University Boulevard, Vancouver, BC V6T 1Z4, Canada.
Correct citation: Huber, D.P.W., and Bohlmann, J. 2004. Terpene synthases and the mediation of plant–insect ecological interactions by terpenoids: a mini-review. *In* Plant Adaptation: Molecular Genetics and Ecology. Proceedings of an International Workshop held December 11–13, 2002, in Vancouver, British Columbia, Canada. *Edited by* Q.C.B. Cronk, J. Whitton, R.H. Ree, and I.E.P. Taylor. NRC Research Press, Ottawa, Ontario. pp. 70–81.

[1]Corresponding author email address: bohlmann@interchange.ubc.ca

Introduction

Of all of the compounds produced by plants that are often referred to as "natural products" or "secondary metabolites", terpenoids are probably the most diverse class.

Terpenoids appear in everyday life as the compounds that are used for such purposes as providing pleasant aromas and flavours in various foods and beverages. Because of their antimicrobial and lipophilic properties, terpenoids imparting pine-like, orange-like, or lemon and lime scents are commonly used in cleaning solutions and many other common household applications. Some terpenoids are important in traditional and modern medicine, such as the anti-malarial drug, artemisinin, or the potent anti-cancer drug, taxol. A large suite of terpenoids, derived from fruit- and vegetable-rich diets and considered to be "nutriceuticals", seem to function as antioxidants or act as otherwise protective substances against a range of common maladies. While these compounds have thus become important to us in a variety of industrial and commercial roles, they are synthesized and used by plants, fungi, and metazoans for many biologically-important purposes. Of particular interest is the use of terpenoids in interactions between individuals of the same or different species.

Terpenoids are all ultimately derived from a five-carbon precursor, isopentenyl diphosphate (IPP) and its isomer dimethylallyl diphosphate (DMAPP). Condensation of two, three, or four five-carbon units results in the formation of the 10-carbon geranyl diphosphate (GPP), the 15-carbon farnesyl diphosphate (FPP), and the 20-carbon geranyl geranyl diphosphate (GGPP) (Koyama and Ogura 1999) (Fig. 1). GPP, FPP, and GGPP are substrates for terpene synthases (TPS), which produce 10-carbon monoterpenes, 15-carbon sesquiterpenes, and 20-carbon diterpenes. While terpenoids containing 20, 30, 40, or more carbon atoms (diterpenes, triterpenes, tetraterpenes, polyterpenes) are common and play important roles in nature in plant biochemistry and ecological interactions, the smaller terpenoids are of particular importance for plant volatile signaling.

The simple structures of the substrates for TPS belie the diversity of the structures that are produced from their catalytic activity (Fig. 2). Two features of the biosynthesis of terpenoids, in combination, provide the fuel for this diversity. First the structures of the substrates are such that many possible carbon-carbon bonds, and combinations of bonds may be formed. Second, the formation of a bond or bonds during catalysis often produces one or more chiral centers in

Fig. 1. The mevalonate pathway, in the cytosol, leading to the synthesis of sesquiterpenoids (A) and the 2-C-methyl-D-erythritol 4-phosphate (MEP) pathway, in the plastids, leading to the synthesis of monoterpenoids and diterpenoids (B). Other abbreviations: dimethylallyl diphosphate (DMAPP); 1-deoxy-D-xylulose 5-phosphate (DXP); endoplasmic reticulum (ER); farnesyl diphosphate (FPP); glyceraldehyde 3-phosphate (GAP); 3-hydroxy-3-methyl-glutaryl coenzyme A (HMG-CoA); isopentenyl diphosphate (IPP); geranyl geranyl diphosphate (GGPP); geranyl diphosphate (GPP); terpene synthase (TPS).

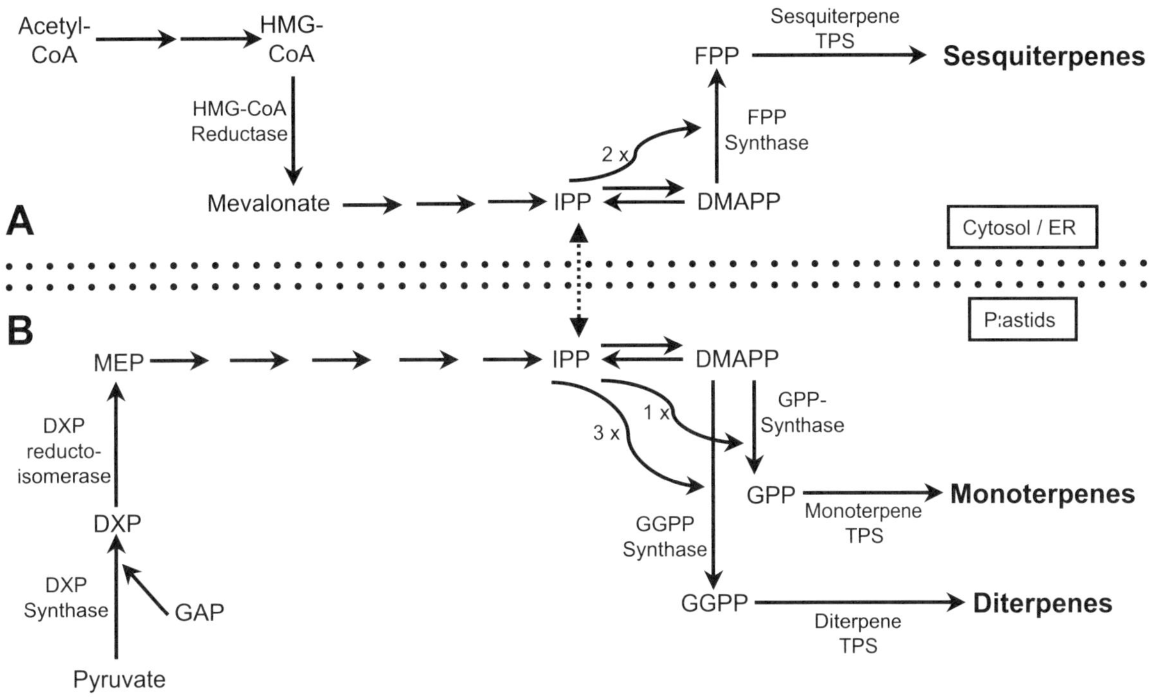

Fig. 2. Representative monoterpenoid structures (A), sesquiterpenoids (B), and a diterpenoid (C).

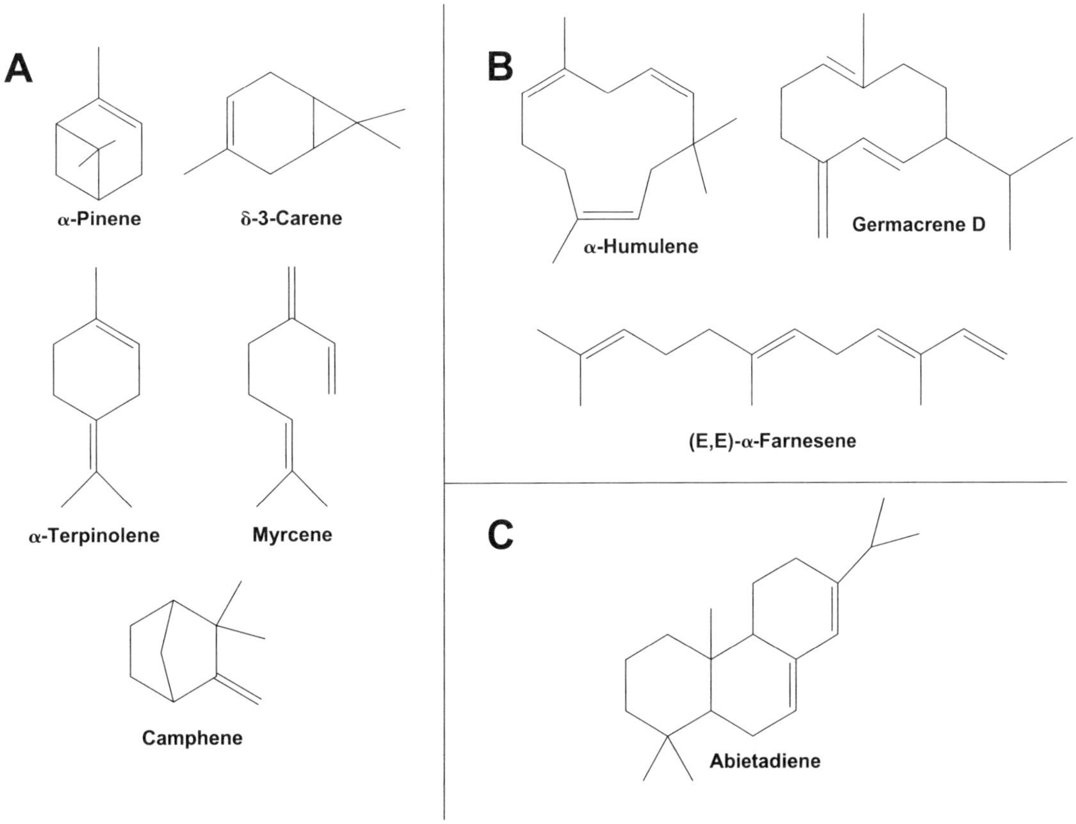

the resulting product (Cane 1999; Wise and Croteau 1999; MacMillan and Beale 1999; Greenhagen and Chappell 2001). The combination of the chemical diversity of the terpenoids with their differing physical characteristics (monoterpenoids and sesquiterpenoids are quite volatile while the diterpenoids are not) and physiological effects (e.g. toxicity, activation of animal sensory neurons, etc.) provides for a great deal of diversity in the functions and thus the chemical ecological roles of the different chemicals.

Biosynthesis of terpenoids

Sesquiterpenoids (C15) are synthesized in the cytoplasm of plant cells whereas monoterpenoids (C10) and diterpenoids (C20) are synthesized in the plastids (McGarvey and Croteau 1995). Precursor IPP and DMAPP, however, may be synthesized by one of two separate pathways depending on localization, the mevalonate pathway in the cytoplasm or the 2-C-methyl-D-erythritol 4-phosphate pathway (MEP) (Fig.1) in the plastids (Lange et al. 2001). IPP and DMAPP are converted to GPP, FPP, or GGPP, the substrates for the many TPS, by the activity of prenyl transferases. TPS produce the variety of terpenoid basic skeletons present in plants. Formation of the array of terpenoid structures by TPS is often followed by secondary modifications such as oxidations, addition of functional groups, or truncations, resulting in enormous terpenoid structural diversity.

TPS represent a large and diverse family of genes and can be grouped into seven subfamilies (Fig. 3), based on amino acid sequence relatedness (Bohlmann et al. 1998*b*; Aubourg et al. 2002; Dudareva et al. 2003). The monophyletic origin of plant TPS is supported by common exon/intron structure of TPS in the seven subfamilies (Trapp and Croteau 2001; Aubourg et al. 2002). The synthases group both in relation to the phylogenetic relatedness of the organisms (angiosperms vs. gymnosperms) that express them and by catalytic function. For instance, a (–)-limonene synthase in an angiosperm would likely be in a separate family from (–)-limonene synthase in a gymnosperm but limonene synthases from different gymnosperms would cluster in the same family (Bohlmann et al. 1998*b*). Amino acid sequences may allow an *a priori* prediction of the general function of an uncharacterized TPS, but not its specific function.

Hence a gymnosperm TPS with an amino acid sequence similar to a known limonene synthase in another species is probably a monoterpene synthase, but is not necessarily a limonene synthase. There are a number of conserved sequence motifs (Bohlmann et al. 1998*b*; Aubourg et al. 2002) and also a large similarity in crystal structure and catalytic mechanism (Starks et al. 1997; Rynkiewicz et al. 2001; Peters and Croteau 2002; Whittington et al. 2002) in TPS across species studied to date, supporting the idea that the diversity of TPS that we witness today has its origin in a common ancestor.

The diversity of terpenoid products constitutively present or inducible in plant tissues is directly related to the diverse array of TPS present in any one species. In addition, while some TPS produce exclusively one product (e.g., Bohlmann et al. 1998*a*) many TPS produce one major terpenoid product and a number of other terpenoids in varying amounts (Fig. 4 and e.g., Bohlmann et al. 1997; Steele et al. 1998; Bohlmann et al. 1999; Bohlmann et al. 2000; Fäldt et al. 2003*a*, *b*; Phillips et al. 2003). On the other hand, the production of separate enantiomers of the same compound in a species is usually, but not always, the role of separate TPS (Schmidt et al. 1998; Phillips et al. 2003) and the stereochemistry of terpenoids may have important ramifications for their biological activity (Hobson et al. 1993; White and Hobson 1993; Erbilgin and Raffa 2001). The number of different TPS in gene families, the fact that many TPS produce a group of products, and the ability of plants to regulate the expression of TPS genes provide plants with the capability of responding to biotic pressures by slight or massive modifications of their total terpene content and the constituents of the mixture, as the situation warrants. The combination of the "multiple TPS enzyme" and the "single enzyme/several products" models of terpene biosynthesis will play an important role in the selective breeding or the genetic modification of agricultural and silvicultural plants in regards to the applied up-regulation or down-regulation of desirable or undesirable terpenoids.

Terpenoids in plant defense against insects

Among the most studied roles of terpenoids are their toxic effects against pathogens (Cobb et al. 1968; Paine and Hanlon 1994, Paine et al. 1997) and insects (Raffa et al. 1985; Werner 1995) in plant defense. This review focuses on plant defenses against insects. Because plants are stationary and because some plants, such as trees, are long-lived and have long generation times, they cannot escape their short-lived, short generation natural enemies in space, nor are they able to evolve as rapidly to escape their enemies in time. A single tree must withstand the attacks of multiple generations of pests in a single location. Its responses must be plastic enough to allow it to respond to each new generation of herbivores with enough vigor to allow it also to reproduce. As a result, over time, plants have evolved both physical and toxic chemical means of dealing with pests. Terpenoids act to inhibit pests by both means.

Defenses may be either constitutive or induced. The constitutive resin of gymnosperms contains a complex cocktail of terpenoids (Erdtman et al. 1968; Persson et al. 1996; Sjödin et al. 1996; Fäldt 2000). The constitutive terpenoid odor or taste of tree resin may be instrumental in the preference of insects for their hosts and the rejection of their nonhosts (Chararas et al. 1982; Edwards et al. 1993). One recent study (Wallin and Raffa 2000) has shown that a bark beetle, *Ips pini*, amends its post-landing behavior on simulated hosts in response to total monoterpene content, and in particular to the concentrations of the monoterpenes β-pinene and limonene.

Induced defenses, on the other hand, become active upon attack by an invader. Recent work in our lab and by others has focused on the induced defense of spruce, *Picea* spp., against bark beetles, insect-associated fungi, and the white pine weevil, *Pissodes strobi.* Feeding and oviposition by *P. strobi* (Alfaro 1995), simulated weevil damage (Tomlin et al. 1998), bark beetle damage (Franceschi et al. 2000), or inoculation of fungi associated with invading insects (Franceschi et al. 2000; Nagy et al. 2000) causes the formation of traumatic resin ducts in the developing xylem and other

Fig. 3. Phylogenetic tree showing representative terpenoid synthases (TPS) from all seven of the TPS subfamilies. Angiosperm sesquiterpene and diterpene synthases are contained in *TPS-a*; angiosperm monoterpene synthases with the RRX_8W motif are contained in *TPS-b*; copalyl diphosphate synthases (CDS) are contained in *TPS-c*; gymnosperm TPS are contained in *TPS-d*; kaurene synthases (KS) are contained in *TPS-e*, a linalool synthase from *Clarkia breweri* and the *Arabidopsis* AtTPS04 are contained in *TPS-f*; the known *Antirrhinum majus* monoterpene synthases (two of three shown in this tree) and *Arabidopsis* AtTPS14 are contained in *TPS-g*. Sequences available from GenBank and from other published material were used in the construction of the phylogenetic tree. The tree was derived using ClustalX and was visualized with TreeView.

Taxon abbreviations are as follows: *Artemisia annua* (*A.a*); *Abies grandis* (*A.g.*); *Antirrhinum majus* (*A.m.*); *Arabidopsis thaliana* (*A.t.*); *Clarkia breweri* (*C.b.*); *Citrus junos* (*C.j.*); *Citrus limon* (*C.l.*); *Cucurbita maxima* (*C.m.*); *Cucumis sativus* (*C.sa.*); *Croton sublyratus* (*C.su.*); *Gossypium arboretum* (*G.a.*); *Lycopersicon esculentum* (*L.e.*); *Lactuca sativa* (*L.s.*); *Mentha aquatica* (*M.a.*); *Mentha longifolia* (*M.l.*); *Mentha* × *piperita* (*M.* × *p.*); *Nicotiana tabacum* (*N.t.*); *Populus* × *canescens* (*P.* × *c.*); *Quercus ilex* (*Q.i.*); *Rosa hybrida* (*R.h.*); *Scoparia dulcis* (*S.d.*); *Salvia officinalis* (*S.o.*); *Stevia rebaudiana* (*S.r.*).

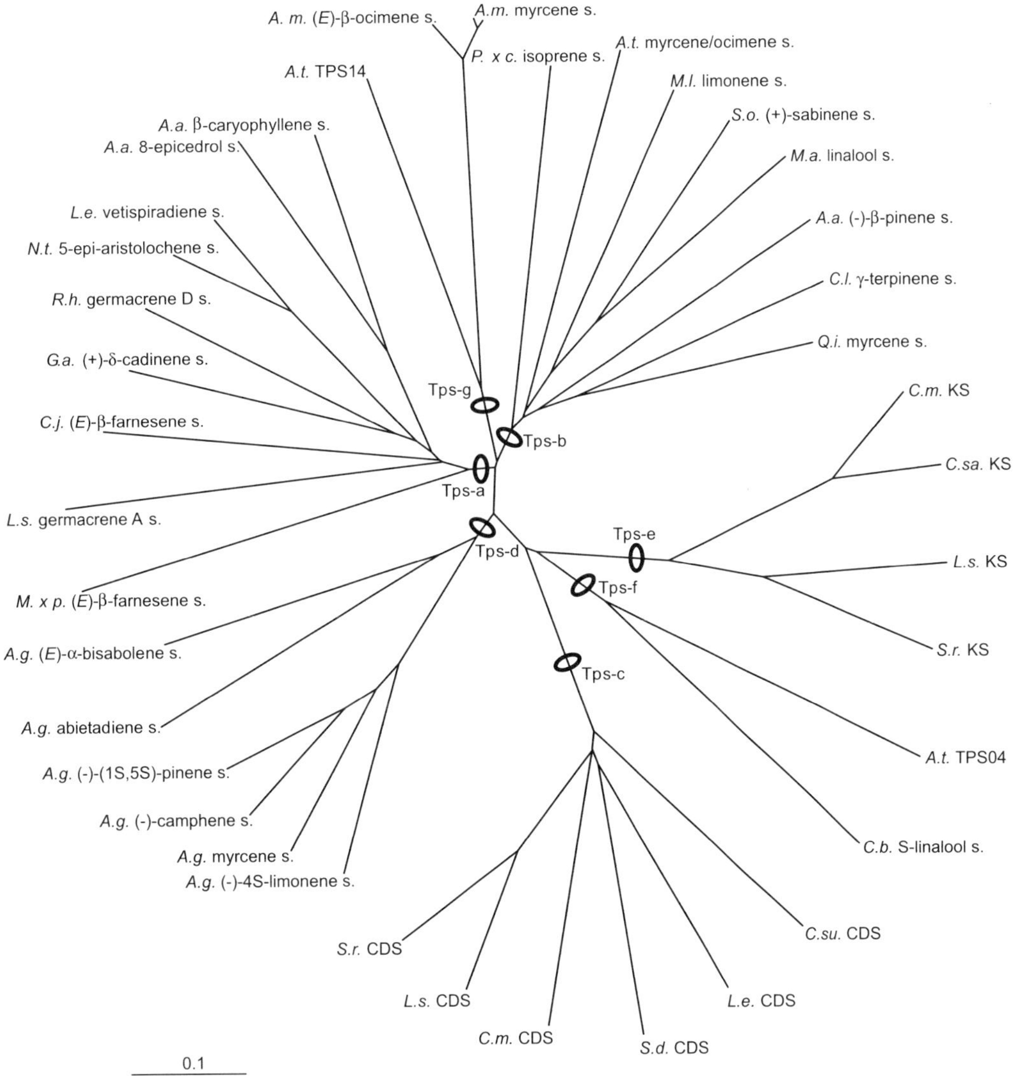

histological changes in the tissues of spruce stems. These ducts bring resin to the site of phloem-excavating adults or newly hatched larvae. When the ducts are severed by insect activity resin flows into the galleries, acting as both a physical and a chemical defense. In addition, the induced resin contained in the traumatic resin ducts is different in its chemical composition from constitutive resin. Specifically, the content of monoterpenes and diterpene resin acids in the induced resin of white spruce, *Picea glauca*, increases in response to simulated weevil damage (Tomlin et al. 2000). The increased monoterpene content results in a less viscous resin and allows the resin to flow more freely into larval feeding galleries where toxic and physical effects may kill the insects. Topical treatment of Norway spruce, *Picea abies*, with methyl jasmonate (MeJA) mimics the effect of insect activity or simulated wounding both histologically and chemically (Franceschi et al. 2002; Martin et al. 2002). In particular, exogenous MeJA treatment of *P. abies* induces new resin duct formation in the developing xylem, increases the terpenoid content in that region (Fig. 5), causes increased activity of prenyl transferases and terpene synthases (Fig. 6), and increases the level of mRNA transcripts of both

Fig. 4. A proposed mechanism for monoterpene biosynthesis by a (+)-3-carene synthase from Norway spruce, *Picea abies* (Fäldt et al. 2003*b*). While (+)-3-carene is the major product, a number of other monoterpene products are synthesized by a single enzyme via alternate intermediates and different carbon-carbon bonds.

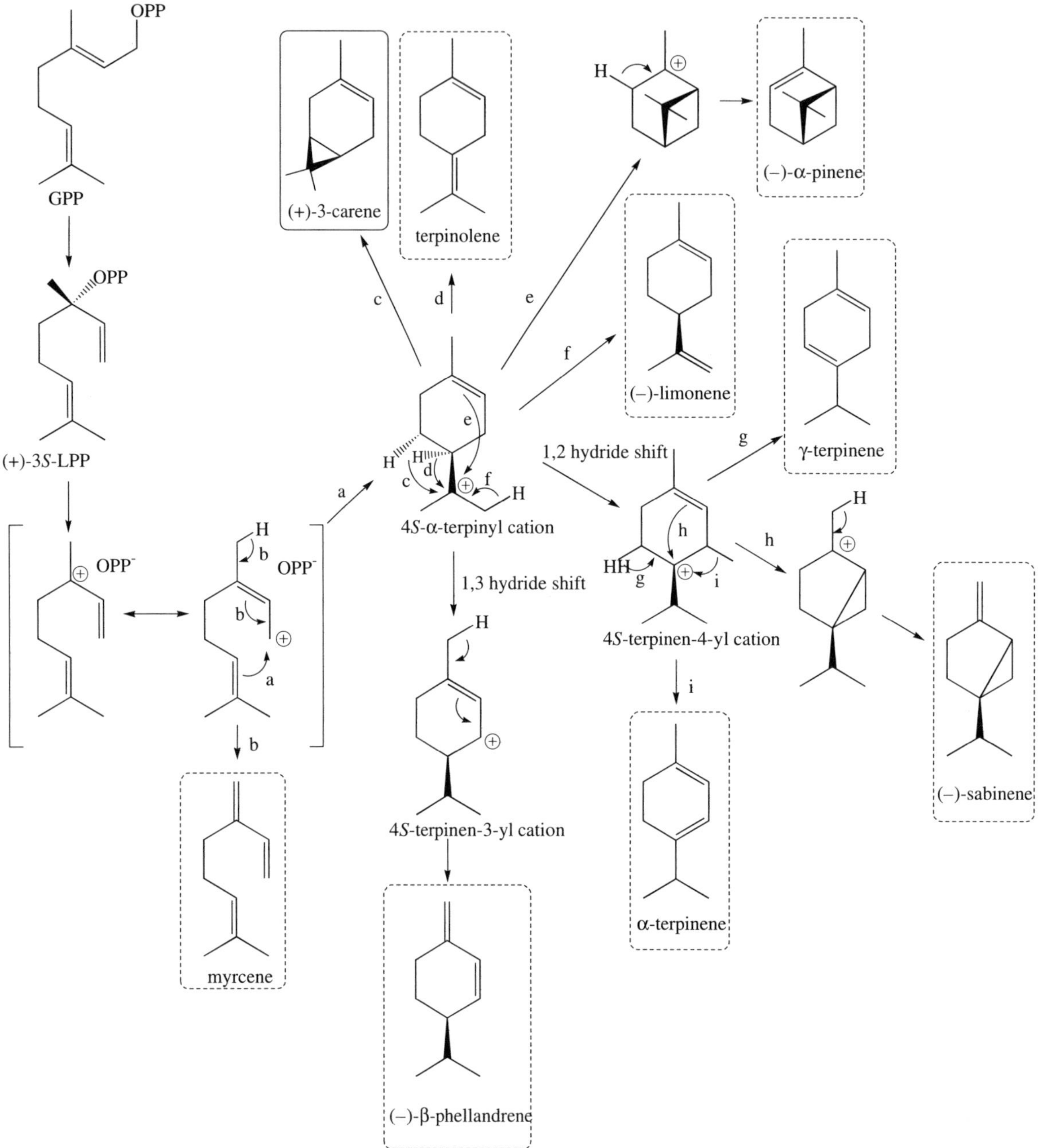

monoterpene synthases and diterpene synthases (Martin et al. 2002; Fäldt et al. 2003*b*). The jasmonate (octadecanoid) signal pathway (Farmer and Ryan 1992; Wasternack and Parthier 1997; Turner et al. 2002) thus seems to be involved in regulating the expression of terpene synthases and the resulting quantity and quality of the accumulation of terpenoids in conifer tissues (Martin et al. 2002; Fäldt et al. 2003*b*).

Changes in terpenoid volatile release in response to MeJA or jasmonic acid application have been reported previously in angiosperms (Hopke et al. 1994; Dicke et al. 1999; Gols et al. 1999; Koch et al. 1999; Halitschke et al. 2000; Ozawa et al. 2000; Schmelz et al. 2001) and more recently in the gymnosperm Norway spruce (Martin et al. 2003*b*). The octadecanoid-induced terpenoid volatiles are often released in a diurnal pattern (Rodriguez-Saona et al. 2001; Martin et al. 2003*b*). Stimuli, such as mechanical wounding or arthropod feeding (mechanical wounding plus oral secretions), result in similar volatile release and buildup of secondary metabolites (Ozawa et al. 2000; Tomlin et al. 2000) and thus are likely instrumental in turning on the jasmonate signal pathway. The expression of TPS and the ultimate production of terpenoids in response to octadecanoid signals should focus attention on manipulation of the signaling pathway for understanding plant resistance to herbivory and other terpenoid-related traits. The combination of manipulation of the

Fig. 5. Methyl jasmonate-induced TPS enzyme activities and terpenoid accumulation in the developing xylem of Norway spruce, *Picea abies* L. Karst. Monoterpene accumulation (left axis, bars) and monoterpene synthase activity (right axis, line graph) is shown in A. Diterpene accumulation (left axis, bars) and diterpene synthase activity (right axis, line graph) is shown in B. In both graphs dark bars and ♦—♦ correspond to methyl jasmonate-treated spruce and light bars and ▲—▲ correspond to untreated spruce. One replicate is shown for each data type. The authors thank D. Martin for her contribution of this figure.

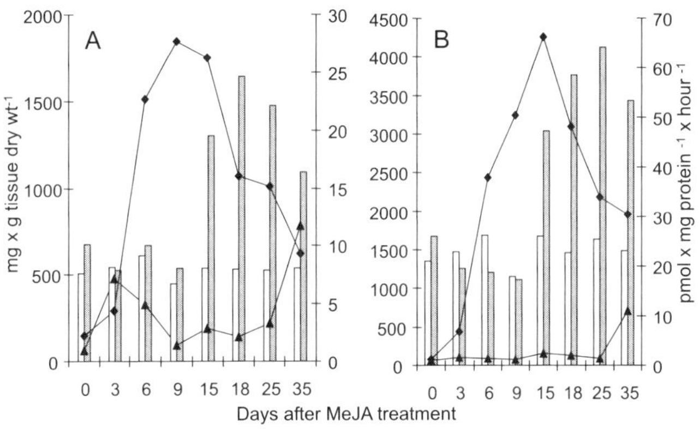

octadecanoid signaling pathway, the direct application of octadecanoids or other oxylipins to plants, and the modification of terpene synthases or their promoters has the potential to provide powerful insights into the responses of plants to herbivory and the plasticity of plant defensive responses in the face of numerous and changing herbivore populations. In addition, a better understanding of adaptive characteristics of plant defense against herbivores such as plant responses to elicitors, the details of the signaling pathways, the anatomical and physiological characteristics that affect signal translocation, the extent and speed of the response in different plants, and the effect of the induced responses on the invading insects or pathogens may provide means to produce better and higher quality yields for many agricultural and silvicultural crops.

Terpenoids in indirect defense

Effects of terpenoids, such as toxicity, that act immediately upon the attacking insect herbivore are often termed direct defenses. Plants, however, have also co-evolved indirect defenses with predators and parasitoids in which the plants signal antagonists of the attacking herbivores (Turlings et al. 1990; Takabayashi and Dicke 1996; Dicke and Vet 1999; Paré and Tumlinson 1999; Kessler and Baldwin 2001; Pichersky and Gershenzon 2002). During feeding by an herbivore, elicitors from the oral secretions of the insect (Mattiacci et al. 1995; Alborn et al. 1997; Halitschke et al. 2001) are deposited on damaged areas of leaves. The elicitors act as signals and induce the plant to release volatiles that serve to attract enemies of the herbivores. Many of the volatiles released from damaged plants, which are known to attract predators and parasitoids, are terpenoids or terpenoid-derived compounds (Takabayashi and Dicke 1996; Alborn et al. 1997; Dicke and Vet 1999; Paré and Tumlinson 1999; Kessler and Baldwin 2001; Pichersky and Gershenzon 2002). In addition, recent work (Arimura et al. 2000, 2002; Karban et al. 2000) has confirmed earlier suggestions (Baldwin and Schultz 1983) that plants are able to respond to neighboring, signaling plants resulting in induced defense responses in the receiving plant. Once again terpenoids are involved as signal molecules (Arimura et al. 2000, 2002).

Recent work on indirect defense in our lab (Arimura et al. 2004) has provided evidence that hybrid poplar, *Populus trichocarpa* × *deltoides*, responds to herbivory by the forest tent caterpillar, *Malacosoma disstria*, by releasing a number of compounds, including terpenoids, from leaves located away from the site of caterpillar feeding. One induced sesquiterpenoid volatile, germacrene D, is released from leaves in a rhythmic, diurnal fashion and is tightly linked to the presence or absence of caterpillars (Fig 6). In addition, we have identified and isolated a functional germacrene D synthase cDNA, *Ptd*TPS1, from the hybrid poplar and we have shown that its transcript accumulation is correlated with insect feeding and with germacrene D volatile release from the infested plant.

Interestingly, while tent caterpillars are known to feed mainly at dusk, night, and dawn (Fitzgerald 1980; Casey et al. 1988; Fitzgerald et al. 1988), *Ptd*TPS1 expression and germacrene D emission was highest during the day. It is still unclear why this is the case, but it may be due to caterpillar oral secretions that remain on the leaves and act as elicitors during the daytime. Alternatively some factor may contribute to a particularly strong effect on *Ptd*TPS1 transcript levels by insect-produced elicitors during the dawn feeding period or during occasional daytime feedings, compared to other times. Daytime opening of stomata may also play a role in any of the above possible scenarios.

Work on the molecular and biochemical regulation of tritrophic interactions in various plants is still in its early stages. Future research will continue to reveal the important role of terpenoids and their associated TPS in this and other forms of plant indirect defense.

Terpenoids in floral scents and other plant emissions

Terpenoids are also common volatiles emanating from flowers (Knudsen et al. 1993) and serve to attract beneficial insect pollinators (Dobson 1994). To this point the main body of work on floral scents has been driven by the cataloguing of compounds for the fragrance and flavour industry and by studies concerned with the dynamics of insect and plant interactions. However, little work has sought to understand the molecular genetic and biochemical bases of floral scent emissions (Dudareva and Pichersky 2000). Two terpenoid synthases have been identified and isolated, and characterized from flowers, a linalool synthase from *Clarkia breweri* (Pichersky et al. 1995; Dudareva et al. 1996) and a germacrene D synthase from *Rosa hybridia* (Guterman et al. 2002). Recent work has led to the discovery of three new terpenoid synthases from snapdragon, *Antirrhinum majus*, specifically two myrcene synthases and an (*E*)-β-ocimene synthase (Dudareva et al. 2003). In *A. majus* the emission of monoterpenoids is rhythmic with highest emission occurring during daylight hours. Monoterpenoid emission seems mainly unaffected by the presence or absence of light, but is

Fig. 6. Systemic release of the volatile, (–)-germacrene D, from leaves of *Populus trichocarpa* × *deltoides* clone H11 during and following feeding by the forest tent caterpillar (FTC), *Malacosoma disstria*, and following mechanical damage (leaves cut in half with scissors), one replicate is shown here. Treatments were applied to the lower leaves of small, greenhouse-grown trees and volatiles were collected from the upper leaves over a time course using equipment similar to that described by Röse et al. 1996. Dark shaded bars correspond to FTC-infested trees; light shaded bars correspond to mechanical wounding treatment. Volatile collections were done for periods of 4 hours, shown along the x-axis. Alternate light and dark sections along the graph correspond to alternate 12-hour light and dark periods. FTCs were on the tree during the period denoted by the double-headed arrow. Samples in which germacrene-D was not detected by gas chromatography-mass spectroscopy are denoted by very small bars.

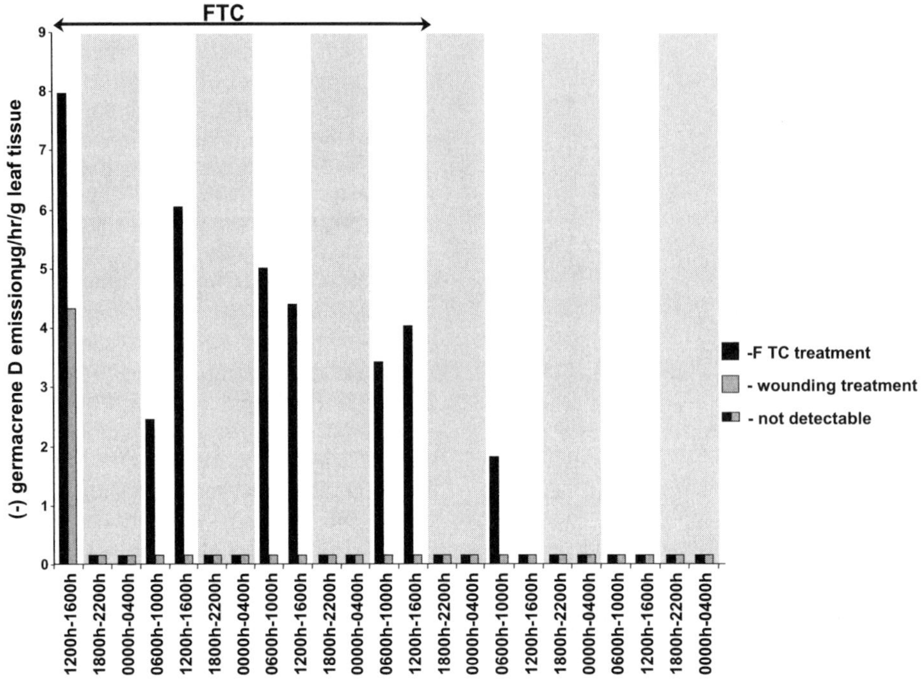

rather tied to the time of day and so is probably regulated by a circadian rhythm. Other studies also reveal the existence of variable floral volatile release at different times of the day and also indicate that certain volatiles may be preferentially released at night to attract nocturnal pollinators (e.g. Pettersson and Knudsen 2001). Daylight emission of floral volatiles is probably an adaptation to attract daytime-active pollinators. In addition, by limiting volatile production and release to one part of the day, the plant can reduce energy expenditures incurred by continual terpenoid biosynthesis.

Floral scent TPS gene expression is directly related to volatile emission in *A. majus* flowers. Furthermore, gene expression analysis has revealed that monoterpene synthase mRNA is most prevalent in the upper lobes of the *A. majus* flower followed by the lower lobes and the tube, meaning that floral volatile production is closely targeted to the area where pollinator activity is most beneficial to the plant's reproductive success (Dudareva et al. 2003). Other work on floral terpenoid volatiles has provided similar results (Dudareva et al. 1996; Raguso and Pichersky 1999). Continued research with a focus on the regulation and emission of floral volatiles in plants, in combination with consideration of the biology of their specific pollinators, should shed more light on the adaptive significance of terpenoid mixtures and TPS gene regulation in pollination biology.

Herbivorous insects also use terpenoid plant volatiles as cues to locate their host plants (Rembold et al. 1991; Kalberer et al. 2001; Mozuraitis et al. 2002) and to avoid toxic or otherwise unsuitable hosts (Werner 1995; Erbilgin and Raffa 2000). Because of the large amount of terpenoids synthesized by their hosts, coniferophagous bark beetle host selection behaviour is predominantly mediated by terpenoids (Byers 1995). Some bark beetles respond positively to specific volatile monoterpenes, but not to others (Miller and Borden 2000; Miller and Lindgren 2000). The specificity of the response may even be related to the chirality of the volatile monoterpenes in some cases (Hobson et al. 1993; White and Hobson 1993). Insects that feed on more than one host at different times of the year may respond to different mixtures of host-derived terpene volatiles during their life cycle (Kalberer et al. 2001). Thus host terpenoid chemistry affects not only the post-alighting success of insects, but also may serve to attract or repel herbivores before they come into contact with the plant.

Terpenoid biosynthesis and use by animals

While the presence of terpenoids in animals, particularly insects, has been known for some time, only recently has attention begun to focus on the role and potential biosynthesis of these compounds in metazoans. Of particular interest in this regard are the bark beetles that use a number of terpenoid-derived compounds as powerful pheromones that mediate their aggregation on host trees (Borden 1985). Because conifer-infesting bark beetles sometimes literally swim in terpenoid resin secretions, it was long thought that pheromone production simply involved the ability of either the insect or their symbiotic microorganisms to modify sequestered terpenes (Hughes 1974; Renwick et al. 1976), evolved from the need to detoxify these compounds. While this may be the case to some extent, it is becoming clearer that some bark beetles at least are able to synthesize pheromone components from simple precursors (Seybold and Tittiger 2003) after feeding or in response to treatments with juvenile hormone. In *Ips pini* and *Dendroctonus jeffreyi* the site of biosynthesis of terpenoid pheromone components has been localized to the tissue of the midgut (Hall et al. 2002; Nardi et al. 2002). Recent work, has revealed an inducible terpene synthase in whole-body extracts of male *I. pini* that converts GPP to myrcene after treatment with juvenile hormone III (Martin et al. 2001; Martin et al. 2003*a*).

If ancestral *I. pini* were reliant upon their host tree to supply precursors for pheromone production, their pheromone emissions would have consisted of an enantiomeric ratio that was dictated by precursors supplied by their host tree. The development, by the beetles, of the ability to produce pheromone components would uncouple this reliance on the chemistry of their host tree. In the case of *I. pini*, different populations respond to different ratios of (+)- and (–)-ipsdienol (Miller et al. 1989, 1996). Such geographic variation in behavioral responses to enantiomeric ratios of pheromones would only have become possible once pheromone production was separated from host tree chemistry to at least some extent. It must be noted, though, that while pheromone production in some bark beetles may be partly or mostly uncoupled from tree chemistry, host tree terpenoids still affect the life of bark beetles (Byers 1995) in many ways that contribute to the ability of trees to survive and reproduce.

Further characterization of bark beetle terpenoid synthases and of other enzymes involved in terpenoid pheromone production and a better understanding of the control of pheromone biosynthesis in these insects will potentially provide better methods for controlling these insects. For instance, an understanding of the internal and external controls of pheromone biosynthesis may provide information pertaining to attack dynamics and timing. Juvenile hormone analogues, other regulators of the synthesis pathway, or inhibitors of various enzymes may provide methods for operational modification of insect behavior and potentially reduce the damaging impact of these insects.

Other insects are known to synthesize terpenoids (Feld et al. 2001) or sequester them from their food sources (Codella and Raffa 1995; Wheeler et al. 2002), often for defensive purposes. Ants and termites, two unrelated groups of social insects, are known to possess a fairly complex array of terpenoids in their glandular tissues (Lloyd et al. 1989; Billen et al. 2000) that may be used by the insects as components of trail pheromones (Janssen et al. 1997) or in defense (Valterová et al. 1989; Roisin et al. 1990).

Recent work has shown that the wood ants, *Formica paralugubris*, which collect and store conifer resin in their nests, do not fare as well in artificial nests without resin due to increased growth of microorganisms (Christe et al. 2003). Presumably the terpenoids in the collected resin act to reduce the growth of pathogens. It can be hypothesized, then, that the various species of ants and termites that have evolved the capacity to secrete terpenoids (Lloyd et al. 1989; Valterová et al. 1989; Roisin et al. 1990; Janssen et al. 1997; Billen et al. 2000) may have done so under selective pressure to destroy any nest-specific microorganisms that might otherwise invade the colony. After the development of the ability to biosynthesize and secrete terpenoids, the compounds would have been available for use in other colony-specific roles (Blum 1970). While it is not clear at this point if the terpenoid secretions of ants and termites are a result of the release of previously sequestered material or are synthesized *de novo*, it seems reasonable to hypothesize that these taxa could potentially provide a rich source of metazoan-derived terpene synthases. In particular it would be valuable to investigate the origin of terpenoids in species of ants that do not have regular interaction with terpenoid-rich plants. If we consider that bark beetles have evolved the capacity to synthesize terpenoids *de novo* (Martin et al. 2001; Martin et al. 2003*a*; Seybold and Tittiger 2003), and thus have been at least partially released from complete reliance on the chemistry of their host trees, it seems quite reasonable that ants may have been under selective pressure to do the same when resin resources were low. Because ants are often quite easy to rear in laboratory conditions, because they are available in large quantities around the world in numerous and varied habitats, and have quite diverse terpenoid secretions (Lloyd et al. 1989; Billen et al. 2000) they would provide good material for such investigations.

Conclusion

The diversity of terpenoids and their ubiquitous presence in nature have made the study of terpene synthases very profitable. Many researchers in this field have limited themselves to certain aspects of terpene-mediated biology such as the discovery and characterization of genes and enzymes, the regulation of gene expression, the regulatory steps on the mevalonate and non-mevalonate pathways, the behavior of plants in response to herbivory or to pollinators, or the behavior of insects encountering the terpenoid secretions of plants. While research in such limited areas is bound to prove fruitful, the most robust answers to questions about the adaptive significance of the plethora of terpenoids in plants will be found when researchers combine their expertise and tackle problems on more than one of these levels.

The field of terpene biology is quickly growing. Each aspect of terpenoid biology, as presented briefly in this review, reveals many unanswered questions and opportunities to turn new answers into operational practice in agriculture and silviculture. In addition, considering the prevalence and importance of terpenoids in nature that has thus far been discovered, it is likely that we are only now skimming the surface. Many more exciting and useful discoveries are waiting to be made.

References

Alborn H.T., Turlings T.C.J., Jones T.H., Stenhagen G., Loughrin J.H., and Tumlinson J.H. 1997. An elicitor of plant volatiles from beet armyworm oral secretion. Science (Washington, D.C.) **276**: 945–949.

Alfaro, R.I. 1995. An induced defense reaction in white spruce to attack by the white pine weevil, *Pissodes strobi*. Can. J. For. Res. **25**: 1725–1730.

Arimura, G., Ozawa, R., Shimoda, T., Nishioka, T., Boland, W., and Takabayashi, J. 2000. Herbivory-induced volatiles elicit defence genes in lima bean leaves. Nature **406**: 512–515.

Arimura, G., Ozawa, R., Nishioka ,T., Boland, W., Koch, T., Kühnemann, F., and Takabayashi, J. 2002. Herbivore-induced volatiles induce the emission of ethylene in neighbouring lima bean plants. Plant J. **29**: 87–98.

Arimura, G., Huber, D.P.W., and Bohlmann, J. 2004. Forest tent caterpillars (*Malacosoma disstria*) induce systemic and diurnal emissions of terpenoid volatiles in hybrid poplar (*Populus trichocarpa* × *deltoides*): cDNA cloning, functional characterization, and patterns of gene expression of (–)-germacrene D synthase, *PtdTPS1*. Plant J. **37**: 603–616.

Aubourg, S., Lecharny, A., and Bohlmann, J. 2002. Genomic analysis of the terpenoid synthase (*AtTPS*) gene family of *Arabidopsis thaliana*. Mol. Genet. Genomics **267**: 730–745.

Baldwin I.T., and Schultz J.C. 1983. Rapid changes in tree leaf chemistry induced by damage: evidence for communication between plants. Science (Washington, D.C.) **221**: 277–279.

Billen, J., Ito, F., Tsuji, K., Schoeters, E., Maile, R., and Morgan, E.D. 2000. Structure and chemistry of the Dufour gland in *Pristomyrmex* ants (Hymenoptera, Formicidae). Acta Zool. **81**: 159–166.

Blum, M.S. 1970. The chemical basis of insect sociality. *In* Chemicals controlling insect behavior. *Edited by* M. Beroza. Academic Press. New York, N.Y. pp. 61–94.

Bohlmann, J., Steele, C.L., and Croteau R. 1997. Monoterpene synthases from grand fir (*Abies grandis*): cDNA isolation, characterization, and functional expression of myrcene synthase, (–)-4S-limonene synthase, and (–)-(1*S*,5*S*)-pinene synthase. J. Biol. Chem. **35**: 21784–21792.

Bohlmann, J., Crock, J., Jetter, R., and Croteau, R. 1998*a*. Terpenoid-based defenses in conifers: cDNA cloning, characterization, and functional expression of wound-inducible (*E*)-α-bisabolene synthase from grand fir (*Abies grandis*). Proc. Natl. Acad. Sci. U.S.A. **95**: 6756–6761.

Bohlmann, J., Gilbert Meyer-Gauen, G., and Croteau, R. 1998*b*. Plant terpenoid synthases: molecular biology and phylogenetic analysis. Proc. Natl. Acad. Sci. U.S.A. **95**: 4126–4133.

Bohlmann, J., Phillips, M., Ramachandiran, V., Katoh, S., and Croteau R. 1999. cDNA cloning, characterization, and functional expression of four new monoterpene synthase members of the *Tpsd* gene family from grand fir (*Abies grandis*). Arch. Biochem. Biophys. **368**: 232–243.

Bohlmann, J., Martin, D., Oldham, N.J., and Gershenzon, J. 2000. Terpenoid secondary metabolism in *Arabidopsis thaliana*: cDNA cloning, characterization, and functional expression of a myrcene/(*E*)-β-ocimene synthase. Arch. Biochem. Biophys. **375**: 261–169.

Borden, J.H. 1985. Aggregation pheromones. *In* Comprehensive insect physiology, biochemistry and pharmacology, Volume 9. *Edited by* G.A. Kerkut and L.I. Glibert. Pergamon, Oxford, U.K. pp. 257–285.

Byers, J.A. 1995. Host-tree chemistry affecting colonization in bark beetles. *In* Chemical ecology of insects. *Edited by* Cardé, R.T. and Bell, W.J. Chapman and Hall, New York, N.Y. pp. 154–213.

Cane, D.E. 1999. Sesquiterpene biosynthesis: cyclization mechanisms. *In* Comprehensive natural products chemistry, Volume 2. *Edited by* D.E. Cane. Pergamon, Oxford, U.K. pp. 155–200.

Casey, T.M., Joos, B., Fitzgerald, T.D., Yurlina, M.E., and Young, P.A. 1988. Synchronized group foraging, thermoregulation, and growth of eastern tent caterpillars in relation to microclimate. Physiol. Zool. **61**: 372–377.

Chararas, C., Revolon, C., Feinberg, M., and Ducauze, C. 1982. Preference of certain Scolytidae for different conifers: a statistical approach. J. Chem. Ecol. **8**: 1093–1110.

Christe, P., Opplinger, A., Bancalà, F., Castella, G., and Chapuisat, M. 2003. Evidence for collective medication in ants. Ecol. Lett. **6**: 19–22.

Cobb, J.F.W., Krstic, M., Zavarin, E., and Barber, J.H.W. 1968. Inhibitory effects of volatile oleoresin components on *Fomes annosum* and four *Ceratocystis* species. Phytopathology **58**: 1327–1335.

Codella, S., and Raffa, K.F. 1995. Host plant influence on chemical defense in conifer sawflies (Hymenoptera: Diprionidae). Oecologia **104**: 1–11.

Dicke, M., and Vet, L. 1999. Plant-carnivore interactions: evolutionary and ecological consequences for plant, herbivore, and carnivore. *In* Herbivores: between plant and predators. *Edited by* H. Olff, V. Brown, and R. Drent. Blackwell Science, Oxford, U.K. pp. 483–520.

Dicke, M., Gols, R., Ludeking, D., and Posthumus, M.A. 1999. Jasmonic acid and herbivory differentially induce carnivore- attracting plant volatiles in lima bean plants. J. Chem. Ecol. **25**: 1907–1922.

Dobson, H.E.M. 1994. Floral volatiles in insect biology. *In* Insect-plant interactions. Volume 5. *Edited by* E. Bernays. CRC Press, Boca Raton, USA. pp. 47–81.

Dudareva, N., and Pichersky, E. 2000. Biochemical and molecular genetic aspects of floral scents. Plant Physiol. **122**: 627–633.

Dudareva, N., Cseke, L., Blanc, V.M., and Pichersky, E. 1996. Evolution of floral scent in *Clarkia*: novel patterns of S-linalool synthase gene expression in the *C. breweri* flower. Plant Cell **8**: 1137–1148.

Dudareva, N., Martin, D., Kish, C.M., Kolosova, N., Gorenstein, N., Fäldt, J, Miller, B., and Bohlmann, J. 2003. (*E*)-β-ocimene synthase and myrcene synthase genes of floral scent biosynthesis in snapdragon: function and expression of three terpene synthase genes of a new TPS-subfamily. Plant Cell **15**: 1227–1241.

Edwards, P.B., Wanjura, W.J., and Brown, W.V. 1993. Selective herbivory by Christmas beetles in response to intraspecific variation in *Eucalyptus* terpenoids. Oecologia **95**: 551–557.

Erbilgin, N., and Raffa, K.F. 2000. Opposing effects of host monoterpenes on responses by two sympatric species of bark beetles to their aggregation pheromones. J. Chem. Ecol. **26**: 2527–2548.

Erbilgin N., and Raffa K.F. 2001. Modulation of predator attraction to pheromones of two prey species by stereochemistry of plant volatiles. Oecologia **127**: 444–453.

Erdtman, H., Kimland, B., Norin, T., and (in part) Daniels P.J.L. 1968. The constituents of the "pocket resin" from Douglas fir *Pseudotsuga menziesii* (Mirb.) Franco. Acta Chem. Scand. **22**: 938–942.

Fäldt, J. 2000. Volatile constituents in conifers and conifer-related wood decaying fungi. Ph.D. thesis. Royal Institute of Technology, Department of Chemistry, Organic Chemistry. Stockholm, Sweden.

Fäldt, J., Arimura, G-I., Gershenzon, J., Takabayashi, J., and Bohlmann, J. 2003*a*. Functional identification of *AtTPS03* as (*E*)-β-ocimene synthase: a monoterpene synthase catalyzing jasmonate- and wound-induced volatile formation in *Arabidopsis thaliana*. Planta **216**: 745–751.

Fäldt, J., Martin, D., Miller, B., Rawat, S., and Bohlmann, J. 2003*b*. Traumatic resin defense in Norway Spruce (*Picea abies*): methyl jasmonate-induced terpene synthase gene expression, and cDNA cloning and functional characterization of (+)-3-carene synthase. Plant Mol. Biol. **51**: 119–133.

Farmer, E.E., and Ryan C.A. 1992. Octadecanoid precursors of jasmonic acid activate the synthesis of wound inducible proteinase inhibitors. Plant Cell **4**: 129–134.

Feld, B.K., Pasteels, J.M., and Boland, W. 2001. *Phaedon cochleariae* and *Gastrophysa viridula* (Coleoptera: Chrysomelidae) produce defensive iridoid monoterpenes *de novo* and are able to sequester glycosidically bound terpenoid precursors. Chemoecology **11**: 191–198.

Fitzgerald, T.D. 1980. An analysis of daily foraging patterns of laboratory colonies of the eastern tent caterpillar, *Malacosoma americanum* (Lepidoptera: Lasiocampidae), recorded photoelectronically. Can. Entomol. **112**: 731–738.

Fitzgerald, T.D., Casey, T., and Joos, B. 1988. Daily foraging schedule of field colonies of the eastern tent caterpillar, *Malacosoma americanum*. *Oecologia* 76: 574–578.

Franceschi, V.R., Krokene, P., Krekling, T., and Christiansen, E. 2000. Phloem parenchyma cells are involved in local and distant defense responses to fungal inoculation or bark beetle attack in Norway spruce (Pinaceae). Am. J. Bot. **87**: 314–326.

Franceschi, V.R., Krekling, T., and Christiansen, E. 2002. Application of methyl jasmonate on *Picea abies* (Pinaceae) stems induces defense-related responses in phloem and xylem. Am. J. Bot. **89**: 578–586.

Gols, R., Posthumus, M.A., and Dicke, M. 1999. Jasmonic acid induces the production of gerbera volatiles that attract the biological control agent *Phytoseiulus persimilis*. Entomol. Exp. Appl. **93**: 77–86.

Greenhagen, B., and Chappell, J. 2001. Molecular scaffolds for chemical wizardry: Learning nature's rules for terpene cyclases. Proc. Natl. Acad. Sci. U.S.A. **98**: 13479–13481.

Guterman, I., Sahlit, M., Menda, N., Piestun, D., Dafny-Yelin, M., Shalev, G., Bar, E., Davydov, O., Ovadis, M., Emanuel, M., Wang, J., Adam, Z., Pichersky, E., Lewinsohn, E., Zamir, D., Vainstein, A., and Weiss, D. 2002. Rose scent: genomics approach to discovering novel floral fragrance-related genes. Plant Cell **14**: 2325–2338.

Halitschke, R., Kessler, A., Kahl, J., Lorenz, A., and Baldwin, I.T. 2000. Ecophysiological comparison of direct and indirect defenses in *Nicotiana attenuata*. Oecologia **124**: 408–417.

Halitschke R., Schittko U., Pohnert G., Boland W, and Baldwin I.T. 2001. Molecular interactions between the specialist herbivore *Manduca sexta* (Lepidoptera, Sphingidae) and its natural host *Nicotiana attenuata*. III. Fatty acid-amino acid conjugates in herbivore oral secretions are necessary and sufficient for herbivore-specific plant responses. Plant Physiol. **125**: 711–717.

Hall, G.M., Tittiger, C., Andrews, G.L., Mastick, G.S., Kuenzli, M., Luo, X., Seybold, S.J., and Blomquist, G.J. 2002. Midgut tissue of male pine engraver, *Ips pini*, synthesized monoterpenoids pheromone component ipsdienol *de novo*. Naturwissenschaften **89**: 79–83.

Hobson K.R., Wood D.L., Cool L.G., White P.R., Ohtsuka T., Kubo I., and Zavarin E. 1993. Chiral specificity in responses by the bark beetle *Dendroctonus valens* to host kairomones. J. Chem. Ecol. **19**: 1837–1846.

Hopke, J., Donath, J., Blechert, S., and Boland, W. 1994. Herbivore-induced volatiles: the emission of acyclic homoterpenes from leaves of *Phaseolus lunatus* and *Zea mays* can be triggered by a β-glucosidase and jasmonic acid. FEBS Letters **352**: 146–150.

Hughes, P.R. 1974. Myrcene: a precursor of pheromones in Ips beetles. J. Insect Physiol. **20**: 1271–1275.

Janssen, E., Hölldobler, B., Kern, F., Bestmann, H.J., and Tsuji, K. 1997. Trail pheromone of myrmicine ant *Prystomyrmex pugens*. J. Chem. Ecol. **23**: 1025–1034.

Kalberer, N.M., Turlings T.C.J., and Rahier, M. 2001. Attraction of a leaf beetle (*Oreina cacaliae*) to damaged host plants. J. Chem. Ecol. **27**: 647–661.

Karban, R., Baldwin, I.T., Baxter, K.J., Laue, G., and Felton, G.W. 2000. Communication between plants: induced resistance in wild tobacco plants following clipping of neighboring sagebrush. Oecologia **125**: 66–71.

Kessler, A., and Baldwin, I.T. 2001. Defensive function of herbivore-induced plant volatile emissions in nature. Science (Washington, D.C.) **291**: 2141–2144.

Knudsen, J.T., Tollsten, L., and Bergström, L.G. 1993. Floral scents — a checklist of volatile compounds isolated by headspace techniques. Phytochemistry **33**: 253–280.

Koch, T., Krumm, T., Jung, V., Engelberth, J, and Boland, W. 1999. Differential induction of plant volatile biosynthesis in the lima bean by early and late intermediates of the octadecanoid-signaling pathway. Plant Physiol. **121**: 153–162.

Koyama, T., and Ogura K. 1999. Isopentenyl diphosphate isomerase and prenyl transferases. *In* Comprehensive Natural Products Chemistry, Volume 2. *Edited* by D.E. Cane. Pergamon, Oxford, UK. pp. 69–96.

Lange, B.M., Ketchum, R.E.B., and Croteau, R.B. 2001. Isoprenoid biosynthesis. Metabolite profiling of peppermint oil gland secretory cells and application to herbicide target analysis. Plant Physiol. **127**: 305–314.

Lloyd, H.A., Blum, M.S., Snelling, R.R., and Evans, S.L. 1989. Chemistry of mandibular and Dufour's gland secretions of ants in genus *Myrmecocystus*. J. Chem. Ecol. **15**: 2589–2600.

MacMillan, J., and Beale, M.H. 1999. Diterpene biosynthesis. *In* Comprehensive Natural Products Chemistry, Volume 2. Edited by D.E. Cane. Pergamon, Oxford. pp. 217–244.

Martin, D., Bohlmann, J., Gershenzon, J., Francke W., and Seybold S.J. 2001. A novel sexspecific and inducible monoterpene synthase activity associated with the bark beetle, *Ips pini* (Say). Proc. 18th Annual Meeting of the International Society of Chemical Ecology. Lake Tahoe, USA. pp.

Martin, D. Tholl, D., Gershenzon, J., and Bohlmann, J. 2002. Methyl jasmonate induces traumatic resin ducts, terpenoid resin biosynthesis, and terpenoid accumulation in developing xylem of Norway spruce stems. Plant Physiol. **129**: 1003–1018.

Martin, D., Bohlmann, J., Gershenzon, J, Francke W., and Seybold, S.J. 2003*a*. A novel sexspecific and inducible monoterpene synthase activity associated with a metazoan. Naturwissenschaften **90**: 173–179.

Martin, D., Gershenzon, J., and Bohlmann, J. 2003*b*. Induction of volatile terpene biosynthesis and diurnal emission by methyl jasmonate in foliage of Norway spruce (*Picea abies*). Plant Physiology **132**: 1586–1599.

Mattiacci, L, Dicke, M., and Posthumus, M.A. 1995. β-Glucosidase: an elicitor of herbivore induced plant odor that attracts host-searching parasitic wasps. Proc. Natl. Acad. Sci. U.S.A. **92**: 2036–2040.

McGarvey, D.J. and Croteau, R. 1995. Terpenoid metabolism. Plant Cell **7**: 1015–1026.

Miller, D.R. and Borden, J.H. 2000. Dose-dependent and species-specific responses of pine bark beetles (Coleoptera: Scolytidae) to monoterpenes in association with pheromones. Can. Entomol. **132**: 183–195.

Miller, D.R. and Lindgren, B.S. 2000. Comparison of alpha-pinene and myrcene on attraction of mountain pine beetle, *Dendroctonus ponderosae* (Coleoptera: Scolytidae) to pheromones in stands of western white pine. J. Entomol. Soc. British Columbia **97**: 41–46.

Miller, D.R., Borden, J.H., and Slessor, K.N. 1989. Interpopulation and intrapopulation variation of the pheromone ipsdienol produced by male pine engravers, *Ips pini* (Say) (Coleoptera: Scolytidae). J. Chem. Ecol. **15**: 233–248.

Miller, D.R., Borden, J.H., and Slessor, K.N. 1996. Enantiospecific pheromone production and response profiles for populations of pine engraver, *Ips pini* (Say) (Coleoptera: Scolytidae), in British Columbia. J. Chem. Ecol. **22**: 2157–2172.

Mozuraitis, R., Stranden, M., Ramirez, M.I., Borg-Karlson, A.-K., and Mustaparta, H. 2002. (–)-Germacrene D increases attraction and oviposition by the tobacco budworm moth *Heliothis virescens*. Chem. Senses **27**: 505–509.

Nagy, N.E., Franceschi, V.R., Solheim, H., Krekling, T., and Christiansen, E. 2000. Wound induced traumatic resin duct development in stems of Norway spruce (Pinaceae): anatomy and cytochemical traits. Am. J. Bot. **87**: 302–3113.

Nardi, J.G., Young, A.G., Ujhelyi, E., Tittiger, C., Lehane, M.J., and Blomquist, G.J. 2002. Specialization of midgut cells for synthesis of male isoprenoid pheromone components in two scolytid beetles, *Dendroctonus jeffreyi* and *Ips pini*. Tiss. Cell 34: 221–231.

Ozawa, R., Arimura G-I., Takabayashi, J., Shimoda, T., and Takaaki, N. 2000. Involvement of jasmonate- and salicylate-related signaling pathways for the roduction of specific herbivore-induced volatiles in plants. Plant Cell Physiol. **41**: 391–398.

Paine, T.D., and Hanlon, C.C. 1994. Influence of oleoresin constituents from *Pinus ponderosa* and *Pinus jeffreyi* on growth of mycangial fungi from *Dendroctonus jeffreyi*. J. Chem. Ecol. **20**: 2551–2563.

Paine T.D., Raffa, K.F., and Harrington, T.C. 1997. Interactions among scolytid bark beetles, their associated fungi, and live host conifers. Annu. Rev. Entomol. **42**: 179–206.

Paré, P.W., and Tumlinson, J.H. 1999. Plant volatiles as a defense against insect herbivores. Plant Physiol. **121**: 325–331.

Persson, M., Sjödin, K., Borg-Karlson, A-K., Norin, T., and Ekberg, I. 1996. Relative amounts and enantiomeric compositions of monoterpene hydrocarbons in xylem and needles of *Picea abies*. Phytochemistry **42**: 1289–1297.

Peters, R.J., and Croteau, R.B. 2002. Abietadiene synthase catalysis: mutational analysis of a prenyl diphosphate ionization-initiated cyclization and rearrangement. Proc. Natl. Acad. Sci. U.S.A. **99**: 580–584.

Pettersson, S. and Knudsen J.T. 2001. Floral scent and nectar production in *Parkia biglobosa* Jacq. (Luguminosae: Mimosoideae). Bot. J. Linn. Soc. **135**: 97–106.

Phillips, M.A., Wildung, M.R., Williams, D.C., Hyatt, D.C., and Croteau, R. 2003. cDNA isolation, functional expression, and characterization of (+)-α-pinene synthase and (–)-α-pinene synthase from loblolly pine (*Pinus taeda*): stereocontrol in pinene biosynthesis. Arch. Biochem. Biophys. **411**: 267–276.

Pichersky E., and Gershenzon J. 2002. The formation and function of plant volatiles: Perfumes for pollinator attraction and defense. Curr. Opin. Plant Biol. **5**: 237–243.

Pichersky, E., Lewinsohn, E., and Croteau, R. 1995. Purification and characterization of S-linalool synthase, an enzyme involved in the production of floral scent in *Clarkia breweri*. Arch. Biochem. Biophys. **316**: 803–807.

Raffa, K.F., Berryman, A.A., Simasko J., Teal W., and Wong, B.L. 1985. Effects of grand fir monoterpenes on the fir engraver, *Scolytus ventralis* (Coleoptera: Scolytidae), and its symbiotic fungus. Environ. Entomol. **14**: 552–556.

Raguso, R.A., and Pichersky, E. 1999. A day in the life of a linalool molecule: chemical communication in a plant-pollinator system. Part 1: Linalool biosynthesis in flowering plants. Plant Sp. Biol. **14**: 95–120.

Rembold, H., Köhne, A.C., and Schroth, A. 1991. Behavioral response of *Heliothis armigera* Hb. (Lep., Noctuidae) moths on a synthetic chickpea (*Cicer arietinum*) kairomone. J. Appl. Entomol. **112**: 254–262.

Renwick, J.A.A., Hughes, P.R., and Krull, I.S. 1976. Selective production of *cis*- and *trans*verbenol from (–)- and (+)-α-pinene by a bark beetle. Science (Washington, D.C.) **191**: 199–201.

Rodriguez-Saona, C., Crafts-Brandner, S.J., Paré, P.W., and Henneberry, T.J. 2001. Exogenous methyl jasmonate induces volatile emissions in cotton plants. J. Chem. Ecol. **27**: 679–695.

Roisin, Y., Everaerts, C., Pasteels, J.M., and Bonnard, O. 1990. Caste dependent reactions to soldier defensive secretion and chiral alarm-recruitment pheromone in *Nasutitermes princes*. J. Chem. Ecol. **16**: 2865–2876.

Röse, U.S.R., Manukian, A., Heath, R.R., and Tumlinson, J.H. 1996. Volatile semiochemicals released from undamaged cotton leaves: a systematic response of living plants to caterpillar damage. Plant Physiol. **111**: 487–495.

Rynkiewicz, M.J., Cane, D.E., and Christianson, D.W. 2001. Structure of trichodiene synthase from *Fusarium sporotrichioides* provides mechanistic inferences on the terpene cyclization cascade. Proc. Natl. Acad. Sci. U.S.A. **98**: 13543–13548.

Schmelz, E.A., Alborn, H.T., and Tumlinson, J.H. 2001. The influence of intact-plant and excised-leaf bioassay designs on volicitin- and jasmonic acid-induced sesquiterpene volatile release in *Zea mays*. Planta **214**: 171–179.

Schmidt, C.O., Mouwmeester, H.J., de Kraker, J-W., and König, W.A. 1998. Biosynthesis of (+)- and (–)-germacrene D in *Solidago canadensis*: isolation of two enantioselective germacrene D synthases. Angew. Chem. Int. Ed. **37**: 1400–1402.

Seybold, S.J., and Tittiger, C. 2003. Biochemistry and molecular biology of *de novo* isoprenoid pheromone production in the Scolytidae. Annu. Rev. Entomol. **48**: 425–453.

Sjödin, K., Persson, M., Borg-Karlson, A-K., and Torbjörn H. 1996. Enantiomeric compositions of monoterpene hydrocarbons in different tissues of four individuals of *Pinus sylvestris*. Phytochemistry **41**: 439–445.

Starks C.M., Back K., Chappell J., and Noel J.P. 1997. Structural basis for cyclic terpene biosynthesis by tobacco 5-epi-aristolochene synthase. Science (Washington, D.C.) **277**: 1815–1820.

Steele, C.L., Crock, J., Bohlmann, J., and Croteau, R. 1998. Sesquiterpene synthases from grand fir (*Abies randis*): comparison of constitutive and wound-induced activities, and cDNA isolation, characterization, and bacterial expression of d selinene synthase and g-humulene synthase. J. Biol. Chem. **273**: 2078–2089.

Takabayashi, J., and Dicke, M. 1996. Plant-carnivore mutualism through herbivore-induced carnivore attractants. Trends Plant Sci. **1**: 109–113.

Tomlin, E.S., Alfaro, R.I., Borden, J.H., and Fangliang, H. 1998. Histological response of resistant and susceptible white spruce to simulated white pine weevil damage. Tree Physiol. **18**: 21–28.

Tomlin, E.S., Antonejevic, E., Alfaro, R.I., and Borden J.H. 2000. Changes in volatile terpene and diterpene resin acid composition of resistant and susceptible white spruce leaders exposed to simulated white pine weevil damage. Tree Physiol. **20**: 1087–1095.

Trapp, S.C., and Croteau, R.B. 2001. Genomic organization of plant terpene synthases and molecular evolutionary implications. Genetics **158**: 811–832.

Turlings, T.C.J., Tumlinson, J.H., and Lewis, W.J. 1990. Exploitation of herbivore induced odors by host-seeking parasitic wasps. Science (Washington, D.C.) **250**: 1251–1253.

Turner, J.G., Ellis, C., and Devoto, A. 2002. The jasmonate signal pathway. Plant Cell **14**: S153–S164.

Valterová, I., Køeèek J., and Vrkoè, J. 1989. Intraspecific variation in the defense secretions of *Nasutitermes ephratae* soldiers and the biological activity of some of their components. Biochem. Syst. Ecol. **17**: 327–323.

Wallin, K.F., and Raffa, K.F. 2000. Influences of host chemical and internal physiology on the multiple steps of postlanding host acceptance behavior of *Ips pini* (Coleoptera: Scolytidae). Environ. Entomol. **29**: 442–453.

Wasternack, C. and Parthier, B. 1997. Jasmonate-signalled plant gene expression. Trends Plant Sci. **2**: 302–307.

Werner, R.A. 1995. Toxicity and repellency of 4-allylanisole and monoterpenes from white spruce and tamarack to the spruce beetle and eastern larch beetle (Coleoptera: Scolytidae). Environ. Entomol. **24**: 372–379.

Wheeler, G.S., Massey, L.M., and Southwell, I.A. 2002. Antipredator defense of biological control agent *Oxyops vitiosa* is mediated by plant volatiles sequestered from the host plant *Melaleuca quinquenervia*. J. Chem. Ecol. **28**: 297–315.

White, P.R. and Hobson, K.R. 1993. Stereospecific antennal response by red turpentine beetle, *Dendroctonus valens* to chiral monoterpenes from ponderosa pine resin. J. Chem. Ecol. **19**: 2193–2202.

Whittington, D.A., Wise, M.L., Urbansky, M., Coates, R.M., Croteau, R.B., and Christianson, D.W. 2002. Bornyl diphosphate synthase: structure and strategy for carbocation manipulation by a terpenoid cyclase. Proc. Natl. Acad. Sci. U.S.A. **99**: 15375–15380.

Wise, M.L. and Croteau, R. 1999. Monoterpene biosynthesis. *In* Comprehensive Natural Products Chemistry, Volume 2. *Edited by* D.E. Cane, Pergamon, Oxford. pp. 97–154.

11
Adaptation in plant speciation: evidence for the role of selection in the evolution of isolating barriers between plant species

Leonie C. Moyle

Abstract: Ascertaining the role of adaptation in speciation presents unique difficulties because not only must the adaptive function of a trait be established but this trait must also be associated with the operation of an isolating barrier between taxonomic groups. Empirical evidence for the role of adaptation in the development of reproductive isolation between plant species is outlined, focusing on experimental and observational studies that identify an explicit connection between putatively adaptive traits and reproductive isolating barriers. Three preliminary patterns emerge: First, there is considerably more evidence that reproductive isolation arises as an indirect by-product of selected changes, rather than as the direct target of selection. Second, adaptation is most commonly associated with the development of pre-mating isolating barriers and/or (environment-dependent) hybrid maladaptation; evidence for the involvement of selection in the development of post-mating pre-zygotic, and post-zygotic 'intrinsic' stages of isolation is weak or absent. Third, the majority of relevant trait changes involve floral traits, and/or traits associated with edaphic adaptation. Current limitations on the available data (e.g., most of the available studies are in herbaceous temperate taxa) mean that the applicability of these conclusions to all plant taxa, including those with diverse life histories and mating systems, remains to be determined. In addition to increasing taxonomic breadth and examining the adaptive effects of polyploidy and hybridization, future research into the role of adaptation in plant speciation should also tackle the potential roles of sexual selection and/or intra- and inter-genomic conflict in the evolution of reproductive isolating barriers.

Introduction

> *"In nature there is a continuum from adaptation to speciation, and the line between them is sometimes more imagined than real."* (Schemske 2000, p. 1073)

Which forces drive the evolution of new species is a fundamental question in contemporary speciation research. Frequently this question is recast in terms of the relative importance of genetic drift versus natural selection in the formation of isolating barriers between diverging taxa (e.g., Barrett 1996; Barton 1996). Whether drift or selection is the predominant force in speciation influences the conditions under which we expect speciation, as well as our expectation of the types of traits involved, and the biotic and abiotic conditions likely to accelerate or retard taxonomic diversification. Nonetheless, despite considerable interest in speciation processes over the last 30 years, many basic questions remain: Do traits under selection act as the basis of reproductive isolation? Does this selection primarily act directly or indirectly to form species barriers? What conditions and agents of selection are most frequently associated with the development of isolation? These questions underscore the extent to which the intimate links between adaptation (i.e., the product of natural selection) and speciation (Schneider 2000) have yet to be fully investigated.

Selection versus drift in speciation

Because species are frequently characterized both by breaks in phenotypic similarity and by breaks in reproductive compatibility, it is tempting to assume a direct connection between forces responsible for the establishment of both. The temptation to invoke selection in the formation of new species is, understandably, even greater when speciation appears to occur very rapidly. In this regard, the evolutionary forces driving angiosperm speciation are especially intriguing because of the spectacularly rapid rate of diversification observed within this group: ~240,000 extant species in under 140 million years of evolution (Ricklefs and Renner 1994) (for comparison, Drosophilids are thought to be ~80 million years old, with ~3200 extant taxa; Powell 1997). Currently several hypotheses appeal to different angiosperm adaptations (e.g., animal-vectored pollination or seed dispersal) in an effort to explain this rapid diversification (e.g., Ricklefs and Renner 1994; Bawa 1995). Nonetheless, just as substantial phenotypic differences between taxa may result from random stochastic forces (e.g., Lande 1976), so too isolating barriers could be the result of neutral or stochastic genetic changes that incidentally produce species barriers between

Leonie C. Moyle. Center for Population Biology, Section of Evolution and Ecology, University of California, Davis, CA 95616, USA. e-mail address: lcmoyle@ucdavis.edu
Correct citation: Moyle, L.C. 2004. Adaptation in plant speciation: evidence for the role of selection in the evolution of isolating barriers between plant species. *In* Plant Adaptation: Molecular Genetics and Ecology. Proceedings of an International Workshop held December 11–13, 2002, in Vancouver, British Columbia, Canada. *Edited by* Q.C.B. Cronk, J. Whitton, R.H. Ree, and I.E.P. Taylor. NRC Research Press, Ottawa, Ontario. pp. 82–93.

differentiating lineages. The 'neutral expectation' for rates of speciation due to stochastic processes alone is not well–defined (although see Barton 1996); in principle, however, accelerated diversification in angiosperms could be consistent with increased susceptibility to reproductive isolation via stochastic forces (e.g., founder events, bottlenecks, and/or inbreeding). The relative importance of drift and selection is presently debated both theoretically (e.g., Turelli et al. 2001) and with respect to observed empirical patterns. For example, it has frequently been suggested that stochastic forces play a large role in diversification processes on islands (e.g., Barrett 1996; Weller et al. 2001), although novel selective environments might also account for island diversification (e.g., Barton 1996).

One root cause for ongoing debates over the role of selection in species diversification is that empirical evidence of selectively-driven speciation is difficult to obtain. Not only must the adaptive function and selective history of a trait or set of traits be identified, but this trait(s) must also be explicitly associated with the operation of an isolating barrier between groups. Accordingly, the evidence for direct, explicit links between adaptation and reproductive isolation is sparse, even in (arguably) better studied animal groups. Indirect evidence has been used to make a case for the relative importance of selection in animal speciation (e.g., Rice and Hostert 1993, reviewing evidence from manipulative laboratory experiments in *Drosophila*). In addition, Schluter (2001) has identified a number of candidate empirical examples (including 2 in plants) that provide solid evidence for the role of divergent natural selection in the formation of species barriers. In this paper, my aim is to outline the available empirical evidence for the role of selection in the formation of new plant species. This outline is not intended to be an exhaustive survey but rather an overview of the patterns that appear to predominate in the experimental and observational literature to date. All of the examples described are drawn from flowering plants (angiosperms), largely because this is where the bulk of data has been collected; for the same reason, almost all are from herbaceous taxa. My goal is to provide some insight into our current understanding of the potential role of adaptation in plant reproductive isolation and speciation. The general questions that motivate this overview are those posed at the outset of this paper: Do traits under selection act as the basis of reproductive isolation? Does this selection primarily act directly or indirectly to form species barriers? What conditions and agents of selection are most frequently associated with the development of isolation?

Adaptation in plant populations

Clearly, our understanding of how adaptation may be implicated in plant speciation is necessarily informed (and limited) by our understanding of plant adaptation itself. The state of play in contemporary plant adaptation is a central theme of this entire volume, and a comprehensive review is inappropriate here (see also Stebbins 1950; Grant 1981; Levin 2000). Nonetheless classical experimental works (e.g., Clausen 1951) through to contemporary QTL analyses (e.g., Mitchell-Olds 1995) provide abundant, though frequently indirect, data concerning the general conditions and factors that appear to facilitate adaptive responses in plants. Of these, primarily biotic factors include interactions with four groups: predators, competitors, pollinators (and other mutualists), and pathogens. Abiotic factors associated with adaptive plant responses include soil, water, light, temperature, and nutrients. Adaptive plant responses addressed in this volume alone include, for example, drought, heavy metal, and salt tolerance (Rajakaruna and Whitton, this volume, Chapter 13), as well as adaptive responses to local light levels (Borevitz, this volume, Chapter 8; Matthews, this volume, Chapter 18), predominant pollinators, and herbivore attack (Miranda et al, this volume, Chapter 19). Adaptive responses to soil and associated conditions (edaphic adaptations) are perhaps the clearest current examples of rapid adaptation in plant populations (Kruckeberg 1986; Macnair and Gardner 1998; Rajakaruna and Whitton, 2004). Serpentine soils, which display a combination of unique (poor) soil structural characteristics, heavy metal content, unusual mineral ratios, nutrient poverty, and rapid water loss (Macnair and Gardner 1998) are especially ripe with conditions that stimulate natural selection (Kruckeberg 1986).

Nonetheless, while studies frequently document a general association between particular phenotypic traits, and a specific set of environmental conditions, it is considerably less common to evaluate adaptive function directly (e.g., through reciprocal transplants), or to test an hypothesis of selection against appropriate null models (Lande 1976; Schluter 2000). In addition, although occasionally the genetic basis is known, in general the genes underlying some of our best examples of plant adaptation remain to be discovered. As a result, our ability to draw definitive conclusions about the process of adaptation in nature and the conditions and traits predominantly involved, remains limited (Schluter 2000). These limitations in turn extend to our current understanding of the role of adaptation in plant speciation.

Selection, adaptation, and the generation of isolation among plant species

Here I outline examples from observational and experimental studies that provide some evidence for the role of adaptation in the development of isolating barriers between species. Where it is thought to be known, the genetic basis of these traits and associated isolating barriers is also noted. In principle, any adaptive change that leads to complete spatial isolation could be considered an adaptation that produces speciation. However, when considering very young taxa, the permanence of spatial barriers alone is uncertain. My aim is to draw somewhat stronger connections between adaptive change and the evolution of species isolating barriers, therefore all the examples discussed here include at least some explicit evaluation of barriers to reproduction or gene flow (including changes in reproductive/mating system) beyond simple spatial separation. As such, my focus is on the link between adaptation and the component of speciation involving the evolution of biological isolating barriers. The framework for the following discussion owes much to Schluter's (2000) characterization of models of ecological speciation (see chapter 8 in Schluter 2000), although that account focuses on the role of ecologically-based divergent natural selection in generating reproductive isolation whereas the discussion here does not require that selection be divergent. A distinction is drawn between isolating barriers that are the by-product of selection for adaptive changes in populations, versus isolation that is the direct target and product of selection; each is discussed in turn. Note that examples discussed here do not involve any polyploid (or hybrid) speciation processes. The possible role of adaptation in polyploid (and/or hybrid) speciation — especially in the persistence and spread of newly formed polyploid/hybrid lineages — is an important topic in it's own right but beyond the scope of this paper (see, for example, Levin 2002; Husband, this volume, Chapter 15).

A. Isolation as a by-product of selection

Reproductive isolation can evolve as a by-product of genetic changes favored by selection in different lineages, either as a true pleiotropic effect or because of genetic correlations or linkage between genes under direct selection and genes that cause reproductive isolation. This process is most likely in allopatry, both because spatially separated populations may be more likely to experience different selective conditions, and because any potential for homogenizing gene flow is limited simply by physical distance alone. Nonetheless, by-product isolation can still operate between taxa occurring in sympatry. For example, selection for character variants (e.g., flower color) that are simultaneously important in ecological function (e.g., attracting new pollinators) and in mating and gene flow (e.g., assortative mating) could lead to evolution of reproductive isolation in sympatry, without direct selection for isolation itself. Note also that by-product isolation does not require divergent environmental conditions for selection to fix 'complementary' genes (i.e., loci that function in their native genetic background but are dysfunctional on foreign genetic backgrounds — Dobzhansky 1936) in different lineages. Identical selective pressures could produce alternative genetic 'solutions' in different populations, as determined solely by the order in which favorable mutations arise; the products of these alternative solutions could, nonetheless, be mutually incompatible when placed in the same genetic background. Finally, while reproductive isolation can evolve at any number of stages, ranging from prevention of mating through to hybrid inviability and sterility, for convenience examples outlined below are organized according to discrete isolation stages; as noted, some individual cases involve simultaneous or sequential changes at multiple stages of isolation.

Pre-mating isolation

Pre-mating isolation barriers may result indirectly if selection favors changes that lead to positive assortative mating within types, thereby reducing gene flow between types. In angiosperms, pre-mating barriers to isolation (apart from spatial isolation) frequently involve changes in modes of pollination, in kinds/classes of pollinators, or in mating systems, roughly captured by three (not necessarily mutually exclusive) categories of phenotypic change: shifts in flowering time, changes in floral morphology; and changes in selfing rate.

Flowering time shifts: Selectively-favored changes in developmental phenology can indirectly lead to isolation if as a consequence organisms are not reproductively receptive at the same time. Ecologically-induced shifts in flowering time appear to be one of the most common ways in which adaptive phenological changes simultaneously produce isolating barriers. Frequently local adaptation has been implicated in these shifts, particularly adaptation to new edaphic environments, which impose strong selection for accelerated phenology, including flowering time, because of limited or ephemeral water availability. For example, in four soil-specialized *Mimulus* species and one serpentine ecotype, early drying conditions appear to have selected for early flowering on serpentine (Macnair and Gardner 1998) or xeric (Kiang and Hamrick 1978) sites. These flowering time shifts produce considerable (though incomplete) isolating barriers between these taxa and their progenitor species; in 2 of these species, analyses indicate that early flowering is partially dominant and partially under major gene control (Macnair and Gardner 1998). In a number of cases, flowering time shifts are accompanied by other changes with isolating effects (e.g., *Salvia* — Grant and Grant 1964). For example, in multiple species of *Clarkia* (e.g., Lewis 1973) and in *Leavenworthia* (Lloyd 1979), selection for early flowering in new, water-ephemeral, habitats appears to have lead to strong selection for the evolution of self-pollination where early flowering outpaces the emergence of pollinators (see further below, and Wyatt 1988 for additional examples). Shifts in daily flowering schedules have also been associated with reproductive isolation in otherwise sympatric and crossable species. For example, *Petunia axillaris* and *Petunia integrifolia* flower nocturnally and diurnally, respectively; in combination with divergent flower colors, these different flowering schedules are synchronous with activity periods of different pollinators (moths and bees, respectively) and appear to effectively prevent gene flow between the two species (Ando et al. 2001).

Changes in floral morphology: Changes in flower color and other aspects of floral morphology (shape, size, symmetry, etc.) may lead to altered pollinator assemblages or modes (ethological isolation), or to changes in placement of pollen on pollinators (mechanical isolation) or to some combination of such factors (e.g., *Pedicularis* — Sprague 1962; *Salvia* — Grant and Grant 1964; *Stylidium* — Armbruster et al. 1994; see Stebbins 1950, Chapter 6, and Waser 1993 for other examples). Although floral changes might have a variety of adaptive functions, in most cases the inferred selective force is the exploitation of new pollinator assemblages (or reproductive assurance via increased selfing rates, see below). For example, it is hypothesized that new species of *Aquilegia* differentiated in flower color and form (especially nectar spurs) to exploit different pollinator assemblages; these shifts indirectly act to reduce or eliminate gene flow between different types, by changing relative attractiveness to different suites of pollinators (Hodges and Arnold 1995). Pollinator-induced shifts in floral form are likely responsible for rapid diversification over very short divergence times within *Aquilegia* (Hodges and Arnold 1995). Floral morphological differences between *Mimulus guttatus* and *Mimulus nudatus* also produce partial pre-mating barriers due to pollinator differences; in particular, reduced flower size and changed shape in *M. nudatus* may be due to a shift from honeybees to sweat bees, favored because of early flowering on drier poorer soils (Macnair and Gardner 1998). Approximately five genes are thought to control these flower size changes (Macnair and Gardner 1998).

Isolating floral changes may involve biochemical rather than structural modifications in morphology. For example, quantitative and qualitative differentiation in floral scent biochemistry among eight morphologically similar, sympatric, and co-flowering *Geonoma* (tropical palm) species is thought to be associated with attraction of different pollinators and therefore reproductive isolation (Knudsen 1999; Knudsen et al. 1999). Similarly, floral scent biochemistry appears to differentiate floral visitation to two sympatric *Ophyrs* orchids (Schiestl and Ayassi 2002). Floral changes may also produce isolating barriers at later stages. For example, changes in style length may lead to post-mating mechanical isolation between species (see section on 'Postmating prezygotic isolation'); selected changes in floral morphology may also result in hybrid maladaptation, where floral intermediates are discriminated against by pollinators (see section on 'Hybrid maladaptation').

Changes in selfing rates: Shifts in selfing rate, particularly from primarily outcrossing to primarily selfing, can act to reduce gene flow with other taxa as a pleiotropic effect. In particular, increased selfing or autogamy may frequently be stimulated by the need for reproductive assurance in new habitats where appropriate pollinators are absent (Jain 1976). Selection for autogamy in northern races of *Gilia achilleifolia*, for example, appears to be a response to absence of normal pollinators; while there is no unambiguous link between selfing changes and reproductive isolation in this case, the southern outcrossing and northern autogamous races do fall into two distinct genetic clusters (Schoen 1982). As indicated above, changes in selfing may frequently accompany flowering time shifts due to ecological factors, especially when this shift places the new flowering period outside the range of activity or presence of traditional pollinators (e.g., *Leavenworthia* — Lloyd 1979; *Clarkia* — Lewis 1973). For example, early flowering times favored on water-poor, heavy metal soils in *Mimulus cupriphilus* appears to have selected for increased selfing to guarantee seed set in the absence of pollinators (Macnair and Gardner 1998); the genetic basis of this reduced flower size is three or more recessive factors. Similarly, the evolution of xeric-tolerant *Mimulus nasutus* from intolerant *M. guttatus* appears to have involved accelerated development and flowering times leading to selection for selfing because of reduced availability of pollinators; reduced visitation might also be due to the relative unattractiveness of small *M. nasutus* plants with few flowers (Kiang and Hamrick 1978; see also above and Wyatt 1988 for other examples.) In a number of these cases, therefore, the evolution of selfing may be a second order effect of initial selective forces acting to differentiate populations. Pollinator absence may also impose selection for shifts in pollination mode. In Hawaiian *Schiedea*, for example, habitat shifts to exposed cliffs and ridges where there are few pollinators may have selected for wind-pollination as well as dioecy (Sakai et al. 1997). Many other examples of pre-pollination barriers that appear to accompany habitat shifts have been identified in island endemics (see Wagner and Funk 1995).

Post-mating, prezygotic isolation

Barriers that act after mating (i.e., after pollen is deposited on the stigma) but before fertilization, could be important in isolating closely related taxa. Incompatible pollen-stigma and pollen-style interactions have been implicated in barriers to artificial crosses between species (de Nettancourt 2001), as has the relative competitive ability of conspecific versus heterospecific pollen in the form of conspecific gamete precedence (e.g., Smith 1968; Arnold et al. 1993; Rieseberg et al. 1995; Marshall et al. 2002 and references therein). Nonetheless, the general significance of post-pollination barriers in bringing about and maintaining isolation between taxa remains unknown, perhaps partly because many studies do not examine factors such as gamete competition when looking at isolating barriers (Howard et al. 1998), but also because isolation at this stage is often accompanied by substantial barriers to gene flow at other stages (e.g., Buchholz et al. 1935; Carney et al. 1994; Rieseberg et al. 1995). Moreover, empirical evidence that such barriers are the by-products of selected changes is rare.

In principle, mismatched interactions between pollen-stigma, -style, or -ovule could result as the by-product of selection for different morphological or physiological trait changes, especially in floral and reproductive characters. For example, successful fertilization may require coordinated timing of mitotic cell cycles in the maturation of male and female gametophytes (Friedman 1999; Zhang and O'Neill 1993). In addition, numerous observational studies suggest that pollen tube lengths and/or growth rates are to some degree matched to style length within individual species (e.g., Buchholz et al. 1935; Smith 1970; Williams and Rouse 1988) presumably as the result of co-adaptation between pollen and style tissues. For example, in *Rhododendron* species and subspecies, taxa with large differences in style length are reciprocally isolated because pollen from short-styled species fails to reach long-styled ovules, whereas pollen from long-styled species frequently overgrows the ovules of short-styled species (Williams et al. 1986; Williams and Rouse 1988). It is hypothesized that this behavior is due to mismatched fertilization signaling between interspecific pollen and stigmata/ovules, possibly associated with mismatched timing of physiological maturity of the male and female gametophytes following mating (Williams et al. 1986); the selective forces responsible for these differences have not, however, been examined. In such cases, it is possible that floral changes selected by pollinators, especially changes in style length, might act to increase post-mating 'mechanical' isolation between taxa by disrupting this match between pollen tube and style length. Nonetheless, isolation due to differences in style lengths alone is likely to be an incomplete, and asymmetric, barrier among most species because (unlike in *Rhododendron*) pollen from most short-styled species is generally out competed in long-styled species but not vice versa (e.g., Buchholz et al. 1935). From the perspective of short-styled species, therefore, adaptively-favored changes that have shortened the style are unlikely to increase post-mating barriers and may actually act to enhance the fertilization success of heterospecific pollen from longer-styled species (e.g., *Datura* — Buchholz et al. 1935, *Mimulus* — Kiang and Hamrick 1978).

Postzygotic isolation

Ecological intermediacy/Environment-dependent 'hybrid maladaptation': Taxa can be effectively isolated, even if they form hybrids, if those hybrids are unable to persist in one or both parental habitats because of unsuitable ecological or environmental conditions (Anderson and Anderson 1954). Such maladapted hybrids have been used as evidence for 'ecological speciation' (Schluter 2000). Environment-dependent hybrid maladaptation is frequently assessed by evaluating the performance of parental and hybrid types in each of the parental environments, using reciprocal transplants. Following the classical studies of Clausen, Keck and Hiesey in *Achillea* and other species (see Clausen 1951), many current examples of hybrid intermediacy or maladaptation in plants come from studies of local adaptation to soil and/or elevation differences, or pollinator differentiation, between plant subspecies or species. For example, elevationally differentiated *Artemisia tridentata* subspecies are found on soils with different structural, mineral, nutrient and heavy metal characteristics; parental types show evidence of home site advantage whereas hybrids are less fit in either parental environment and most fit in an intermediate environment (Wang et al. 1997). Similarly, hybrids of *Mimulus cardinalis* and *Mimulus lewisii* evaluated in parental environments show high seedling mortality and viability selection for traits (including flower color) that match the local parental type (Nobs and Hiesey 1958). Similar patterns of viability selection against hybrids in parental environments have been seen in *Gossypium*, *Lycopersicon*, and *Phaseolus* interspecific crosses, and in an interracial cross in *Potentilla glandulosa* (see Grant 1967 for review and references).

Selectively favored changes in floral morphology may also result in hybrid maladaptation if floral intermediates are discriminated against by pollinators, i.e., are maladaptated to parental 'pollinator environments'. For example, *Ipomopsis* hybrids with intermediate floral phenotypes are disadvantaged (under-visited compared to parental types) in the presence of the two major parental pollinators (Campbell et al. 1997). Similarly, sympatric populations of *Salvia mellifera* and *Salvia apiana* have different sets of principal pollinators and these pollinators rarely visit F1 hybrids (Grant and Grant 1964). In hybrids of *Mimulus cardinalis* and *M. lewisii* (Schemske and Bradshaw 1999) pollinators preferentially used F2 hybrids with greater proportions of genes from the corresponding preferred parental type (Schemske and Bradshaw 1999); the genetic basis of flower color difference appears to be few genes of large effect (Bradshaw et al. 1995). Apart from reciprocal transplant studies, other weaker lines of evidence may also suggest hybrid maladaptation. For example, sympatric species that do not generally hybridize under natural conditions but that can be found hybridizing in artificially disturbed habitats (e.g., *Salvia* — Anderson and Anderson 1954; *Iris* — Lenz 1959) indirectly suggest that under normal circumstances ecological differentiation maintains isolation between parental types (Stebbins 1950; Grant 1971).

Hybrid inviability and sterility (*environment independent*): Isolation due to the presence of 'intrinsic' barriers — such as hybrid inviability, sterility, and breakdowni — is an effective, and most likely permanent, barrier to gene flow between diverging taxa. There are extensive data on species crossability, and hybrid viability and fertility in interracial and interspecific crosses from multiple diverse angiosperm groups, much of it collected as part of classical experimental taxonomic studies (see e.g., Stebbins 1950; Clausen 1951; Grant 1981). Nonetheless, although many such studies demonstrate a general relationship between morphological differentiation and degree of hybrid inviability or sterility, in almost all cases the forces responsible for fixing genes that cause reduced hybrid fitness (and even the nature of the genes themselves) remain unknown.

The best empirical evidence for the role of selection in fixing genes that produce post-zygotic isolation has been found in *Mimulus guttatus*. The major gene for copper tolerance in the tolerant ecotype of *M. guttatus* cosegregates with a complementary gene involved in hybrid inviability in crosses with the intolerant ecotype (Macnair and Christie 1983). Although it is not yet known whether hybrid inviability is a real pleiotropic effect of copper tolerance, or results from close linkage between tolerance and inviability alleles, this finding strongly suggests that selection for tolerance was responsible for the spread of a locus that directly influences F1 hybrid inviability. Forces responsible for the spread of the other factor involved in the incompatibility reaction are, however, unknown (Christie and Macnair 1987), as are those underlying the spread of a second set of complementary factors that appear to be widespread and polymorphic in natural *M. guttatus* populations (Christie and Macnair 1987).

Apart from this example, the paucity of evidence for direct links between genes underlying adaptive traits and those responsible for intrinsic post-zygotic isolation is remarkable. Much of this is likely due to the difficulty of identifying both the genetic basis of hybrid inviability or sterility and the adaptive force responsible for fixing these loci. In many individual cases the adaptive value of traits is thought to be reasonably well understood but the link to hybrid inviability and sterility is correlative only. For example, in *Salvia apiana* and *Salvia mellifera* hybrids, partial incompatibility barriers (crossing incompatibility and F2 breakdown) accompany adaptive species differences in floral morphology, but explicit links between morphological and viability traits has not been established (Grant and Grant 1964). *Mimulus guttatus* and *Mimulus nudatus* are separated by considerable seed inviability (possibly due to genes of large effect) but it is not yet known whether this barrier is directly associated with adaptive (e.g., floral and edaphic) changes between the two species (Macnair and Gardner 1998). Similarly, in *Delphinium nelsonii*, local adaptation (as measured with reciprocal transplants) appears to covary with seed set (close versus distant crosses) and viability and fecundity of F1 offspring under field conditions, suggesting that ecologically-based 'outbreeding depression' might be linked with crossing success and F1 viability (Waser 1993).

Conversely, in some cases the genetic basis of hybrid failure is reasonably well-understood but the link to selection and adaptation is unclear. In *Crepis*, for example, a complementary gene acts as a dominant semi-lethal in hybrids between *C. tectorum* and *C. capiliaris*, leading to death of the seedling carrier, however this locus does not have a known phenotypic function in its native background (Hollingshead 1930). Similarly, a pair of interacting loci of unknown function appear to be responsible for male sterility among *Mimulus nasutus* and *M. guttatus* (Fishman and Willis 2001). More generally, cytoplasmic male sterility — the abortive development of anthers and/or pollen due to negative interactions between cytoplasmic and nuclear genes — is a common observation in interspecific crosses, and is well-understood genetically in specific cases (Kaul 1988; Schnable and Wise 1998). Nonetheless, we do not presently know whether selective forces are commonly responsible for fixing incompatible cyto-nuclear differences in diverging lineages, even in cases where local adaptation has been documented (e.g., Burke et al. 1998; Galloway and Fenster 1999; Campbell and Waser 2001). Finally, selective elimination of certain gene combinations in hybrid or backcross genetic backgrounds — as inferred from patterns of marker segregation distortion in hybrids — could indicate loci that are genically incompatible between species (e.g., Fishman et al. 2001). Grant (1967) outlines several examples of apparent viability selection against specific hybrid genotypes and marker introgressions in *Lycopersicon*, *Phaseolus*, *Rubus*, and *Tragopogon*; introgressions that appear to specifically produce pollen sterility have been demonstrated in a number of species including *Chrysanthemum*, *Lycopersicon*, *Nicotiana*, and *Phaseolus* (Grant 1967). The role of selection in fixing these apparently incompatible genic differences is, however, unknown. In principle, if altered segregation ratios could be associated with specific morphological or other phenotypic traits (e.g., Rick 1963) this might suggest that selection for such morphological differences was important in fixing genes contributing to inviability between species. Unfortunately, to date, many of these classical and more contemporary (e.g., Bernacchi and Tanksley 1997; Harushima et al. 2002) genetic studies involve cultivated species, so that their findings could be substantially influenced by the effects of artificial selection for domestication.

B. Direct selection for isolation

Barriers between species might also be generated by direct selection for isolation itself, if reproductive interactions impose some fitness cost on one or either parental species in sympatry. These costs potentially include reproductive interference that can be both direct (e.g., competition for pollinators) and indirect (e.g., stigmatic contamination with heterotypic pollen and/or gamete wastage) (Grant 1966; Waser 1983, Armbruster et al. 1994). If the selective disadvantage of competition for pollinators, for example, is sufficiently high, selection should act to change traits that influence pollinator visitation in one or both sympatric taxa, resulting in reproductive character displacement (Waser 1983). For example, in *Stylidium* variation in three floral morphological traits determines the attractiveness of flowers to particular pollinating species and the specific placement of pollen on the visitor, thereby defining the 'pollinator niche' for each species (Armbruster et al. 1994). Patterns of morphological differentiation between 31 *Stylidium* species at 25 sympatric sites are consistent with the operation of reproductive character displacement to avoid reproductive interference between taxa. Nonetheless, in this case it is unclear whether the observed character displacement was involved in sympatric speciation or merely in allowing coexistence of already well-defined species (Armbruster et al. 1994). Waser (1983) outlines several similar cases where reproductive character differences between sympatric congeners might be the result of character displacement due to pollinator competition.

Direct selection for isolation can also result in cases where hybrids are formed but are less fit than their parental types due to the build-up of partial incompatibility during a preceding period of allopatry. Under such circumstances, selection is expected to favor the evolution of pre-mating isolating barriers to prevent the formation of unfit hybrids, leading to speciation via reinforcement. The evidence for reinforcement in plant species is limited (c.f. Butlin 1989), although there are a small number of cases which exhibit patterns consistent with the operation of reinforcement. For example, there is evidence for flower color displacement and an increase in self-compatibility in the presence of congeneric species in *Phlox* (Levin and Kerster 1967; Levin 1985). Similarly, in *Arenaria uniflora*, selfing may have evolved as a selected response to reduce detrimental gene flow from the closely related *Arenaria glabra* (Fishman and Wyatt 1999). Populations of *A. uniflora* found in sympatry with *A. glabra* are predominantly selfing; non-selfing *A. uniflora* planted in sympatry with *A. glabra* experience a strong fitness disadvantage due to heterospecific pollen transfer usurping *A. uniflora* ovules (Fishman and Wyatt 1999).

Other examples from non-manipulative, observational studies may be consistent with the operation of reinforcement. For example, character displacement in flowering time in *Anthoxanthum* and *Agrostis* is hypothesized to be an adaptation for reduced gene flow between heavy-metal tolerant and non-tolerant types (McNeilly and Antonovics 1968) although this hypothesis has not been directly evaluated. Similarly Grant (1971) hypothesized that increased post-mating barriers between sympatric versus allopatric species of *Gilia* was due to reinforcement, although other alternative explanations are not excluded by the current data, including the possibility that sympatric species are more distantly related than allopatric species.

Summary: empirical evidence for selection/adaptation in speciation

The examples outlined here provide some evidence that traits under selection might act as the basis of reproductive isolation between newly forming plant species (Table 1). In the main, however, the hypothesized associations between adaptations and species isolating barriers are based on indirect or correlative evidence, and even less is known about the genetic basis of adaptive changes that produce reproductive isolation. Accordingly, it is clearly premature to draw definitive conclusions about factors that predominate in the evolution of isolating barriers in plants (even solely in herbaceous angiosperms). Nevertheless a number of preliminary patterns emerge from the preceding overview. First, there is considerably more evidence that isolation arises as an indirect by-product of selected changes, rather than as the direct target of selection. Second, adaptation appears to be most commonly associated with the development of pre-mating isolating barriers and/or (environment-dependent) hybrid maladaptation. Evidence for the involvement of selection in the development of post-mating pre-zygotic, and post-

Table 1. Empirical evidence for the role of adaptation in plant reproductive isolation: illustrative case studies.

Stage of Isolation	Observed change/species difference	Example species (reference)	Hypothesized selective force
A. Indirect selection			
Prezygotic			
Pre-mating	Flowering phenology		
	• seasonal shift	*Mimulus* (Kiang and Hamrick 1987, Macnair and Gardner 1998)	Edaphic/water limitation
	• diurnal shift	*Petunia* (Ando et al. 2001)	New pollinators
	Floral morphology		
	• structural change	*Aquilegia* (Hodges and Arnold 1995)	New pollinators
		Mimulus (Macnair and Gardner 1998)	
		Pedicularis (Sprague 1962)	
		Salvia (Grant and Grant 1963)	
	• biochemical change	*Geonoma* (Knudsen 1999)	New pollinators
		Orphyrs (Schiestl and Ayassi 2002)	
	Increased selfing rate	*Clarkia* (Lewis 1973)	Absence of pollinators/ reproductive assurance
		Leavenworthia (Lloyd 1979)	
		Gilia (Schoen 1982)	
		Mimulus (Kiang and Hamrick 1987)	
Post-mating	Floral morphology	Unknown	New pollinators? Absence of pollinators?
	Timing of gametophyte maturity (hypothesized)	*Rhododendron* (Williams et al. 1986)	Unknown
Postzygotic			
Environment-dependent "hybrid maladaptation"	Poor hybrid performance in parental conditions	*Achillea* (Clausen 1951)	Multiple
		Artemisia (Wang et al. 1998)	
		Mimulus (Nobs and Hiesey 1958)	
	Pollinator discrimination against hybrids	*Ipomopsis* (Campbell et al. 1997)	New pollinators
		Salvia (Grant and Grant 1964)	
		Mimulus (Schemske and Bradshaw 1999)	
Environment-independent hybrid inviability/ sterility	Complementary gene that cosegregates with adaptive trait	*Mimulus* (Macnair and Christie 1983)	Edaphic/copper tolerance
B. Direct selection			
Prezygotic			
Pre-mating	Floral morphology (color)	*Phlox* (Levin and Kerster 1967, Levin 1985)	Reproductive interference from conspecific
	Increased selfing rate	*Arenaria* (Fishman and Wyatt 1999) *Phlox* (Levin and Kerster 1967, Levin 1985)	Reproductive interference from conspecific

zygotic 'intrinsic' stages of isolation is currently limited. Third, the vast majority of relevant trait changes appear to involve floral characters, and/or traits associated with local soil adaptation (especially local adaptation to heavy metals and water scarcity). Therefore, while pollinators and some abiotic drivers of adaptation are well-represented, other conditions that generally stimulate plant adaptation — especially other biotic factors such as competitors, predators/herbivores, and pathogens — have been rarely implicated in the development of isolating barriers.

Some of these patterns may primarily be due to logistical and experimental constraints on studies to date, especially the relative difficulty of establishing a causal connection between the action of selection and the development of barriers at various different stages of isolation. For example, differentiation in floral traits — and the consequent behavioral effects on pollinators — can be relatively easily assayed, generally without a direct analysis of the specific genetic basis of the trait(s) involved. Quantifying hybrid maladaptation is also relatively tractable in plants, and reciprocal transplantation experiments are a traditional component of studies of local adaptation (although, given this, there is a surprisingly limited number of analyses of hybrid maladaptation in wild plant species under natural conditions; Campbell and Waser 2001). In contrast, identifying a selective basis for changes responsible for 'intrinsic' hybrid inviability or fertility requires intensive segregation analyses of associations between specific phenotypic traits and the loci responsible for inviable or sterile hybrids, in addition to establishing an adaptive role for the phenotypic traits of interest.

Another relevant factor is the genetic complexity of the traits underlying adaptation and associated species barriers. It is easier to identify associations between adaptive characters and isolation loci for traits with a simple genetic basis, in particular – systems in which single sets of complementary genes are associated with adaptive traits based on one or few genes of large effect (e.g., copper tolerance and hybrid inviability in *Mimulus guttatus* — Macnair and Christie 1983). It is possible that part of the general bias towards detecting isolation in association with local soil adaptation (including heavy metals and water scarcity) and/or changes in floral traits can be explained by the relative genetic complexity of these particular adaptive traits. Although considerably more data are needed to determine whether such traits are typically genetically complex or simple, it is interesting to note that, for example, multiple studies suggest that only one or few major genes confer metal tolerance in higher plants (see Macnair 1993). Similarly, it has been suggested that floral differences between species are typically based on few genes of large effect (Gottlieb 1984). In contrast, the influence of such factors as interactions with competitors, herbivores, and pathogens on species isolation may simply be more difficult to quantify empirically. The involvement of, for example, seed predation (*Polemonium foliosissimum*, *Ipomopsis aggregata* — Brody 1997) or herbivore defenses (*Dalechampia* — Armbruster 1997) in the evolution of floral traits appears to be complex and difficult to dissect without intensive experimental analysis. The connection between interspecific competition and isolating barriers may be similarly diffuse and difficult to establish, even though physical exclusion due to reduced competitive ability may also be operating in cases of edaphic adaptation (e.g., Macnair and Gardner 1998), for example.

Finally, it is important to note that the weight of current evidence is likely influenced by relative experimental effort expended in specific empirical systems. Many of the studies outlined here focus on temperate herbaceous species (especially from western U.S.A.), which likely reflects a bias towards experimental and observational tractability in these groups (as well as — to a lesser extent — an interest in local edaphic adaptation). As such they may not be representative of the predominant factors and traits involved in the evolution of reproductive isolation in plants as a whole.

Adaptation in plant speciation: future work

We presently do not know how much of a role selection, versus other evolutionary forces, plays in the generation of new species. Nonetheless, there is some evidence that climatic, edaphic, and biotic factors may play a role in the evolution of plant isolating barriers, particularly when changes such as flowering time, floral morphology, patterns of selfing, and ecological tolerance, are involved (Table 1). The limitations evident in the available evidence in turn suggest multiple avenues for further research on the role of selection in plant speciation:

How frequently are isolating barriers associated with adaptive changes and which agents of selection are most frequently involved?

The relative importance of selection versus drift in the evolution of isolating barriers, and selective conditions and target traits predominantly responsible for the evolution of isolating barriers, are likely to be strongly dependent on the specific life history and reproductive characteristics of the particular group of interest. As such, establishing more direct links between plant adaptation and isolating barriers must include a diversity plant groups, spanning a range of population structures, life history characters, and breeding systems. Such efforts will undoubtedly be facilitated by the increasing variety of molecular genetic tools available to plant evolutionary biologists. Perhaps the most promising contemporary method (at least in genetically tractable taxa) for identifying loci underlying adaptive traits and reproductive isolation, as well as the links between them, are QTL mapping studies (Via and Hawthorne 1998). Most such analyses to date have involved cultivated species (e.g., Bernacchi and Tanksley 1997; Harushima et al. 2002) and so might reflect the influence of artificial selection for domestication. Nonetheless, QTL analyses in natural systems have been very successful in identifying loci associated with adaptive phenotypic traits (e.g., Fishman et al. 2001) and promise to be equally successful in identifying associations (if any) with loci underlying species isolating barriers. These techniques are also increasingly accommodating of large species with longer life spans (e.g., Bradshaw and Stettler 1995). Ultimately, by identifying the specific factors underlying reproductive isolation, molecular population genetic analyses of these loci in combination with functional analyses of the phenotypic and adaptive effect of molecular differences, will provide definitive evidence for or against the role of selection in the evolution of such factors between species. Although very few such analyses are currently available for 'isolation genes', and all are in animals (e.g., Ting et al. 1998; Barbash et al. 2003), they indicate that selection may indeed be responsible for the rapid evolution of genes underlying interspecific reproductive isolation.

In contrast, in systems without substantial molecular genetic development, alternative (albeit indirect) approaches can be used to more explicitly evaluate the association between divergence in adaptive traits and the development of isolating barriers. For example, multivariate regression or path analysis of multiple molecular and morphological trait differences, including reproductive barriers, between taxa can evaluate the independent associations between isolating barriers, and putatively neutral and adaptive differentiation (e.g., Tilley et al. 1990). This approach allows a simultaneous assessment of the relative causal importance of different genetic traits in the evolution of reproductive isolation, and the identification of adaptive traits that are potentially associated with these barriers, without the development of extensive molecular genetic tools beyond simple genetic markers (Moyle, unpublished).

The generation of these kinds of data in a more biologically diverse set of plant groups will better enable us to evaluate the general importance of different forces in driving the evolution of isolating barriers. For example, one factor unexamined here is the possible influence of population structure (and the potential concomitant effects of breeding system) on the rate at which isolating barriers evolve, and the degree to which adaptively driven changes are involved. Some studies suggest that species with smaller population sizes or more inbreeding may exhibit stronger crossing barriers (e.g., Weller et al. 2001) or more transgressive segregation (Rieseberg et al. 1999), for example. Conversely, limited dispersal distances, and local geographical structuring could also enhance linkage disequilibrium between loci responsible for local adaptation and those involved in isolation, such as pollen recognition/mate discrimination (Waser 1993). In both cases, species with smaller, more fragmented populations might be expected to exhibit higher rates of diversification; the challenge, however, is to distinguish the diversifying effects of drift from those of selection, under this population structure. Alternative modes of speciation may also generate conflicting expectations regarding the role of adaptation. For example, speciation in peripheral isolates may be more likely to involve adaptation to new, previously marginal habitats, but could also be initiated by drift acting in small, low-density populations at the edge of a species range (Antonovics 1976). Naturally, the evolution of reproductive isolation may also be a complex interaction between drift and selective forces (e.g., Fenster and Galloway 2000) especially where population structure and mating system could act to limit local population sizes and amplify opportunities for drift, even within a context of local adaptation. Clearly, more studies in taxa with contrasting mating systems and population structures, for example, may help to resolve the potential influence of these factors.

Similarly, in cases where selection does play a role in the development of species isolating barriers, the trait changes predominantly involved will likely also vary between groups. Grant (1981), for example, hypothesizes that vegetative and physiological trait changes will be most important in early differentiation among predominantly wind-pollinated taxa, whereas early divergence in floral traits will be of more importance in animal-pollinated taxa (Grant 1981). In addition, as indicated above, it remains to be tested whether adaptive traits involved in the evolution of isolation are typically genetically complex or simple.

Is isolation a direct or an indirect effect of adaptive changes?

Partly because of the paucity of evidence supporting its operation, the general importance of reinforcement in the evolution of new species remains much debated (Marshall et al. 2002). On the basis of studies reviewed here, it is clear that evidence for direct selection for reproductive isolation is rare in plants. Whether this is due to biological factors that make reproductive character displacement and reinforcement genuinely less likely in specific plant systems (Levin 1970; Grant 1971) or merely the lack of direct studies of reinforcement speciation in plants remains to be determined. For example, it has been argued that selection to reinforce characters that favor homotypic matings will be relatively ineffective in plant species that rely on insect pollinators, especially if pollinators are insufficiently constant in discriminating between different floral types (Grant 1981; 1994). If this is generally the case in plants, wind-pollinated species should be expected to show much higher frequencies of speciation via reinforcement than insect-pollinated groups. Alternatively, it has been proposed that heterospecific matings may be insufficiently costly to drive the process of reinforcement in plants that can manipulate resource provisioning of developing embryos to minimize investment in hybrid offspring (Levin 1970). If this is the case, reinforcement speciation should occur more frequently in plant systems where females are unable to redirect resource investment from hybrid to intraspecific offspring (for example, gymnosperms), or from hybrid offspring to future reproductive opportunities (for example, ephemeral and annual plants). Clearly, considerably more data are needed to evaluate such broad scale expectations. Fortunately, field-based studies of pre-zygotic stages of reproductive isolation are particularly tractable in plant species (see, for example, Ramsey et al. 2003). In addition, one fruitful alternative approach that has been applied in animal taxa (e.g., Coyne and Orr 1989) is the broad comparison of patterns of isolation among sympatric versus allopatric species pairs within well-studied genera (see also Moyle et al. 2004). If reinforcement is operating in these systems, an enhanced frequency of character displacement, such as local habitat differentiation or non-overlapping reproductive (e.g., flowering) timing, would be expected among sympatric species pairs as a result of past selective pressure for avoidance of heterospecific matings. This comparative approach would also be valuable in assessing the additional influences of pollination mode, and reproductive and life history strategies, on the frequency of reinforcement.

What other forces are involved in the evolution of new plant species?

The focus in this paper has been on studies that provide evidence for an explicit connection between presumptively adaptive changes and reproductive isolating barriers. I have not addressed the forces involved in polyploid and hybrid speciation although these could also substantially influence predominant patterns of plant speciation. Hybrid and polyploid speciation are likely to involve their own, somewhat unique role for adaptation. In particular, newly formed polyploid or hybrid lineages must overcome considerable potential challenges from competition and hybridization in order to persist and become established in a population (e.g., Felber 1991; Rieseberg 1997; Levin 2002). This suggests that there might frequently be other selectively important advantages to polyploid/hybrid lineages that allow their persistence in populations of diploids/parental types, when they are the minority type. More research attention to the adaptive significance of hybridization and polyploidy, especially the comparative ecological tolerance of hybrids and polyploids in comparison to their progenitors, will provide valuable insight into these issues.

Finally, the potential for additional phenomena such as sexual selection and intra- and inter-genomic conflict to act as diversifying forces has been addressed in the context of animal speciation (Rice 1998; Arnqvist et al. 2001; Panhuis et al. 2001) but is largely unexamined in plants (Skogsmyr and Lankinen 2002). Numerous authors have pointed out that plants (angiosperms) exhibit many biological features (e.g., complex pollen-stigma, pollen-style and maternal-embryo-endosperm interactions) that may facilitate the operation of sexual selection, male-male competition, and/or male-female and parent-offspring genetic conflicts in plant systems (Westoby and Rice 1982; Queller 1987; Charlesworth et al. 1987; Lyons et al. 1989; Walsh and Charlesworth 1992; Waser 1993). Nonetheless, appropriate experimental analyses have been limited to date (Walsh and Charlesworth 1992; Howard 1999; Skogsmyr and Lankinen 2002); as such, the potential role of sexual selection and genetic conflicts in the evolution of species barriers — especially in comparison to more conventional biotic and abiotic selective factors — also remains to be fully investigated in plants.

Acknowledgements

Thanks to J. Whitton, L. D. Gottlieb, and two anonymous reviewers for helpful comments, and to the Center for Population Biology, UC Davis, for financial support.

References

Anderson, E., and Anderson, B. 1954. Introgression of *Salvia apiana* and *Salvia mellifera*. Ann. Mo. Bot. Gard. **41**: 329–338.

Ando, T., Nomura, M., Tsukahara, J., Watanabe, H., Kokubun, H., Tsukamoto, T., Hashimoto, G., Marchesi, E., and Kitching, I.J. 2001. Reproductive isolation in a native population of *Petunia* sensu Jussieu (Solanaceae). Ann. Bot. **88**: 403–413.

Antonovics, J. 1976. The nature of limits to natural selection. Ann. Mo. Bot. Gard. **63**: 224–247.

Armbruster, W.S. 1997. Exaptations link evolution of plant-herbivore and plant pollinator interactions: a phylogenetic inquiry. Ecology **78**: 1661–1672.

Armbruster, W.S., Edwards, M.E., and Debevec, E.M. 1994. Floral character placement generates assemblage structure of western Australian triggerplants (*Stylidium*). Ecology **75**: 315–329.

Arnold, M.L., Hamrick, J.L., and Bennett, B.D. 1993. Interspecific pollen competition and reproductive isolation in iris. J. Hered. **84**: 13–16.

Arnqvist, G., Edvardsson, M., Friberg, U., and Nilsson, T. 2000. Sexual conflict promotes speciation in insects. Proc. Natl. Acad. Sci. U.S.A. **97**: 10460–10464.

Barbash, D.A., Siino, D.F., Tarone, A.M., and Roote, J. 2003. A rapidly evolving MYB-related protein causes species isolation in *Drosophila*. Proc. Natl. Acad. Sci. U.S.A. **100**: 5302–5307.

Barrett, S.C.H. 1996. The reproductive biology and genetics of island plants. Phil. Trans. Roy. Soc. Lond. B Biol. Sci. **351**: 725–733.

Barton, N.H. 1996. Natural selection and random genetic drift as causes of evolution on islands. Phil. Trans. Roy. Soc. Lond. B Biol. Sci. **351**: 785–794.

Bawa, K.S. 1995. Pollination, seed dispersal, and diversification of angiosperms. Trends Ecol. Evol. **10**: 311–312.

Bernacchi, D., and S. D. Tanksley. 1997. Interspecific backcross of *Lycopersicon esculentum* × *L. hirsutum*: linkage analysis and a QTL study of sexual compatibility factors and floral traits. Genetics **147**: 861–877.

Bradshaw, H.D., and Stettler, R.F. 1995. Molecular genetics of growth and development in *Populus*. IV. mapping QTLs with large effects on growth, form, and phenology traits in a forest tree. Genetics **139**: 963–973.

Bradshaw, H.D., Wilbert, S.M., Otto, K.G., and Schemske, D.W. 1995. Genetic mapping of flower traits associated with reproductive isolation in monkeyflowers (*Mimulus*). Nature **376**: 762–765.

Brody, A.K. 1997. Effects of pollinators, herbivores, and seed predators on flowering phenology. Ecology **78**: 1624–1631.

Buchholz, J.T., Williams, L.F., and Blakeslee, A.F. 1935. Pollen-tube growth of ten species of *Datura* in interspecific pollinations. Proc. Natl. Acad. Sci. U.S.A. **21**: 651–656.

Burke, J.M., Voss, T.J., and Arnold, M.L. 1998. Genetic interactions and natural selection in Louisiana *Iris* hybrids. Evolution **52**: 1304–1310.

Butlin, R.K. 1989. Reinforcement of premating isolation. *In* Speciation and its consequences. *Edited by* D. Otte and J.A. Endler. Sinauer, Cambridge, MA. pp. 158–179.

Campbell, D.R., and Waser, N.M. 2001. Genotype-by-environment interaction and the fitness of plant hybrids in the wild. Evolution **55**: 669–676.

Campbell, D.R., Waser, N.M., and Melendez-Ackerman, E.J. 1997. Analyzing pollinator mediated selection in a plant hybrid zone: hummingbird visitation patterns on three spatial scales. Am. Nat. **149**: 295–315.

Carney, S.E., Cruzan, M.B., and Arnold, M.L. 1994. Reproductive interactions between hybridizing irises: analyses of pollen-tube growth and fertilization success. Am. J. Bot. **81**: 1169–1175.

Charlesworth, D., Schemske, D.W., and Sork, V.L. 1987. The evolution of plant reproductive characters: sexual versus natural selection. *In* Evolution of Sex. *Edited by* S.C. Stearns. Birkhauser, Basel. pp. 317–335.

Christie, P., and Macnair, M.R. 1987. The distribution of postmating reproductive isolating genes in populations of the yellow monkey flower, *Mimulus guttatus*. Evolution **41**: 571–578.

Clausen, J. 1951. Stages in the evolution of plant species. Hafner, New York, NY.

Coyne, J.A., and Orr, H.A. 1989. Patterns of speciation in *Drosophila*. Evolution **43**: 362–381.

deNettancourt, D. 2001. Incompatibility and incongruity in wild and cultivated plants. Springer, Berlin.

Dobzhansky, T. 1936. Studies of hybrid sterility. II. Localization of sterility factors in *Drosophila pseudoobscura* hybrids. Genetics **21**: 113–135.

Felber, F. 1991. Establishment of a tetraploid cytotype in a diploid population: effect of the relative fitness of cytotypes. J. Evol. Biol. **4**: 195–207.

Fenster, C.B., and Galloway, L.F. 2000. Population differentiation in an annual legume: genetic architecture. Evolution **54**: 1157–1172.

Fishman, L., and Wyatt, R. 1999. Pollinator-mediated competition, reproductive character displacement, and the evolution of selfing in *Arenaria uniflora* (Caryophyllaceae). Evolution **53**: 1723–1733.

Fishman, L., and Willis, J.H. 2001. Evidence for Dobzhansky-Muller incompatibilities contributing to the sterility of hybrids between *Mimulus guttatus* and *M. nasutus*. Evolution **55**: 1932–1942.

Fishman, L., Kelly, A.J., Morgan, E., and Willis, J.H. 2001. A genetic map in the *Mimulus guttatus* species complex reveals transmission ratio distortion due to heterospecific interactions. Genetics **159**: 1701–1716.

Friedman, W.E. 1999. Expression of the cell cycle in sperm of *Arabidopsis*: implications for understanding patterns of gametogenesis and fertilization in plants and other eukaryotes. Development **126**: 1065–1075.

Galloway, L.F. and Fenster, C.B. 1999. The effect of nuclear and cytoplasmic genes on fitness and local adaptation in an annual legume *Chamaecrista fasciculata*. Evolution **53**: 1734–1743.

Gottlieb, L.D. 1984. Genetics and morphological evolution in plants. Am. Nat. **123**: 687–709.

Grant, K.A. and Grant, V. 1964. Mechanical isolation of *Salvia apiana* and *Salvia mellifera* (Labiatae). Evolution **18**: 196–212.

Grant, V. 1966. The selective origin of incompatibility barriers in the plant genus *Gilia*. Am. Nat. **100**: 99–118.

Grant, V. 1967. Linkage between morphology and viability in plant species. Am. Nat. **101**: 125–139.

Grant, V. 1971. Plant speciation. Columbia University Press, New York, NY.

Grant, V. 1981. Plant speciation, Second Edition. Columbia University Press, New York, NY.

Harushima, Y., Nakagahra, M., Yano, M., Sasaki, T., and Kurata, N. 2002. Diverse variation of reproductive barriers in three interspecific rice crosses. Genetics **160**: 313–322.

Hodges, S.A., and Arnold, M.L. 1995. Spurring plant diversification: are nectar spurs a key innovation? Proc. R. Soc. Lond. B Biol. Sci. **262**: 343–348.

Hollingshead, L. 1930. A lethal factor in *Crepis* effective only in an interspecific hybrid. Genetics **15**: 114–140.

Howard, D.J. 1999. Conspecific sperm and pollen precedence and speciation. Annu. Rev. Ecol. Syst. **30**: 109–132.

Howard, D.J., Reece, M., Gregory, P.G., Chu, J., and Cain, M.L. 1998. The evolution of barriers to fertilization between closely related organisms. *In*: Endless forms: species and speciation. *Edited by* D.J. Howard and S.H. Berlocher. Oxford Univ. Press, New York, NY. pp. 279–288.

Jain, S.K. 1976. The evolution of inbreeding in plants. Annu. Rev. Ecol. Syst. **7**: 469–495.

Kaul, M.L.H. 1988. Male sterility in higher plants. Springer-Verlag, Berlin.

Kiang, Y.T., and Hamrick, J.L. 1978. Reproductive isolation in the *Mimulus guttatus-M. nasutus* complex. Am. Midl. Nat. **100**: 269–276.

Knudsen, J.T. 1999. Floral scent differentiation among co-flowering, sympatric species of *Geonoma* (Aracaceae). Plant Species Biol. **14**: 137–142.

Knudsen, J.T., Andersson, S., and Bergman, P. 1999. Floral scent attraction in *Geonoma macrostachys*, an understory palm of the Amazonian rainforest. Oikos **85**: 409–418.

Kruckeberg, A.R. 1986. The stimulus of unusual geologies for plant speciation. Syst. Bot. **11**: 455–463.

Lande, R. 1976. Natural selection and random genetic drift in phenotypic evolution. Evolution **30**: 314–334.

Lenz, L.W. 1959. Hybridization and speciation in the pacific coast irises. Aliso **4**: 237–309.

Levin, D.A. 1970. Reinforcement of reproductive isolation: plants versus animals. Am. Nat **104**: 571–581.

Levin, D.A. 1985. Reproductive character displacement in *Phlox*. Evolution **39**: 1275–1281.

Levin, D.A. 2000. The origin, expansion, and demise of plant species. Oxford University Press, New York, NY.

Levin, D.A. 2002. The role of chromosomal change in plant evolution. Oxford University Press, New York, NY.

Levin, D.A. and H.W. Kerster. 1967. Natural selection for reproductive isolation in *Phlox*. Evolution **21**: 679–687.

Lewis, H. 1973. The origin of diploid neospecies in *Clarkia*. Am. Nat **107**: 161–170.

Lloyd, D.G. 1979. Some reproductive factors affecting the selection of self-fertilization in plants. Am. Nat **113**: 67–79.

Lyons, E.E., Waser, N.M., Price, M.V., Antonovics, J., and Motten, A.F. 1989. Sources of variation in plant reproductive success and implications for concepts of sexual selection. Am. Nat **134**: 409–433.

Macnair, M.R. 1993. The genetics of metal tolerance in vascular plants. New Phytol. **124**: 541–559.

Macnair, M.R., and Christie, P. 1983. Reproductive isolation as a pleiotropic effect of copper tolerance in *Mimulus guttatus*? Heredity **50**: 295–302.

Macnair, M.R., and Gardner, M. 1998. The evolution of edaphic endemics. *In* Endless forms: species and speciation. *Edited by* D.J. Howard and S.H. Berlocher. Oxford University Press, New York, NY. pp. 157–171.

Marshall, J.L., Arnold, M.L., and Howard, D.J. 2002. Reinforcement: the road not taken. Trends Ecol. Evol. **17**: 558–562.

McNeilly, T., and Antonovics, J. 1968. Evolution in closely adjacent plant populations. IV. Barriers to gene flow. Heredity **23**: 205–218.

Mitchell-Olds, T. 1995. The molecular basis of quantitative genetic variation in natural populations. Trends Ecol. Evol. **10**: 324–328.

Moyle, L.C., Olson, M.S., and Tiffin, P. 2004. Patterns of reproductive isolation in three angiosperm genera. Evolution **58**: 1195–1208.

Nobs, M.A., and. Hiesey, W.M. 1958. Performance of *Mimulus* races and their hybrids in contrasting environments. Carnegie Inst. Washington Yearbook **57**: 270–272.

Panhuis, T.M., Butin, R., Zuk, M., and Tregenza, T. 2001. Sexual selection and speciation. Trends Ecol. Evol. **16**: 364–371.

Powell, J.R. 1997. Progress and prospects in evolutionary biology: the *Drosophila* model. Oxford University Press, New York, NY.

Queller, D.C. 1987. Sexual selection in flowering plants. *In* Sexual selection: testing the alternatives. *Edited by* J.W. Bradbury and M.W. Anderson. John Wiley and Sons, New York, NY. pp. 165–179.

Ramsey, J., Bradshaw, H.D., and Schemske, D.W. 2003. Components of reproductive isolation between the monkeyflowers *Mimulus lewisii* and *M. cardinalis* (Pyrmaceae). Evolution **57**: 1520–1534.

Rice, W.R. 1998. Intergenomic conflict, interlocus antagonistic coevolution, and the evolution of reproductive isolation. *In* Endless forms: species and speciation. *Edited by* D.J. Howard, and S.H. Berlocher. Oxford University Press, New York, NY. pp. 261–270.

Rice, W.R., and Hostert, E.E. 1993. Laboratory experiments on speciation: what have we learned in 40 years? Evolution **47**: 1637–1653.

Rick, C.M. 1963. Barriers to interbreeding in *Lycopersicon peruvianum*. Evolution **17**: 216–232.

Ricklefs, R.E., and Renner, S. S. 1994. Species richness within families of flowering plants. Evolution **48**: 1619–1636.

Rieseberg, L.H. 1997. Hybrid origins of plant species. Annu. Rev. Ecol. Syst. **28**: 359–389.

Rieseberg, L.H., Archer, M.A., and Wayne, R.K. 1999. Transgressive segregation, adaptation and speciation. Heredity **83**: 363–372.

Rieseberg, L.H., Desrochers, A.M., and Youn, S.J. 1995. Interspecific pollen competition as a reproductive barriers between sympatric species of *Helianthus* (Asteraceae). Am. J. Bot. **82**: 515–519.

Sakai, A.K., Weller, S.G., Wager, W.L., Soltis, P.S., and Soltis, D.E. 1997. Phylogenetic perspectives on the evolution of dioecy: Adaptive radiation in the endemic Hawaiian genera *Schiedea* and *Alsinidendron* (Caryophyllaceae: Alsinoideae). *In* Molecular evolution and adaptive radiation. *Edited by* T.J. Givnish and K.J. Sytsma. Cambridge University Press, Cambridge, UK. pp. 455–473.

Schemske, D.W. 2000. Understanding the origin of species. Evolution **54**: 1069–1073.

Schemske, D.W. and Bradshaw, H.D. 1999. Pollinator preference and the evolution of floral traits in monkey flowers (*Mimulus*). Proc. Natl. Acad. Sci. U.S.A. **96**: 11910–11915.

Schiestl, F.P., and Ayasse, M. 2002. Do changes in floral odor cause speciation in sexually deceptive orchids? Plant Syst. Evol. **234**: 11–119.

Schluter, D. 2000. The ecology of adaptive radiation. Oxford University Press, Oxford, UK.

Schluter, D. 2001. Ecology and the origin of species. Trends Ecol. Evol. **16**: 372–380.

Schnable, P.S., and Wise, R.P. 1998. The molecular basis of cytoplasmic male sterility and fertility restoration. Trends Plant Sci. **3**: 175–180.

Schneider, C.J. 2000. Commentary: natural selection and speciation. Proc. Natl. Acad. Sci. U.S.A. **97**: 12398–12399.

Schoen, D.J. 1982. Genetic variation and the breeding system of *Gilia achilleifolia*. Evolution **36**: 361–370.

Skogsmyr, I., and Lankinen, A. 2002. Sexual selection: an evolutionary force in plants? Biol. Rev. Camb. Philos. Soc. **77**: 537–562.

Smith, E.B. 1968. Pollen competition and relatedness in *Haplopappus* section *Isopappus*. Bot. Gaz. **129**: 371–373.

Smith, E.B. 1970. Pollen competition and relatedness in *Haplopappus* section *Isopappus* (Compositae).II. Am. J. Bot. **57**: 874–880.

Sprague, E.F. 1962. Pollination and evolution in *Pedicularis* (Scrophulariaceae). Aliso **5**: 181–209.

Stebbins, G.L. 1950. Variation and evolution in plants. Columbia University Press, New York, NY.

Tilley, S.G., Verrell, P.A., and Arnold, S.J. 1990. Correspondence between sexual isolation and allozyme differentiation: a test in the salamander *Desmognathus ochrophaeus*. Proc. Natl. Acad. Sci. U.S.A. **87**: 2715–2719.

Ting, C.T., Tsaur, S.C., Wu, M.L., and Wu, C.I. 1998. A rapidly evolving homeobox at the site of a hybrid sterility gene. Science (Washington, D.C.) **282**: 1501–1504.

Turelli, M., Barton, N.H., and Coyne, J.A. 2001. Theory and speciation. Trends Ecol. Evol. **16**: 330–343.

Via, S., and Hawthorne, D.J. 1998. The genetics of speciation: promises and prospects of quantitative trait locus mapping. *In* Endless forms: species and speciation. *Edited by* D.J. Howard and S.H. Berlocher. Oxford University Press, New York, NY. pp. 352–364.

Wagner, W.L., and Funk, V.A. 1995. Hawaiian biogeography: evolution on a hot spot archipelago. Smithsonian Institution Press, Washington, DC.

Walsh, N.E., and Charlesworth, D. 1992. Evolutionary interpretations of differences in pollen tube growth rates. Quart. Rev. Biol. **67**: 19–37.

Wang, H., McArthur, E.D., Sanderson, S.C., Graham, J.H., and Freeman, D.C. 1997. A narrow hybrid zone between two subspecies of big sagebrush (*Artemisia tridentata*: Asteraceae). IV. Reciprocal transplant experiments. Evolution **51**: 95–102.

Waser, N.M. 1983. Competition for pollination and floral character differences among sympatric plant species: a review of evidence. *In* Handbook of experimental pollination biology. *Edited by* C.E. Jones and R.J. Little. Scientific and Academic, New York, NY. pp. 277–293.

Waser, N.M. 1993. Population structure, optimal outbreeding, and assortative mating in angiosperms. *In* The natural history of inbreeding and outbreeding: theoretical and empirical perspectives. *Edited by* N.M. Thornhill. University of Chicago Press, Chicago, IL. pp. 173–199.

Weller, S.G., Sakai, A.K., and Wagner, W.L. 2001. Artificial and natural hybridization in *Schiedea* and *Alsinodendron* (Caryophyllaceae: Alsinoideae): the importance of phylogeny, genetic divergence, breeding system, and population size. Syst. Bot. **26**: 571–584.

Westoby, M., and Rice, B. 1982. Evolution of seed plants and inclusive fitness of plant tissues. Evolution **36**: 713–724.

Williams, E.G., and Rouse, J.L. 1988. Disparate style lengths contribute to isolation of species in *Rhododendron*. Aust. J. Bot. **36**: 183–191.

Williams, E.G., Kaul, V., Rouse, J.L., and Palser, B.F. 1986. Overgrowth of pollen tubes in embryo sacs of *Rhododendron* following interspecific pollinations. Aust. J. Bot. **34**: 413–423.

Wyatt, R. 1988. Phylogenetic aspects of the evolution of self-pollination. *In* Plant evolutionary biology. *Edited by* L.D. Gottlieb and S.K. Jain. Chapman and Hall, London, UK. pp. 109–131.

Zhang, X.S., and O'Neill, S.D. 1993. Ovary and gametophyte development are coordinately regulated by auxin and ethylene following pollination. Plant Cell **5**: 403–418.

12
Discussion report: New methods and tools for plant adaptation — what do we need?

Barbara K. Mable

Abstract: The final discussion of the workshop on "Molecular Genetics and Ecology of Plant Adaptation" was based on the question: "New methods and tools for plant adaptation — what do we need?" The major ideas discussed can be synthesized into three categories: (1) biological tools — approaches for deciding which species to propose as candidates for genome sequencing projects, approaches for choosing which genes or gene regions to focus on that can be applied to a broad range of species, the importance of extending methods such as transformation to a wider array of species, and the importance of combining traditional methods such as reciprocal transplants and direct assessment of differences in expression of candidate "adaptation" genes under natural conditions; (2) hard tools — the power of combining data from different sources about a given species or type of adaptation, development of better tools for assessing phenotypic changes, development of more sensitive methods of looking at polymorphisms to include rare changes that are not detectable using current methods, and the utility of using cDNA libraries as a starting point to study phenotypic variation in new species and then advancing to more extensive genomics approaches; and (3) theoretical/analytical tools — development of new analytical methods and statistical approaches designed to reflect our changing views of genomic structure and development of better methods for directly relating phenotypes to genotypes.

Introduction

The 3-day workshop on "Molecular Genetics and Ecology of Plant Adaptation" ended, rather appropriately, with a discussion of the tools necessary to attain the vision of an "ecomolecular synthesis" proposed by Quentin Cronk (UBC Botanical Garden) in the opening talk of the conference. One of the great strengths of the workshop was that it provided plant ecologists, evolutionary geneticists, and molecular geneticists with a rare opportunity to share their ideas with each other in a structured but relatively informal setting. The common goal was to assess the most efficient approach to integrating information from all areas, to better understand how organisms adapt to the changing environments in which they find themselves.

The final brainstorming session, chaired by Carl Douglas (Department of Botany, UBC) focused on the question: "New Methods and Tools for Plant Adaptation — what do we need?" With the opportunities opened up by the completion of various genome sequencing projects, it had been difficult to keep this topic out of the discussions on the previous days, on 'The Intellectual Framework for a Plant Adaptation Science" (chaired by Quentin Cronk), and "Answered and Unanswered Questions in Plant Adaptation" (chaired by Dolph Schluter, Department of Zoology, UBC). This reflects the group's underlying excitement at the prospect of finally being able to genetically dissect the phenotypic changes that result in differential survival and adaptation, now that we have (or will soon have) the molecular tools to determine the factors causing these changes at the level of genes and gene interactions. This enthusiasm was apparent throughout the three days of the workshop.

Carl Douglas opened the final discussion by drawing a distinction between the kinds of "biological" (i.e., which model systems, which adaptive changes, which genes/genomes to study) and "hard" (i.e., approaches to genome sequencing, the utility of expression arrays, quantitative methods for going from quantitative trait loci to actual genes, methods for directly relating phenotypes to adaptive changes) tools that we should be focusing on to achieve the proposed "ecomolecular synthesis". For the purposes of this overview "theoretical and analytical" tools (i.e., analytical approaches for handling data from new technological advances, theoretical framework in which to synthesize growing databases) are discussed separately from "hard tools" in an effort to lay out more clearly the progress needed in both areas.

The ideas proposed during the final discussion fell under three headings and drew on information presented in talks throughout the meeting as a sample of the progress that is currently being made by this community in each of these areas. There was a strong bias in the discussion towards biological tools because this is the area where ecologists and evolutionary biologists (who represented a larger proportion of the participants) have the most to contribute. The main participants in the discussion were Justin Borevitz (Salk Institute), Quentin Cronk (UBC Botanical Garden), Mitchell

Barbara K. Mable[1]. Department of Botany, University of Guelph, Guelph, ON N1G 2W1, Canada.
e-mail address: bmable@uoguelph.ca
Correct citation: Mable, B.K. 2004. Discussion report: New methods and tools for plant adaptation — what do we need? *In* Plant Adaptation: Molecular Genetics and Ecology. Proceedings of an International Workshop held December 11–13, 2002, in Vancouver, British Columbia, Canada. *Edited by* Q.C.B. Cronk, J. Whitton, R.H. Ree, and I.E.P. Taylor. NRC Research Press, Ottawa, Ontario. pp. 94–100.

[1]Current address: Division of Environmental and Evolutionary Biology, University of Glasgow, Glasgow, Scotland, U.K. G12 8QQ. e-mail address: b.mable@bio.gla.ac.uk

Cruzan (Department of Biology, Portland State), Tom Givnish (Kilauea Field Station, University of Wisconsin-Madison), Scott Hodges (Life Sciences, University of California-Santa Barbara), Michael Purugganan (Department of Genetics, North Carolina State University), Mark Rausher (Department of Biology, Duke University), Carl Schlichting (Department of Ecology and Evolutionary Biology, University of Connecticut),and John Willis (Department of Biology, Duke University).

Proposed strategies

Biological tools

The discussion focused predominantly on the following questions: (1) What other genome projects should be initiated and how should we choose which species to concentrate on next?; (2) What is the best approach to choosing which genes to focus on?; and (3) For the species that will have genome sequencing projects completed, what other biological tools should we be preparing to complement the hard tools, to more directly relate genomic attributes to adaptive changes?

For obvious practical and economic reasons, the initial wave of genome sequencing projects focused on organisms that were already being used as model systems for genetic studies, with a bias towards those with small genomes. Current predictions are that on the order of 30 plant genomes will be sequenced within the next 10 years (and some people think that this is a vast underestimate); as an integrated community we should become involved in directing where the next wave of focus lies. To learn more about how molecular changes relate to adaptation, the time is ripe to go beyond the more traditional models and propose new models (or model groups) that encompass a wide range of variation (e.g., variation in morphologies and ecologies) so that molecular variation and molecular responses can be more directly related back to environmental variation and adaptation. One suggestion was that species like *Potentilla glandulosa* (Rosaceae) would be ideal because the problem of adaptation has been approached from a reaction norms perspective, which may be more realistic in terms of how organisms adapt to their environments. Other suggestions for model groups included *Mimulus* (Scrophulariaceae), because it has been used extensively as a model for variation in reproductive strategies and ecological traits and there is quite a large group of people interested in its biology; *Collinsia* (Scrophulariaceae), because it has not yet been used as a model but has very interesting biology; *Arabidopsis lyrata* (Brassicaceae), because it is an outcrossing relative of the widely used model *Arabidopsis thaliana* and shows substantial variation in ecology and morphology across its range; and *Populus trichocarpa* (Salicaceae), because its genome is currently being sequenced in an approach designed to be more integrated with other research efforts.

In choosing models that will shed light on adaptation, it would be best to concentrate on species for which we know something about life history variation and perhaps have preliminary information on phenotype-genotype mapping. However, we could also take a closer look at the ecology of models currently in use and determine if genomic information already available could be used to test predictions about adaptation. It was also suggested that we should be choosing genomes that will allow us to apply genomic variation in a phylogenetic context (i.e., we should be trying to get representatives from lineages in between those already sequenced). People also choose models for different reasons; there are groups who use one organism and ask many questions that can be tested using that species, but there is also a large group who choose models to address particular questions. These are both valid approaches to studying variation but the question was raised: how do we choose models in a way that can satisfy both?

The problem with choosing a model for ecological approaches is that there is such vast variation in habitats and strategies for adaptation among organisms that it is difficult to choose a single representative that encompasses even a small fraction of this variation. It was suggested that perhaps what we really need is to come up with a set of common markers that could be used to make comparisons across a wide range of species. Currently, everyone has their own set of quantitative trait loci (QTLs) that they apply to their own organisms under a particular set of conditions of interest to them. It would be much more productive at this stage to come up with common goals and questions that could be addressed across a wide range of taxa, rather than just focusing on particular models. In order to do this, there would have to be a shift away from anonymous markers to gene-based markers (especially those in relatively low copy number) that could be isolated from a wide range of species, to allow examination of polymorphisms and mapping in relation to ecological variation. These cross-species markers (second generation markers) could provide information on things like conservation of gene order and synteny, which could then be mapped onto an angiosperm phylogeny to allow a historical framework when making inferences about genomic changes and adaptation (i.e., accounting for shared history when interpreting patterns of genomic restructuring). A potential concern was raised that genes conservative enough to amplify across all angiosperms might not be the best candidates for studying adaptation because they may not show an appropriate degree of variation, although the benefit of using this approach to examine the evolution of genome structure seems very powerful. There is also the nontrivial problem of identifying true orthologues using this type of approach — especially considering findings that most genes exist in large gene families rather than the single gene copy model that was prevalent prior to large-scale genomics.

The ultimate biological usage of genomes — relating phenotypes to genotypes — is likely to be a major focus of the "postgenomics" era, but this is not a trivial problem. Looking at phenotypic variation is what ecologists/evolutionary biologists are good at — we should be leading the way on this front – but what is the best approach for doing this? There was a suggestion that perhaps we should be focusing on genes in particular pathways known to be involved in adaptation to particular ecological pressures, or we should be focusing on particular kinds of adaptations so that we can concentrate efforts to search for the elusive "gene × environment interaction" (g × e) in a more narrow arena. To assess the importance of this component of phenotypic variation we need a lot of genes that we think are adaptively important but we don't yet have the tools to test whether they actually are. If we focus initially on a more limited set of adaptive responses, we might be able to develop tools that could then be generalized to new systems. For example, if we gained a lot of information on one type of response in a wide variety of taxa it might accelerate the chance of successfully relating phenotype to genotype. Phenotypes are the hard part. It was pointed out that, despite years of collecting QTLs, no one has managed to clone a QTL of even moderate effect, and only a few of large effect. The genome databases currently available could be used as a starting point to identify candidate sets of genes that might be related to particular responses. However, just by looking at variation across genomes, while we might see overall correlations of changes with environment, epistatic interactions and changes in regulatory elements such as transcription factors are likely to be extremely important but would not necessarily be isolated by this type of approach. By focusing on particular types of responses we might be able to look at a wider range of genetic changes.

In order to efficiently work across a wide range of taxa, it would be helpful to be able to apply some of the more powerful biologically based tools developed for model systems to other species. *Agrobacterium*-mediated transformation, for example, is a method that has been used successfully in only a limited number of taxa but it would be very useful to develop a generalizable transformation method that would work in any species, or at least in a subset with particular attributes. In poplar, for example, success of transformation techniques is genotype dependent. Vacuum infiltration transformation of young ovules will work in about 10% of species and this may be related to variation in the timing of opening of the gynoecium (i.e., it will only work on species for which the ovules are open early in development). Perhaps this is one thing we should look at when choosing new model systems — ones that we predict will be able to be transformed using this type of method because of their developmental profile. Root tissue culture has been used to transform *Arabidopsis lyrata* (which is closely related to *A. thaliana*) even though it has quite a different rate of development, but it would be useful to define what attributes would be required to extend this to other groups. No one really knows how gun transformation works but perhaps it could be adapted to a wider variety of taxa. It was decided that this is the realm of plant molecular biologists — we should challenge the experts in this area to come up with new methods that are more broadly applicable across taxa. It was also pointed out that perhaps it would be most efficient to develop service labs that are good at transforming non-model species — we don't have to do everything ourselves.

Dolph Schluter, an animal evolutionary ecologist in the Zoology Department at UBC, was asked what he most envied plant biologists for — the answer was reciprocal transplant experiments. The sedentary nature of plants, the ability to propagate lines clonally, and the relatively short generation times are biological aspects that make plants very powerful tools to compare phenotypic changes with genetic variation. In this age of advancing technology, it is easy to ignore the basics and get swept up in the latest technological trends. However, the real test of adaptation can only be realized by looking at variation under natural conditions. This type of approach should come back into fashion. For example, it has been noted that often one finds only a subset of the variability in responses under greenhouse conditions that you would find under field conditions. It was pointed out that, even though reciprocal transplant experiments have been performed since the 1930's, we haven't really progressed much beyond that because the critical step of taking what we think are adaptive trait loci back to set up field experiment trials has been quite limited. These early experiments overlooked a lot of factors and current approaches to ecology are continually changing, but the question posed was: will molecular tools help in this area? While this could be a focus of future studies, a lot could be learned by going back to conclusions based on previous molecular studies to test whether or not laboratory-based variation is reflected in natural populations, or whether differences in populations sampled from across a species range can be related to differences in habitat or geography. This combination of approaches is where the true power of an ecomolecular synthesis could be realized.

"Hard" tools

The discussion focused mainly on the following questions: (1) What kinds of genomic information do we need to achieve an ecomolecular synthesis? (i.e., what kinds of tools do we need to relate phenotypic variation to genotypic?); and (2) When developing new biological models, what is the most efficient strategy for beginning a survey of the genome?

It is clear that a large number of whole-genome sequencing projects are currently underway and the database on variation at the genomic level will continue to expand, but it is really variation at the level of phenotypes that is required to contribute to the envisioned ecomolecular synthesis. This work is currently receiving a vast amount of time and money by the molecular biology community in the fields of proteomics and expression arrays but, in order to directly relate these changes to ecological processes, random strategies for linking genes to functions may not be sufficient. There is enough data already available on a number of plants to link together all of the information that we know about particular groups. High resolution genetic maps and ESTs (expressed sequence tags) could be matched up to BAC (bacterial artificial chromosome) libraries, fingerprinting approaches could be taken to link information from various sources, and microarrays using cDNA could be compared to genomic hybridizations to distinguish expression changes from genomic restructuring changes. Fingerprinting can cover about 10% of the genome, requires little costs and could be done in months. Physical maps could be combined with genetic maps potentially on dozens of species. Techniques for mapping and high-throughput approaches to genotyping in general are continuing to improve and it should soon be possible to produce very fine-scale maps on a lot of different species. For example, linkage disequilibrium (LD) mapping ties particular ancestral haplotypes to variation in quantitative traits and provides potentially much finer resolution maps because it relies on historical recombination rather than on experimental crosses. However, substantial disequilibrium and a large number of polymorphisms are required to be able to recognize the ancestral types. Sensitivity of markers is also likely to improve, which will result in higher overall resolution. AFLPs (amplified length polymorphisms) revolutionized mapping techniques when they were proposed but new approaches are constantly being created and high-throughput screening methods are becoming standard laboratory procedures. Fine maps of a particular locus no longer have to rely on transformation but instead, a large-scale screening approach of individuals from second-generation crosses could be taken.

With proteomics one can be confident that a particular protein is involved in a particular pathway but it doesn't really tell you how its expression might change in different contexts. It is this context-dependence that really needs to be evaluated. Knowing that a protein is in a particular tissue doesn't mean that it is processed correctly. The problem with current approaches to microarrays is that people tend to jump to clustering of genes with similar expression patterns before determining whether they have reproducible results. Although spotted arrays are still more affordable than those based on oligonucleotide "features" to detect each gene (Affymetrix chips), reproducibility in the latter is much higher, so there tends to be a trade-off in terms of feasible levels of biological replication and data quality. Ideally, experiments need to be set up so that each gene is treated as a multiple test because adequate controls and replication are essential to meaningful interpretation.

To complement studies of SNPs (single nucleotide polymorphisms), a relatively newer but potentially very powerful approach is to use SFPs (single feature polymorphisms) to look for polymorphisms in expression, using oligonucleotide-based microarrays. Even in nonmodel organisms this is becoming an affordable and straightforward approach. This could be the trend of the future for directly relating phenotypes to genotypes at a whole genome scale, provided that sufficient efforts are taken to ensure reproducibility. For example, this type of array profile can provide a sensitive readout for what types of genes or pathways are involved in adaptation (e.g., to something like shade avoidance) across a large number of species and/or across a wide range of environmental conditions. To enable interpretation of patterns of hybridization of more genetically distant genotypes or species to a given array, DNA could be used as a type of phylogenetic control for comparing expression polymorphisms. Not only could these DNA-based arrays correct for differences in gene copy number but they could also provide an indication of genomic rearrangements or duplications that might cause misinterpretation of apparent expression polymorphisms. They would effectively provide a direct comparison between SNPs and SFPs and could also be compared to data from other sources such as QTLs or LD maps. Soon these genome approaches will be like Eppendorf tubes; the costs will become so trivial that we won't even think about how many need to be used to address a particular question.

One problem with current technologies is that it is very difficult to detect rare differences, which could be very important in at least the initial stages of adaptation to a new environment. Techniques such as micoarrays are improving in reliability but the signal: noise ratio is still not high enough to allow detection of rare or subtle differences in expression patterns. The jury is still out over whether phenotypes are controlled by many genes of small effect or few genes of larger effect but it is realistic to assume that both types of patterns (as well as everything in between) might be important. The limitations of our current tools are that they can only really detect the latter in any kind of statistically confident way so it is difficult to assess the relative importance of each. Many of the genomes currently sequenced are exceptionally small compared to average and we don't yet know how much this will bias our interpretation about the control of gene expression. Will larger genomes be subject to more redundancy or more specialization of functions by duplicated regions? It is simply too soon to tell.

What strategies should be used when initiating work on a new model or using a species for which much is known about natural history or ecological variation but little about genetics? It was suggested that the most efficient first step would be to start with the creation of cDNA libraries. This is getting very simple and relatively inexpensive, but to further reduce costs these libraries could be shared between laboratories or financial resources could be pooled and the work could be farmed out to biotechnology laboratories. These libraries could be generated for a few different strains or closely related species, or could be generated from different populations of the same species adapted to different environments. The libraries then could be used to spot microarrays to identify markers for use in expression studies and to look for polymorphisms. Arrays based on single gene features (Affymetrix chipmaps) could then be used to look for polymorphisms that could be directly related back to adaptive changes, or different species could be hybridized to the same chip using both DNA and cDNA to correct for phylogenetic differences when looking for evidence of selection.

Related to hard tools is the question of funding for projects on new model species. There will be a need to convince funding agencies to sink money into more or different organisms. Perhaps communities working on particular groups could get together to justify the choice of new models (e.g., *Mimulus*). It will also be important to convince funding bodies of the importance of combining efforts from across disciplines. Governments are already rewarding collaborative efforts but tend to focus on collaborations between industry and academics rather than on collaborations among academics with different areas of expertise. We need to proactively work to change these attitudes at the governmental level.

Theoretical/analytical tools

One of the consequences of the extremely rapid accumulation of data provided by genomic approaches to studying genetics is that theoretical and analytical approaches have lagged slightly behind. In order to progress in the difficult task of relating genetic variation directly to phenotypes, this is an area that perhaps will require the greatest attention. Before we move on to new species and more and more sophisticated methods for dissecting genomes, perhaps we should take a step backward and synthesize the information already accumulated to generate new theoretical models in which to frame questions. It is already becoming abundantly clear from genome sequencing studies that many of the well-established views about how molecular evolution proceeds need to be re-evaluated. For example, rather than most genes existing in single copies (which has been the working model since DNA-DNA hybridization studies examined the relative proportion of genes in each fraction in the 1960s), even very small genomes like *A. thaliana* are characterized by an extensive history of gene duplication, leaving few genes in single copies. This presents a major analytical problem because establishing orthology is often essential to interpretation of patterns and correlations between phenotypes and genotypes. For example, QTLs that apparently differ in position or function between species could be an artifact of hybridization to different gene copies in different species, resulting in misleading conclusions. The significance of recombination to these types of studies is also not very well known. We need new analytical methods to detect genes that might occur in low copy numbers as well as ways of more reliably determining orthology, and we need new theoretical models that incorporate our changing views on molecular evolution.

There is currently an enormous world-wide effort concentrated in the field of bioinformatics to extract the maximum amount of information from genomic databases. However, much of this can be seen as a simple "data mining" exercise, without consideration of what kinds of questions could be asked. "Data mining" can be viewed as an equivalent to descriptive natural history studies and perhaps it should be approached in the same way. The ecological, population genetics, and evolutionary models proposed in the first half of the 20th century (many of which are still in wide use today) relied on synthesizing information so that we could progress beyond natural history to predictive models for explaining the forces driving genetic and phenotypic variation. The natural history of genomes could also benefit from this type of approach. Establishing a theoretical framework in which to pose questions about patterns of genome evolution, with analytical and statistical methods designed specifically for these questions, seems like an area that really needs the attention of theoretical ecologists and population geneticists. One current problem with many bioinformatics initiatives is that they are often advanced without much input from biologists (i.e., there is very little biology in the "bio"). To demonstrate, in a recent request for grant proposals under the blanket of the Canadian Bioinformatics Integration Network (CBIN: http://www.cbin.ca) to access large sums of money set aside by the Canadian government for bioinformatics research, "biological sciences" was not included in the options for affiliation and none of the application categories included biological research. This needs to be changed—only by combining biology with high-powered computation can we hope to understand the forces that structure genomes and that lead to differential adaptation. In general, the ecological and evolutionary communities are more used to using quantitative tools than molecular geneticists and the two communities should be working more closely together in the analytical side of things. Early microarray experiments have tended to be received with reservation in the former group because of lack of replication but just because technologies have not been used well in the past doesn't mean that they can't be in the future. By integrating approaches from biologists and computational scientists we have the best chance of exploiting information from the technologies that currently exist.

While it is true that statistical and analytical approaches to relating variation in QTLs to phenotypic changes are continually under improvement, there are also other approaches that might have potential as indicators of the association between genotypes and phenotypes. Although it is clear that the rapid cycling nature of many focal plants has allowed plant biologists to progress rapidly in QTL mapping and information gained from detailed crossing experiments, approaches developed for non-rapid cycling organisms should not be overlooked. Human geneticists, for example, obviously don't have the luxury of controlled crossing experiments but they have been very successful in finding microsatellites and genes responsible for differences in traits by taking a population genetics and pedigree-based approach to the problem. In the study of human diseases for example, linkage disequilibrium mapping techniques have been used to define the limits of regions involved in particular syndromes, and transmission tests using pedigrees have been very powerful in identifying genetic changes resulting in susceptibility. Family-level pedigrees or even deeper level phylogenetics to look at adaptive changes can provide powerful analytical tools for interpreting patterns of variation. One interesting technique that has been proposed is to predict ancestral protein sequences for particular genes based on phylogenetic studies and then use these sequences to transform extant species ("molecular paleontology") for direct tests of how changes in the proteins lead to differences in adaptation.

One thing that studies of linkage disequilibrium have shown rather dramatically is that making causal connections between allelic variation and particular phenotypes may be seriously confounded by much larger blocks of linkage disequilibrium than we use to consider. Large linked regions (sometimes even under the control of single transcription factors) results in the potential for genetic hitchhiking and thus in misidentification of which genes are actually involved in control of a particular phenotype. To account for this we really need analytical and statistical tests that are able to factor out position effects and consider models of pleiotropy and epistasis in the interpretation of phenotype-genotype correlations. This emphasizes the need not only to keep up with technological advances but also to continually revise our theoretical and analytical approaches as new information about microevolutionary processes becomes available.

Progress

One of the goals of the workshop was to synthesize ideas on the approaches we should be taking in the future to most efficiently integrate advances in technology with studies of ecological adaptation. It is clearly evident from the range of talks presented at this meeting that the problem is already being addressed and significant progress is in the works. Many of the issues raised during the final discussion were introduced throughout the conference. While it is not possible to summarize all of this work here, examples will be drawn from these presentations as an introduction to the progress currently being made towards achieving an "eco-molecular synthesis".

Export of genomics techniques to semi-model species and new approaches to analyze the resulting data are both areas that were well covered in the conference. As discussed by Quentin Cronk, Carl Douglas, Peter Constabel, and Kermit

Ritland, a combined mapping and BAC end sequencing approach is being applied to the poplar genome (*Populus trichocarpa*) for comparison with similar databases on spruce and with hybrids with other economically important tree species. These species were chosen both for economic importance and to gain insight into phylogenetic differences in genomes. In this collaborative effort information on ecological variation, variation in particular biochemical pathways, herbivore defenses, QTL mapping, SNPs, ESTs and polymorphisms based on linkage disequlibrium will be combined to generate a synthesized overview of variation, to be studied using new analytical methods for plant population genomics. Research groups working on *Mimulus*, which has long been used as an evolutionary and ecological model, are taking more whole-genome level approaches and are making progress towards proposing members of this genus as candidates for genome studies (Toby Bradshaw, John Willis). Ecotypes of *Arabidopsis thaliana* and closely related species in the Brassicaceae (cultivated *Brassica* species, outcrossing *Ipomoe*a and *A. lyrata*) are being promoted as complementary models to learn more about variation within a large family and in different environments (Justin Borevitz, Heather McKhann, Barbara Mable, Michael Purugganan, Mark Rausher). Several genome sequencing projects in the Brassicaceae are already underway and it is likely that more species will be sequenced in the future. Carl Schlichting suggested that *Potentilla glandulosa* might make a good candidate for genomic studies because it is being studied from a reaction norms perspective and has already been used in microarray experiments.

As pointed out by Quentin Cronk in his introductory address, the power of concentrating on particular types of adaptations is that research can be being conducted at different hierarchical levels by different groups, which is essential for fully understanding the full spectrum of adaptive processes. For example, on the topic of adaptation to the light environment, Justin Borevitz focused on ecotypic differences and polymorphisms within species, Tom Givnish focused on general traits giving rise to differences between species, and Sara Mathews focused on the genes responsible for major transitions and key innovations in major clades. The importance of considering phylogenetic relationships when looking at adaptive changes was emphasized in several talks. For example, adaptation to edaphic environments has been found to have a phylogenetic component, with edaphic specialists tending to be phylogenetically clustered (Nishi Rajakaruna, Jeannette Whitton). The origin of vivipary in mangroves may be phylogenetically rather than adaptively determined (Suhua Shi).

This emphasizes a point raised by Toby Bradshaw that adaptation must be demonstrated, not assumed, because "bad things happen to good genes". Evidence for selection on particular genes is being approached using comparisons of rates in different parts of pathways to distinguish evolutionary constraints from evolutionary opportunism (Mark Rausher), continually improving population genetics tests (Michael Purugganan, Justin Borevitz, Sara Matthews, Peter Tiffin) or comparisons of differences in expression patterns among species (Peter Constabel). Jörg Bohlmann discussed the importance of looking at whole classes of genes and gene families when considering adaptation. Carl Schlichting pointed out that adaptation should be considered from a reaction norms perspective, because we don't yet know how much variation is hidden or released under rare or novel environmental conditions. Brian Husband suggested that both ecological pressure and other sources of genetic variation (e.g., polyploidy) should be considered when defining the limits of adaptation. The problem of establishing clear associations between phenotypes and genotypes is currently receiving a lot of attention. Dolph Schluter pointed out that experiments to achieve comparative analyses of function is relatively rare in plants — for example, despite years of research on the group, a statistical correlation between environment and phenotypes in Hawaiian silverswords is lacking. He suggested four steps to testing hypotheses about divergent selection that are possible in plants but not yet widely used: (1) using interaction variance to quantify effects in reciprocal transplant experiments; (2) transplanting hybrids to compare fitness differences in the parental habitats; (3) direct measurements of selection; and (4) using QTLs compared to neutral expectations.

High throughput methods and genomic approaches such as microarrays are being used more and more in ecological and evolutionary studies and it is important to keep up with the advances in biotechnology, but without losing sight of more traditional methods. Direct application of these techniques to naturally occurring variation that can be tested under field conditions is now a feasible approach that is being applied to ecotypes of the model species *A. thaliana*. Justin Borevitz introduced the ideas about using SFPs to look at variation across ecotypes in *A. thaliana* but also to look at responses across a more phylogenetically distant array of species using a whole-genome genotyping approach. These studies can be used not only to relate variation to geographic or habitat differences, but also as a tool to look at differences in genomic structuring across species. Michael Purugganan discussed using *A. thaliana* as a model to look at variation in shoot architecture and genetic variation across ecotypes, combined with both field and growth chamber transplants to assess the full spectrum of variation (he noted that a lot of genes are missed when plants are grown only in a growth chamber). Heather McKhann discussed approaches for integrating information about *A. thaliana* and discussed current work looking for genes involved in drought and cold tolerance stress across ecotypes by doing association studies relating genotypes to phenotypes.

A number of people discussed the importance of looking at interactions both within and between species when considering processes resulting in speciation, which emphasizes that biological interactions also are important factors in adaptation. Scott Hedges and John Willis both discussed using QTLs to look for evidence and causes of segregation distortion in interspecific hybrids. Amy Bouck discussed using linkage mapping with AFLPs to look at reproductive isolation in Louisiana irises, whose genome is so large that a fine-scale map is spaced as wide as the whole *A. thaliana* genome. Leonie Moyle discussed the development of more sensitive methods of looking at factors affecting reproductive isolation by incorporating methods drawn from path analysis. Don Levin discussed the importance of considering cytoplasmic factors in speciation rather than just nuclear genes. The importance of relating changes to other organisms with which species interacts was emphasized in relation to pollinators (Elizabeth Elle) and host-race formation (Lee Taylor).

Finally, progress is being made on theoretical and analytical approaches that incorporate new insights from genomic approaches to molecular evolution and phylogenetics. Sara Good-Avila discussed methods for achieving more rigorous tests of phylogenetic rates. Sara Mathews discussed powerful tests for detecting selection based on phylogenetic approaches and techniques for reconstructing ancestral gene states to directly test predictions about adaptive responses. Mitch Cruzan suggested that it might be time to return to a view of adaptive gene complexes on complex fitness surfaces and ridges rather than simply looking for adaptive peaks. Philip Awadalla described methods for looking for recombination on a genomic scale using more realistic models than have been used previously and Kermit Ritland discussed more general approaches to defining a new field of plant population genomics by altering population genetics models to incorporate genomic scale data.

Conclusions

Although many specific ideas were raised in the final discussion and throughout the conference about how advances in genomic technology can be applied most efficiently to the understanding of genetic control of ecological variation, the main take home message is that molecular biologists and ecologists/evolutionary biologists should be working together more often to achieve this common goal. By combining efforts and mutually exploiting the expertise of the two groups we will have the best chance of achieving an "eco-molecular synthesis". This conference was a good first step in bringing together researchers from disparate fields and hopefully the trend towards working together will continue to bloom. Large-scale genomics is changing the way we view molecular variation and is also likely to dramatically alter our view of the forces influencing adaptive changes as well. We should be prepared to continually shift our approaches and theoretical models to accommodate new ideas based on the exciting advances in molecular evolution.

Acknowledgements

This summary is a combination of ideas expressed by other people, superimposed by my own interpretations and biases. The lack of referencing to particular individuals is intentional, so that I can take full credit for any misinterpretations without accepting credit for the majority of ideas introduced. Most people discussed ideas that they raised in their individual talks and I have tried to emphasize these points in the "progress" section to give stronger clues as to whose contributions are whose. I thank the conference organizers for being able to participate in such an interesting and fruitful workshop.

TRAITS, POPULATIONS, AND SPECIES: CASE STUDIES IN PLANT ADAPTATION

13
Trends in the evolution of edaphic specialists with an example of parallel evolution in the *Lasthenia californica* complex

Nishanta Rajakaruna and Jeannette Whitton

Abstract: Adaptation to unusual soil conditions is a common phenomenon in plants, and closely related taxa are frequently distinguished by their edaphic tolerances. The biology of edaphic endemics further suggests that parallel evolution of edaphic adaptations is not unusual, and that such adaptations may often have effects on gene flow between derivative populations and their progenitors. These observations imply that the study of edaphic divergence is fertile ground for understanding the role of natural selection in adaptation and perhaps in speciation as well. Our work on the *Lasthenia californica* complex provides a case in point. Two races occupy distinct edaphic habitats throughout the range of the complex: race A occupies ionically extreme habitats, while race C occurs in ionically more moderate but drier sites. Phylogenetic data reveal that neither race is monophyletic, and further suggest that race C populations have originated multiple times from race A. Physiological data indicate consistent differences between the races in sodium tolerance, with race A having 20-fold greater sodium uptake rates than race C. Populations of each race also differ in Ca^{2+}/Mg^{2+} uptake, with race A having 2-fold higher uptake rates. Preliminary data further suggest that reproductive isolation is stronger between populations of different races than between populations of the same race. Taken together, these findings point to the development of parallel changes in reproductive compatibility accompanying parallel evolution of edaphic races in the *L. californica* complex. While this example represents one of the better documented cases of parallel adaptation in plants, our characterization of edaphic specialists suggests that these will likely provide further cases that will reveal the importance of edaphic adaptation in plant speciation.

Nishanta Rajakaruna. College of the Atlantic, 105 Eden Street, Bar Harbor, ME 04609, U.S.A.
e-mail address: nrajakaruna@coa.edu
Jeannette Whitton. Department of Botany, University of British Columbia, 3529-6270 University Boulevard, Vancouver, BC V6T 1Z4, Canada.
Correct citation: Rajakaruna, N. and Whitton, J. 2004. Trends in the evolution of edaphic specialists with an example of parallel evolution in the *Lasthenia californica* complex. *In* Plant Adaptation: Molecular Genetics and Ecology. Proceedings of an International Workshop held December 11–13, 2002, in Vancouver, British Columbia, Canada. *Edited by* Q.C.B. Cronk, J. Whitton, R.H. Ree, and I.E.P. Taylor. NRC Research Press, Ottawa, Ontario. pp. 103–110.

"Recurrent evolution of taxa not only is possible, but may not be unusual in complexes that have undergone ecological diversification." D.A. Levin (2001)

Introduction

The study of adaptation marks a central focus in evolutionary biology, with one major thrust of these endeavours being to determine the relative importance of adaptive evolution in diversification. When putative adaptive traits have a conspicuous effect on reproductive isolation, e.g., traits that affect pollinator preferences (Schemske and Bradshaw 1999) or key innovations such as nectar spurs (Hodges and Arnold 1995), the link between adaptation and speciation is easiest to draw. However, less conspicuous features of organisms that have effects on fitness and are subject to natural selection may also contribute to speciation and diversification of lineages. For example, a number of plant species are known to occur only under specific, typically extreme soil conditions, being excluded from other substrates (Kruckeberg 2002). The range of extreme soils that harbour edaphic endemics includes hyper-saline soils such as vernal pools and alkaline flats, soils rich in heavy metals, such as mine tailings and serpentine soils, and nitrogen-rich sites such as guano deposits. Where edaphic endemics represent derived expansion and specialization onto these substrates, adaptation to specific soil conditions may have played a key role in diversification (Macnair 1987; Macnair and Gardner 1998). The link between edaphic adaptation and reproductive isolation may be achieved through various means, including direct impacts on reproductive isolation, for example via pleiotropy or linkage, and indirect effects, e.g., reinforcement, accumulation of pre- or post-zygotic isolating barriers in allopatry.

We wish to focus attention on the potential significance of edaphic divergence in plant speciation. We begin with a synopsis of some key aspects of the biology of edaphic specialists that contribute to their value as systems for the study of adaptation and speciation. We focus in particular on parallel adaptive shifts because these cases provide key opportunities to study the selective pressures and genetic mechanisms that contribute to the establishment of adaptation. In addition, cases of parallel adaptation provide the chance to study the relationship between adaptation and reproductive isolation with the benefit of natural replication. We then summarize our recent research characterizing edaphic specialization in the *Lasthenia californica* complex, which we believe is among the best documented cases of parallel evolution of edaphic specialization in plants. Although the *Lasthenia* work characterizes parallel adaptation in greater detail than has been done in many other systems, it seems clear that parallel shifts are likely a common outcome of edaphic specialization. Whether parallel adaptation is commonly accompanied by parallel reproductive isolation remains unclear, but further research in this area will likely provide valuable insights into the role of natural selection in diversification.

The natural history of edaphic specialization

Most plant population biologists are familiar with the classic papers titled "*Evolution in closely adjacent plant populations*" of Bradshaw, McNeilly, Antonovics and co-workers (e.g., McNeilly 1968; McNeilly and Antonovics 1968; Antonovics and Bradshaw 1970) that focused primarily on the evolution of metal tolerance in populations on mine tailings. Much of the work was conducted on two species of grasses, *Anthoxanthum odoratum* and *Agrostis tenuis*, and provides a classic demonstration of the power of strong natural selection in maintaining distinct sub-populations despite the potential for gene flow (McNeilly 1968; McNeilly and Bradshaw 1968). These classic papers, along with subsequent work on mine tailings also illustrate two especially intriguing features of the biology of edaphic specialists that are relevant to speciation. First, edaphic specialists seem to have a tendency to evolve tolerance of specific conditions multiple times, that is there is a tendency towards parallel establishment of edaphic specialization. Whether this represents *de novo* evolution of alleles conferring specialization, or parallel increases in frequency of alleles that have arisen a single time remains unclear (see below). Second, edaphic shifts often have direct or indirect effects on patterns of reproductive isolation.

In the early characterization of mine tailing populations in *Agrostis* and *Anthoxanthum*, parallel evolution was suspected because of the geographical separation of various populations on mine tailings, and the widespread occurrence of non-tolerant populations (Gregory and Bradshaw 1965). Furthermore, copper tolerance was shown to be present at low frequencies in seed samples collected in separate populations of *Agrostis tenuis* growing on non-contaminated soils, pointing to the likely parallel establishment of the trait in isolated localities (McNeilly and Bradshaw 1968). The hypothesis of parallel evolution of metal tolerance in *Agrostis tenuis* has since been further supported by analysis of genotype × environment interaction (G × E) in copper tolerant populations from distinct localities, which reveals that copper tolerant populations display significant heterogeneity in their response to levels of copper (Nicholls and McNeilly 1982). While these patterns may reflect distinct origins of copper tolerance, they could also reflect the presence of unique modifiers of a common mechanism of tolerance that in fact had a single origin (Schat et al. 1996).

In much of the literature discussing the origins of metal tolerance, the geographical distribution of metal-tolerant populations again provides the primary indication of parallel origins. While this is a reasonable hypothesis, it is also possible that tolerant populations have been established by long-distance dispersal, perhaps aided by accidental transport by mine workers (Schat et al. 1996). Further evidence of multiple origins comes from crossing studies that examine patterns of segregation of tolerance that would indicate the action of multiple independent loci. The action of independent loci would be revealed either by a breakdown of metal tolerance in the F2 generation in crosses between homozygous tolerant individuals, but might also be suggested by the presence of transgressive segregation in the F1 generation. To date, such tests tend to support a common genetic basis for tolerance, though the presence of modifiers has been noted in some populations. For example, in *Silene vulgaris*, crosses among copper, zinc and cadmium tolerant and non-tolerant populations from Germany and Ireland reveal that a single locus is responsible for tolerance to each metal, though at least one additional locus contributes to additional copper tolerance in one population (Schat et al. 1993). In this case, the modifier could have arisen subsequent to the establishment of the widespread tolerant genotype at this site. While a common genetic basis for tolerance could be regarded as evidence against parallel evolution of tolerance, studies of the genetic basis of metal tolerance suggest that constraints may limit the number of loci that can successfully mutate to yield tolerant genotypes. Therefore it is possible that crossing studies fail to show a breakdown of tolerance because mutation of the same genetic locus has given rise to tolerant genotypes independently. Finally, it should be noted that even if the same allele is responsible for tolerance in distinct populations, this allele may have been present at low frequencies in non-tolerant populations and thus may have become established in parallel at multiple localities (Schat et al. 1996). In such cases, parallel evolution of metal tolerance could be indicated by phylogenetic data that support multiple origins of tolerant populations. Where phylogenetic studies of edaphically distinct populations have been conducted, it is noteworthy that these uniformly point to parallel evolution.

Under the extreme conditions presented by severely metal-contaminated sites, the role of natural selection in establishing tolerance is understood, given that non-tolerant individuals generally do not survive the mine tailing environment. In less dramatic cases, the pattern of parallel evolution itself is suggestive of the action of natural selection, because parallel establishment of ecologically relevant traits is unlikely to occur as a result of stochastic processes (Levin 2001). For example, in previous work on *Lasthenia californica*, analysis of isozyme variation (Desrochers and Bohm 1998) and phylogenetic data suggested the existence of geographically-based subdivisions within the complex. Analysis of chemical characteristics of soils and plant tissue demonstrated the existence of edaphic races and suggested that divergent natural selection might play a role in diversification of the races (Rajakaruna and Bohm 1999). Combining these data has revealed a pattern of parallel evolution of an ecologically relevant suite of traits (Rajakaruna et al. 2003*c*), which serves as evidence for the action of natural selection in establishing racial differences. Given that relatively few systems have been explored and that many of these provide some indication of parallel evolution, this tendency would seem to be a common characteristic of edaphic specialists.

A phenomenon that may be related to parallel evolution of tolerance is the evolution of multiple tolerances within species. In many cases, lineages that have evolved tolerance to one edaphic extreme have also adapted to other extreme factors. For example, tolerance of multiple populations to four elements (copper, nickel, zinc and lead) was assessed (Gregory and Bradshaw 1965) in *Agrostis tenuis*. Although most populations occur on pasture soils that have trace levels of heavy metals and comprise plants that are not tolerant of heavy metal contamination, mine tailing populations variously display tolerance to zinc, copper and lead, generally matching their tolerance to levels present in soils at the collection locality (Gregory and Bradshaw 1965). The authors also demonstrated that tolerance of each of these metals did not confer tolerance to the others, indicating somewhat distinct mechanisms in each case. Still, that this species has evolved tolerance to normally toxic levels of multiple ions suggest that an underlying trait, perhaps involving tolerance of low pH (Gregory and Bradshaw 1965), drought (Hughes et al. 2001) or hyper saline conditions is widespread in the progenitors of edaphic specialists and contributes to their ability to evolve tolerance to specific ionic extremes. It seems plausible that such underlying traits could facilitate either the evolution of multiple tolerances or the parallel evolution of tolerance of a single heavy metal.

The second intriguing feature of edaphic endemics is the relationship between the shift in edaphic tolerance and changes in patterns of reproductive isolation. Such changes may arise as a direct consequence of adaptive shifts, either because reproductive compatibility is a by-product of physiological adaptation to new edaphic conditions or because of linkage or pleiotropy of loci affecting ecological shifts and those affecting reproductive isolation. Alternatively, enhanced reproductive isolation between divergent edaphic specialists may reflect the action of reinforcement, i.e., selection for reduced gene flow to avoid maladaptive hybridization following a period of divergence in allopatry. It is important to note that the two alternatives are not mutually exclusive. For instance, in the classic mine tailing studies, heavy metal tolerant and non-tolerant populations of both *Anthoxanthum odoratum* and *Agrostis tenuis* were shown to have genetically-controlled differences in flowering time (McNeilly and Antonovics 1968), with tolerant populations flowering earlier in both cases. Plants closest to the mine boundary showed the greatest difference in flowering times, which McNeilly and Antonovics (1968) interpret as evidence for the action of reinforcement, though they state that a portion of the flowering time shifts also arises as a by-product of adaptation to local conditions. They noted a relationship between flowering time and soil temperatures to support this claim. It is likely that changes in phenology can also evolve as a direct by-product of shifts in edaphic tolerances. For example, flowering time differences are associated with differences in sodium accumulation in wheat (Taeb et al. 1992). Furthermore, even if the initial divergence in flowering time is entirely environmentally-determined, this pattern can contribute to the accumulation of genetically-based flowering time differences. Stam (1983) demonstrated theoretically that spatially structured differences in flowering time can lead to spatially-coincident, genetically-based differences in flowering time. Of course as different edaphic habitats are almost by necessity spatially isolated from one another, individuals that occur in distinct edaphic environments are likely to experience decreased gene flow relative to individuals that occur in the same habitat (L. H. Rieseberg, pers. comm.). Another tendency that has been repeatedly observed in edaphic endemics is a shift towards increasing self-fertility. Increased selfing rates are thought to have arisen as a mechanisms to prevent gene flow between mine and non-mine populations in both *Anthoxanthum odoratum* and *Agrostis tenuis* (Antonovics 1968), though this explanation was not favored to explain increased self-fertility in *Armeria maritima* populations from zinc mines (Lefèbvre 1970). Lefèbvre (1970) instead interprets increased self-fertility as having provided reproductive assurance during long distance colonization of mine sites, as there are no adjacent non-mine populations in this region.

Examples of changes in post-mating reproductive isolation accompanying edaphic shifts are less common, as might be expected given that these are generally more likely to accumulate as a by-product of divergence. However, post-mating reproductive isolation can include mechanisms that act prior to zygote formation, such as pollen-pistil incompatibilities. Such post-mating, pre-zygotic isolation can be subject to reinforcement, and thus may be an indirect outcome of adaptive divergence. The most widely cited example of post-mating reproductive isolation associated with edaphic shifts comes from populations of *Mimulus guttatus* adapted to copper mine tailings. Macnair and co-workers have documented that the genetic locus that confers copper tolerance is closely linked to or has a pleiotropic effect on viability of hybrids between tolerant and non-tolerant individuals (Christie and Macnair 1983). Searcy and Macnair (1990) also demonstrated that copper uptake may contribute to pollen pistil incompatibilities, suggesting that crossing studies involving extremes should perhaps be conducted.

The case for parallel evolution of edaphic races in the *Lasthenia californica* complex

Lasthenia californica DC. ex Lindl. *sensu* Ornduff (1966; 1993), the common goldfields of California, provides an ideal system for the study of parallel speciation driven by edaphic forces. Common goldfields, as previously delimited, displays the widest range of edaphic tolerances within the genus, occupying diverse habitats within the Californian Floristic Province (Ornduff 1966; 1993). Despite the range of habitats occupied by populations within the complex, a survey of soil features and elemental composition of plants (Rajakaruna and Bohm 1999) concluded that flavonoid races previously described for the complex (Bohm et al. 1989; Desrochers and Bohm 1993) correspond to edaphic races. Race A plants predominate in ionically-extreme habitats such as coastal bluffs, alkaline flats, vernal pools, and serpentine outcrops; these plants are characterized by the presence of sulfated flavonoids (Bohm et al. 1989). Race C plants are found in ionically-benign but drier sites such as roadside pastures and oak-woodlands; these plants lack the sulfated flavonoids characteristic of race A. Interestingly, the two races grow in parapatry on a serpentine outcrop at Jasper Ridge Biological Preserve (Stanford University, San Mateo County, CA), but at this site, each race occupies microhabitats that correspond to the edaphic conditions that distinguish the races throughout the range of the complex.

Given that we suspect that selection has contributed to the establishment and maintenance of the patterns of distribution of the two races, both at Jasper Ridge and throughout the species range, we initiated a study to examine relationships among populations of the complex. The major findings, outlined below substantiate the role of selection and point to parallel origins of at least one of the edaphic races.

Traits and taxa can evolve in parallel by means other than natural selection. For example, multiple independent origins have been documented for numerous plant polyploids and for diploid hybrid taxa. Here, we are concerned with parallel evolution *sensu* Schluter and Nagel (1995) and Levin (2001), implying the action of natural selection. Levin (2001) suggests two criteria that are necessary to demonstrate parallel evolution of a taxon. First, populations of the descendent taxon must be phylogenetically independent. Second, the shared characteristics must be the products of natural selection. Where parallel evolution results in parallel reproductive isolation, parallel speciation is demonstrated. Schluter and Nagel (1995) provide two additional criteria necessary to document parallel speciation. The third criterion states that separate descendant populations that are found in similar environments must be reproductively isolated from the ancestral populations. Finally, the separate descendant populations must not be reproductively isolated from one another. Few cases exist in which all four criteria have been addressed (Schluter and Nagel 1995) and Levin (2001) indicates that there are almost no unassailable examples of parallel speciation in the plant literature. We will present evidence to address each of these four criteria in *Lasthenia californica*. While the evidence is not complete in each case, in our opinion, it clearly demonstrates parallel evolution, and points toward parallel speciation. We will note areas in need of further corroboration and outline our plans for additional studies.

Evidence of parallel origins of edaphic races in the *Lasthenia californica* complex

Based on a comprehensive ITS/ETS/cpDNA phylogenetic study, Chan et al. (2001; 2002) found that *L. californica sensu* Ornduff represents two geographically-based, non-sister clades. Chan (2001) and Chan et al. (2002) recognized the clades as two cryptic taxa, *L. californica* subsp. *californica* representing the northern clade and *L. gracilis* DC. (Greene) representing the southern clade. The molecular phylogeny allowed us to examine the relationship of the edaphic races to the newly recognized phylogenetic taxa. If ecological selection has played a role in the origin of edaphic races, then similar edaphic tolerances may have evolved in parallel within one or both of the phylogenetic taxa. Indeed, a previous study of allozyme variation (Desrochers and Bohm 1995) indicated that single populations of race A and C cluster with sets of populations belonging to the alternative race. We determined the edaphic race and phylogenetic affinities of 33 populations from throughout the range of the complex. Flavonoid profiles were used to assign populations of Chan et al. (2001, 2002) to edaphic races and characterization of nuclear ribosomal ITS regions of several representatives from the two edaphic races were used to determine phylogenetic affinities (Rajakaruna et al. 2003*c*). Examination of flavonoid profiles and ITS sequences revealed that edaphic races are not concordant with the newly circumscribed taxa, *L. californica* subsp. *californica* and *L. gracilis*. Of the 16 populations of *L. gracilis*, 13 were race A and 3 were race C. Of the 17 populations of *L. californica* subsp. *californica*, 11 were race C and 6, race A. Furthermore, a distance-based analysis of RAPD variation (Rajakaruna 2002) rooted using the information from phylogenetic markers indicates that race A is most likely ancestral in the complex, resulting in an inference of parallel origin of race C in each of the two clades. This further implies that the sulfated flavonoid that is diagnostic for race A was lost independently in derived populations of Race C.

Parallel physiological adaptation

Schluter and Nagel (1995) and Levin's (2001) second criterion aims at establishing that natural selection is the cause of parallel origins: an adaptive mechanism must be identified and tested. In the original characterization of the races (Rajakaruna and Bohm 1999) two chemical factors that most strongly distinguished both the habitats and the plant tissue elemental composition of races were the levels of sodium and magnesium. Race A plants occur in soils averaging 60.8 ppm sodium and 1147 ppm magnesium, while race C sites average 19.9 ppm and 280.6 ppm of each element, respectively. Further examination of the relationship between tissue and soil sodium levels from the two races at Jasper Ridge suggested that the two races differ in their sodium and magnesium uptake physiologies (Rajakaruna and Bohm 1999). Because these differences are likely to reflect differential adaptation to edaphic conditions, we conducted a hydroponic study of sodium uptake physiology in two populations of each race, including phylogenetically divergent populations of both races. Mean Na^+ uptake rates as well as tissue accumulation patterns for Na^+ were estimated. Results suggest that race A plants from both lineages are better tolerant of sodium: Na^+ uptake rates of race A plants were 20-fold higher than those of race C plants. Further, race A plants translocated ~50% of absorbed Na^+ to the shoot compared with <30% in race C, and thus, when combined with higher uptake rates, race A plants were able to accumulate ~5 fold greater concentrations of Na^+ in the above-ground tissues than race C plants. Race C plants from each of the distinct phylogenetic lineages restricted translocation of Na^+ to shoots, a mechanism often used by plants intolerant of toxic elements (Fitter and Hay, 1987). Figure 1 shows the total shoot accumulation as well as shoot/root ratio of Na^+ in race A and C populations from both taxa. The higher shoot concentrations of Na^+ found in race A plants suggest that a similar mechanism may be involved in conferring tolerance to this potentially toxic cation. In the case of Na^+, tolerant species are able to maintain higher concentrations in the shoot and sequester the ions in the vacuole (Amtmann and Sanders, 1999) via Na^+/H^+ antiporters (Apse et al. 1999; Blumwald et al. 2000). We plan to investigate whether the presence/absence or level of expression of the Na^+/H^+ antiporter gene is responsible for the differences in uptake and accumulation of Na^+ in the two races. Such a study will be essential to address the underlying biochemical/genetic basis of these traits. Germination, root growth, and survivorship

Fig. 1. Total shoot sodium (left) and shoot/root sodium (right) for race A and C populations belonging to *L. californica* subsp. *californica* (A_C and C_C, respectively) and *L. gracilis* (A_G and C_G, respectively). Race A populations of both species allocate significantly more sodium to shoot than race C populations.

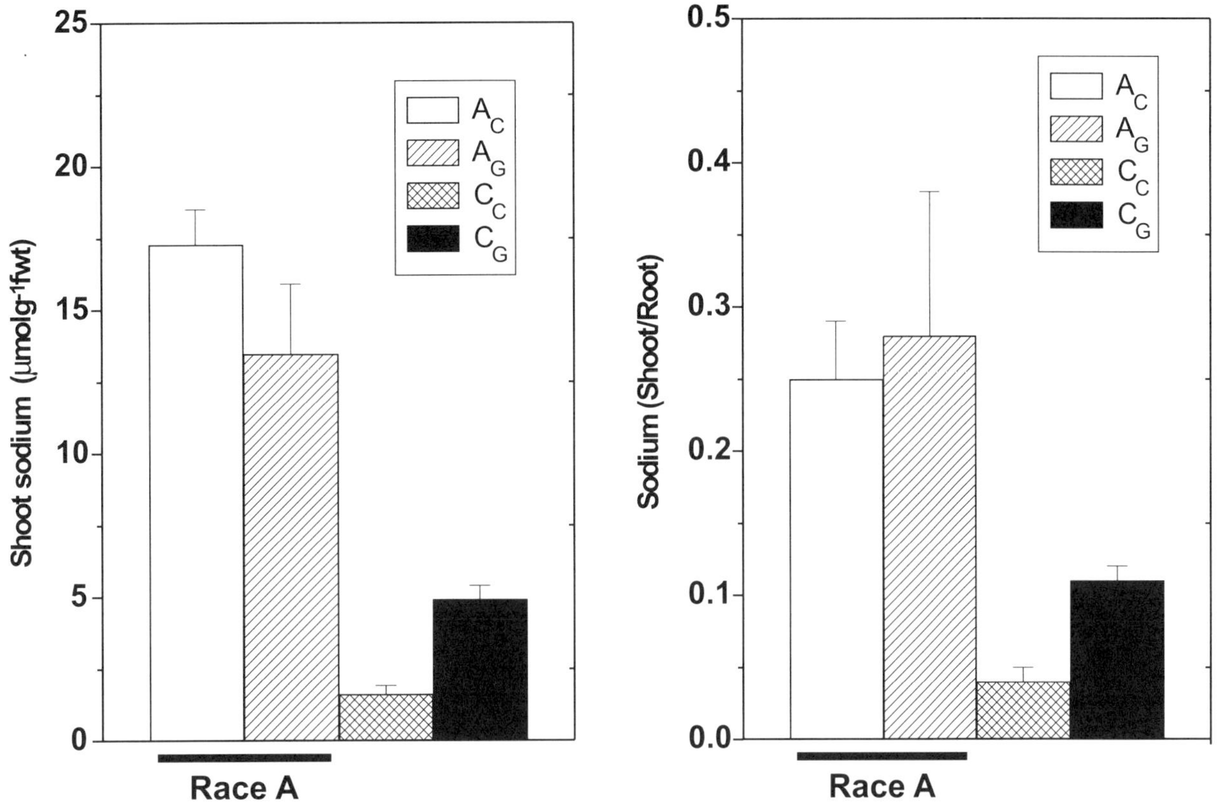

estimates also indicated greater tolerance by race A to Na^+ (Rajakaruna et al. 2003*a*). Further, significant Genotype × Treatment interactions were observed for all tolerance measures, suggesting that these races are genetically differentiated in their tolerance responses.

As described above, results of population genetic studies suggest that race A is ancestral in the complex and therefore both the sulfated flavonoids and Na^+ tolerance have been lost in race C populations. While the genetic basis of these differences and their correlation with habitats of race C suggest that they contribute to differential adaptation, it is not clear how the loss of these traits is adaptive, unless they result in a relative fitness cost under the conditions in which race C occurs. This hypothesized cost of tolerance seems plausible, and seems to agree with the results of a drought stress experiment that we conducted using seed from race A and race C populations at Jasper Ridge (Rajakaruna et al. 2003*b*).

At Jasper Ridge (and throughout the range of the complex), race C populations occur in ionically-benign and water stressed environments (Rajakaruna and Bohm 1999), while race A occurs in ionically challenging but wetter sites. We examined the relative fitness of individuals of the two races under varying levels of drought stress. We conducted an experiment in potting soil in the greenhouse that showed that race C plants from the Jasper Ridge site are more drought tolerant than race A plants from this site (Rajakaruna et al. 2003*b*). Race C plants reach reproductive maturity faster and have greater reproductive fitness under drought conditions. It is unclear at this stage whether drought tolerance is unique to all race C populations, however it seems likely given the water-stressed nature of their edaphic habitats.

The coincident loss of both sulfated flavonoids and traits responsible for Na^+ tolerance is of interest in light of the possible link between these traits (Barron et al. 1988). Sulfated flavonoids are often found in plants growing under saline conditions and we recently hypothesized that there may be an ecologically related adaptive mechanism linking their presence to salt tolerance (Rajakaruna et al. 2003*c*). Hence, it is possible to envision the loss of these potentially linked traits in race C populations not exposed to salinity or other ionic stresses. The loss of these traits combined with a gain in tolerance to water stress is plausible under the known climatic and edaphic shifts that have taken place in California (Howard 1951). Ostensibly, a shift from wet, saline conditions to dry, non-saline habitats could have led to a relaxation of selection on traits conferring salt tolerance while selecting for traits such as early flowering and lower root/shoot ratios that confer drought tolerance. Further, studies have shown that salt tolerance may have a pleiotropic effect on early flowering (Taeb et al. 1992). This study shows a possible link between differential adaptive traits observed for the two races in the *L. californica* complex: early-flowering and Na^+-intolerance in race C and later-flowering and Na^+-tolerance in race A.

The case for parallel evolution of reproductive isolation

The third and fourth criteria of Schluter and Nagel (1995) are critical to demonstrate parallel speciation. Although parallel evolution is common (Schluter and Nagel 1995; Levin 2001), parallel evolution of traits conferring reproductive isolation, i.e., parallel speciation, is uncommon in the plant literature (Levin 2001). While examples of parallel evolution provide ideal setting for the study of the action of natural selection in divergence, a causative link between parallel adaptation and parallel speciation need not exist. In fact, it is easy to imagine that a link between parallel adaptation and reproductive isolation could lead to multiple independently derived units each reproductively isolated from one another. However, where cases of parallel evolution are documented, we have the opportunity to examine the link between adaptation and reproductive isolation, as we have attempted to do in the *Lasthenia californica* complex. The parallel occurrence of physiologically distinct races in divergent lineages allowed us to test for reproductive isolation between the races. In order to determine the extent of reproductive isolation between the races, seven populations were used to estimate levels of crossability between the two races within and between the phylogenetic lineages (Rajakaruna 2002). The three *L. californica* subsp. *californica* populations consisted of 2 race A populations and 1 race C population while the four *L. gracilis* populations consisted of 3 race A populations and 1 race C population. Seed set per cross was estimated by counting the number of cypselae that were full and dark (prior germination studies from field-collected cypselae have shown that the two features are good indications of viability) and then averaging the number from two heads used in reciprocal crosses. Figure 2 shows intra- and inter-racial seed set within and between the two phylogenetic lineages. The column representing the race C × C within-clade cross includes only intra-population crosses (only 1 race C population from each species was used) and thus should be interpreted with caution. Nevertheless, reproductive isolation appears to be strongest in inter-racial crosses both within and between clades. According to theory, if reproductive isolation is evolving as a direct or indirect consequence of edaphic specialization, we would predict greater reproductive compatibility between independently derived populations of race C than between race C populations and their closely related race A counterparts. Our preliminary observations are consistent with this prediction. Examination of pollen tube growth following experimental crossing (Fig. 3) reveals that the mean pollination success of race C populations with race A populations from the same clade is 5%, while the success rate of crosses between race C populations of distinct origins is 40% (Rajakaruna and Whitton, unpublished). We must caution though, that we consider these findings preliminary, and in particular we note that the patterns of interpopulation crossability are highly variable and will require careful study to tease apart the contributions to reproductive isolation of divergent selection on edaphic tolerances versus other effects. Nonetheless, our observed trends corroborate the final two criteria of Schluter and Nagel (1995), i.e., descendent populations must be reproductively isolated from ancestral populations, and separate descendent populations must not be reproductively isolated.

Fig. 2. Mean percent seed set within and between race A and C plants belonging to both phylogenetic species (clades). Interracial seed set is significantly lower than intra-racial crosses both within and between clades.

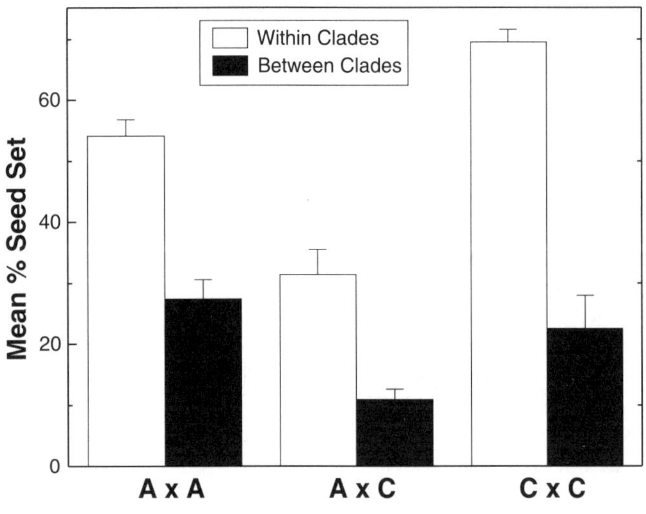

Fig. 3. The relationship between pollination success and seed set in experimental crosses within and between races of *Lasthenia californica*. Most of the isolation comes about by pollen incompatibility reactions, i.e., isolation between races both within and between clades is determined primarily by a prezygotic and postmating mechanism.

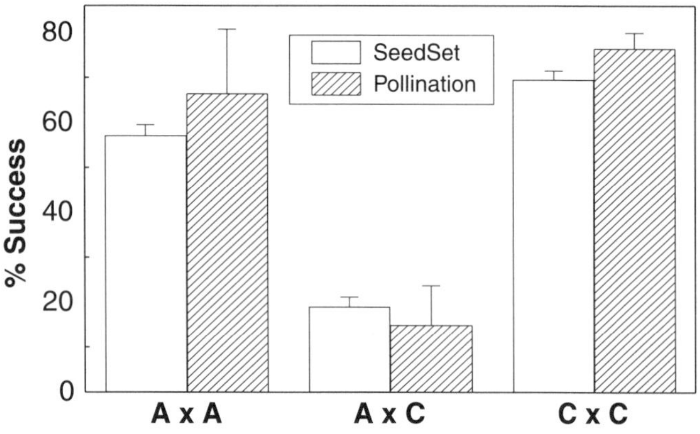

The role of parallel evolution in the origin of species is at an early stage of discovery and we believe that our studies identify a system that can be used as a model for the study of parallel speciation in plants. Ongoing studies are aimed at providing additional supportive evidence to further our hypothesis of parallel speciation of edaphic races in the *L. californica* complex.

Furthermore, our findings from *Lasthenia* californica must not be taken in isolation, as our review of the literature leads to the hypothesis that parallel evolution may be relatively common outcome of edaphic divergence, and that edaphic divergence can have effects on reproductive isolation that could lead to parallel speciation. Thus, these systems are ideal settings in which to study the link between adaptation and speciation.

Acknowledgements

We are grateful for financial support of our research on *Lasthenia californica* from an NSERC Canada Discovery Grant to JW.

References

Allan, W.R., and Sheppard, P.M. 1971.Copper tolerance in some California populations of the monkey flower, *Mimulus guttatus*. Proc. R. Soc. Lond. B Biol. Sci. **177**: 177–196.

Amtmann, A., and Sanders, D. 1999. Mechanisms of Na^+ uptake by plant cells. Adv. Bot. Res. **29**: 76–112.

Antonovics, J. 1968. Evolution in closely adjacent plant populations. V. Evolution of self-fertility. Heredity **23**: 219–238.

Antonovics, J., and Bradshaw, A.D. 1970. Evolution in closely adjacent plant populations. VIII. Clinal patterns at a mine boundary. Heredity **25**: 349–362.

Apse, M.P., Aharon, G.S., Snedden, W.A., and Blumwald, E. 1999. Salt tolerance conferred by overexpression of a vacuolar Na^+/H^+ antiport in *Arabidopsis*. Science (Washington, D.C.) **285**: 1256–1258.

Barron, D., Varin, L., Ibrahim, R.K., Harborne, J.B., and Williams, C.A. 1988. Sulfated flavonoids — an update. Phytochemistry **27**: 2375–2395.

Blumwald, E., Aharon, G.S., and Apse, M.P. 2000. Sodium transport in plant cells. Biochim. Biophys. Acta **1465**: 140–151.

Bohm, B. A., Herring, A., Nicholls, K.W., Bohm, L.R., and Ornduff, R. 1989. A six-year study of flavonoid distribution in a population of *Lasthenia californica* (Asteraceae). Am. J. Bot. **76**: 157–162.

Chan, R., Baldwin, B.G., and Ornduff, R. 2001. Goldfields revisited: A molecular phylogenetic perspective on the evolution of *Lasthenia* (Compositae: Heliantheae *sensu lato*). Int. J. Plant Sci. **162**: 1347–1360.

Chan, R., Baldwin, B.G., and Ornduff, R. 2002. Cryptic goldfields: A molecular phylogenetic re-investigation of *Lasthenia californica sensu lato* and close relatives (Compositae: Heliantheae *sensu lato*). Am. J. Bot. **89**: 1103–1112.

Christie, P., and Macnair, M.R. 1983. Compllimentary lethal factors in two North American populations of the yellow monkeyflower. J. Hered. **75**: 510–511.

Desrochers, A.M., and Bohm, B.A. 1993. Flavonoid variation in *Lasthenia californica*. Biochem. Syst. Ecol. **21**: 449–453.

Desrochers, A.M., and Bohm, B.A. 1995. Biosystematic study of *Lasthenia californica* (Asteraceae). Syst. Bot. **20**: 65–84.

Fitter, A.H., and Hay, R.K.M. 1987. Environmental Physiology of Plants. Second Edition. Academic Press, London.

Gregory, R.P.G., and Bradshaw, A.D. 1965. Heavy metal tolerance in populations of *Agrostis tenuis* Sibth. and other grasses. New Phytol. **64**: 131–143.

Hodges, S.A., and Arnold, M.L. 1995. Spurring plant diversification: are floral nectar spurs a key innovation? Proc. R. Soc. Lond. B Biol. Sci. **262**: 343–348.

Howard, A.D. 1951 The development of the landscape of the San Francisco Bay counties. California Divisions Mines Bull. **154**: 95–106.

Hughes, R., Bachmann, K., Smirnoff, N., and Macnair, M.R. 2001. The role of drought tolerance in serpentine tolerance in the *Mimulus guttatus* Fischer ex DC. complex. S. Afr. J. Sci. **97**: 581–586.

Kruckeberg, A.R. 2002. Geology and plant life: the effects of landforms and rock types on plants. University of Washington Press, Seattle, WA.

Lefèbvre, C. 1970. Self-fertility in maritime and zinc mine populations of *Armeria maritima* (Mill.) Willd. Evolution **24**: 571–577.

Levin, D.A. 2001. The recurrent origin of plant races and species. Syst. Bot. **26**: 197–204.

Macnair, M.R. 1987. Heavy metal tolerance in plants: a model evolutionary system. Trends Ecol. Evol. **2**: 354–359.

Macnair, M.R., and Gardner, M. 1998. The evolution of edaphic endemics. *In* Endless forms: species and speciation. *Edited by* D.J. Howard and S.H. Berlocher. Oxford University Press, New York. pp. 157–171.

McNeilly, T. 1968. Evolution in closely adjacent plant populations. III. *Agrostis tenuis* on a small copper mine. Heredity **23**: 99–108.

McNeilly, T., and Antonovics, J. 1968. Evolution of closely adjacent plant populations. IV. Barriers to gene flow. Heredity **23**: 205–218.

McNeilly, T., and Bradshaw, A.D. 1968. Evolutionary processes in populations of copper tolerant *Agrostis tenuis*. Evolution **22**: 108–118.

Nichols M.K., and McNeilly, T. 1982. The possible polyphyletic origin of copper tolerance in *Agrostis tenuis* (Gramineae). Plant Syst. Evol. **140**: 109–117.

Ornduff, R. 1966. A biosystematic survey of the Goldfield genus *Lasthenia* (Compositae: Helenieae). Univ. California Publ. Bot. **40**: 1–92.

Ornduff, R. 1993. *Lasthenia*. *In* The Jepson Manual: Higher Plants of California. *Edited by* J.C. Hickman. University of California Press, Berkeley, CA. pp. 298–300.

Rajakaruna, N. 2002. Evolution in the *Lasthenia californica* complex (Asteraceae). Ph.D. thesis, Botany, University of British Columbia, Vancouver, BC.

Rajakaruna, N., and Bohm, B.A. 1999. The edaphic factor and patterns of variation in *Lasthenia californica* (Asteraceae). Am. J. Bot. **86**: 1576–1596.

Rajakaruna, N., Siddiqi, M.Y., Whitton, J., Bohm, B.A., and Glass, A.D.M. 2003*a*. Differential responses to Na^+/K^+ and Ca^{2+}/Mg^{2+} in two edaphic races in the *Lasthenia californica* complex (Asteraceae). New Phytol. **157**: 93–103.

Rajakaruna, N., Bradfield, G.E., Bohm, B.A., and Whitton, J. 2003*b*. Adaptive differentiation in response to water stress by edaphic races of *Lasthenia californica* (Asteraceae). Int. J. Plant Sci. **164**: 371–376.

Rajakaruna, N., Baldwin, B.G., Chan, R., Desrochers, A.M., Bohm, B.A., and Whitton, J. 2003*c*. Edaphic races and phylogenetic taxa in the *Lasthenia californica* complex (Asteraceae: Heliantheae): an hypothesis of parallel evolution. Mol. Ecol. **12**: 1675–1679.

Schat, H., Kuiper, E., Ten Bookum, W.M., and Vooijs, R. 1993. A general model for the genetic control of copper tolerance in *Silene vulgaris*: evidence from crosses between plants from different tolerant populations. Heredity **70**: 142–147.

Schat, H., Vooijs, R., and Kuiper, E. 1996. Identical major gene loci for heavy metal tolerances that have independently evolved in different local populations and subspecies of *Silene vulgaris*. Evolution **50**: 1888–1895.

Schemske, D.W., and Bradshaw, H.D., Jr. 1999. Pollinator preference and the evolution of floral traits in monkeyflowers (*Mimulus*). Proc. Natl. Acad. Sci. U.S.A. **96**: 11910–11915.

Schluter, D., and Nagel, L.M. 1995. Parallel speciation by natural selection. Am. Nat. **146**: 292–301.

Searcy, K.B., and Macnair, M.R. 1990. Differential seed production in *Mimulus guttatus* in response to increasing concentrations of copper in the pistil by pollen from copper tolerant and sensitive sources. Evolution **44**: 1424–1435.

Stam, P. 1983. The evolution of reproductive isolation in closely adjacent plant populations through differential flowering time. Heredity **50**: 105–118.

Taeb, M., Koebner, R.M.D., Forster, B.P., and Law, C.N. 1992. Association between genes controlling flowering time and shoot sodium accumulation in the Triticeae. Plant Soil **146**: 117–121.

14 Floral adaptations and biotic and abiotic selection pressures

Elizabeth Elle

Abstract: Explanations for the evolution of floral phenotypes have historically focussed on pollinator choice, but phenotypes also reflect adaptation to abiotic selection pressures. Populations of *Collinsia parviflora* vary significantly in floral phenotype, specifically in corolla size. Traditional explanations for this variation (pollinator choice and the reproductive assurance benefit of selfing) explain some of the phenotypic variation observed. However, genetic correlations between corolla size and time to reproductive maturity indicate that an abiotic factor, pattern of precipitation, may also be an important agent of selection. Rapidly developing populations of *C. parviflora* have small flowers and are found in drier, more ephemeral, ecological settings, where there is limited time to grow large and build large flowers. There has been extensive research on the molecular switch to flowering in model genetic systems; this research on a native plant species suggests that studying natural variation in developmental traits may be a productive way to link molecular genetic and ecological approaches to research on plant adaptation.

"The tubes of the corollas of the common red and incarnate clovers (Trifolium pratense and incarnatum) do not on a hasty glance appear to differ in length; yet the hive-bee can easily suck the nectar out of the incarnate clover, but not out of the common red clover, which is visited by humble-bees alone. I can understand how a flower and a bee might slowly become, either simultaneously or one after the other, modified and adapted in the most perfect manner to each other."

Charles Darwin, *On the Origin of Species*, 1859, pp. 94–95.

Introduction

The classic approach to the study of floral traits has been to assume that they result from reciprocal selection between plants and specific pollinators—so-called pollination syndromes. Pollinators have been invoked to explain the adaptive value of floral colors, nectar rewards, flower shape and more, and so have been considered one of the primary agents of plant speciation (Stebbins 1970; Crepet 1984). Although the importance of genetic drift and evolutionary constraints is acknowledged in the modern literature, the importance of pollinators for divergence in floral traits (and ultimately, local adaptation and speciation) is usually considered paramount (Thomson and Wilson 1996), especially when there is evidence of spatial or temporal variation in the pollinator community (Galen 1989; Galen and Stanton 1989; Schemske and Horvitz 1989). The recent integration of studies of plant-pollinator relationships with QTL analysis of relevant floral traits gives us a more complete understanding of the relationship between pollinator choice and floral trait change (Hodges and Arnold 1994; Bradshaw et al. 1995; Schemske and Bradshaw 1999; Fishman et al. 2002; and see chapters this volume), with the emphasis remaining on the importance of pollinators for the process of plant speciation.

Less attention has been paid to the importance of the abiotic environment for floral evolution. A classic example of the importance of edaphic factors for mating system traits is the difference in phenology and selfing rate between plants growing on mine tailings and nearby plants on normal soil (Antonovics 1967). Populations of *Lasthenia californica* are also partially reproductively isolated due to flowering phenology, related to edaphic differences in soil ions and moisture availability (Rajakaruna and Whitton, this volume, Chapter 13). Evidence also exists for serpentine systems, in which flowering time has been shown to differ for *Collinsia sparsiflora* growing on serpentine and non-serpentine patches (J. W. Wright, M. L. Stanton, and R. Scherson, unpublished data). Moreover, moisture limitation reduces herkogamy and the outcrossing rate in some hermaphroditic species (Schoen 1982; Holtsford and Ellstrand 1992; Elle and Hare 2002), and is also associated with gender dimorphism (Sakai and Weller 1999; Case and Barrett 2001). Some related taxa differ not only in pollination syndrome, but also in local soil nutrient and moisture conditions, e.g., *Aquilegia formosa* and *A. pubescens* (Grant 1952; Chase and Raven 1975; Hodges and Arnold 1994), and *Lobelia cardinalis* and *L. siphilitica* (Caruso et al. 2003); the relative importance of edaphic factors and the pollinator community for divergence of these species pairs is unclear.

Geographic variation in floral traits has been documented for many species, and has been linked to both biotic and abiotic selection pressures. Within species, larger flowers often occur in populations that grow in wetter or less marginal habitats (Rick et al. 1977; Eckhart and Geber 1999). This pattern may be due to the cost of flower maintenance in arid environments; a survey of floral longevity showed that long-lived flowers tended to occur in cool, moist areas, and within

Elizabeth Elle. Department of Biological Sciences, Simon Fraser University, Burnaby, BC V5A 1S6, Canada.
e-mail address: eelle@sfu.ca
Correct citation: Elle, E. 2004. Floral adaptations and biotic and abiotic selection pressures. *In* Plant Adaptation: Molecular Genetics and Ecology. Proceedings of an International Workshop held December 11–13, 2002, in Vancouver, British Columbia, Canada. *Edited by* Q.C.B. Cronk, J. Whitton, R.H. Ree, and I.E.P. Taylor. NRC Research Press, Ottawa, Ontario. pp. 111–118.

variable taxa (though not across all taxa and habitats) large-flowered forms lasted longest (Primack 1985). Alternatively, there may be selection pressures peculiar to the short growing season found arid habitats, such that rapidly developing forms, which tend to be small-flowered, perform well there (Guerrant 1989, and see below). If short growing seasons are important for variation in flower size, one might also expect flower size to decrease with increasing altitude, which has been shown for some species (Grant 1971; Jonas and Geber 1999; Totland 2001). Cases in which flower size increases rather than decreases with altitude have been explained by altitudinal variation in species composition of the pollinator community (Galen et al. 1987; Scobell and Scott 2002), in part because pollinator body size may increase with altitude due to the effect of temperature on energy budgets (Malo and Baonza 2002). Latitudinal variation is more complex; an increase in flower size occurs with more northerly latitudes for *Clarkia unguiculata* (Jonas and Geber 1999), flowering is later and flower number lower at northerly latitudes in *Prunella vulgaris* (Winn and Gross 1993), and complex latitudinal variation in flowering time has been documented for *Beta vulgaris* (Van Dijk et al. 1997). In each case, latitudinal variation is hypothesized to vary with the length of the growing season, and its potential interaction with plant developmental constraints. Overall, then, abiotic explanations for geographic patterns (climatic, altitudinal, and latitudinal) in floral traits are somewhat more common than biotic explanations.

Evidence regarding the importance of abiotic factors for floral trait variation (above) suggests that the study of the evolution of floral traits would be greatly enhanced by broader thinking about the selective pressures that affect flowers. Pollinators are important, but should be only one aspect of a comprehensive research program that seeks to move past the obvious in an effort to understand floral adaptation. Because most research focuses on one or a few populations of plants for logistical reasons, the sort of geographic variation discussed above is often overlooked. I discuss a special case of floral adaptation: adaptation for self-pollination; and use my current research on the evolution of selfing to illustrate how including a geographic component (and the resulting abiotic variation) in research programs can aid in our understanding of floral evolution.

A special case: self-pollination

Biotic selection

Most plant species have mixed mating systems, rather than being strictly selfing or outcrossing (Vogler and Kalisz 2001), and variance in the relative amount of selfing is expected to reflect some balance between its costs and benefits. Historically, there has been an emphasis on genetic explanations for mixed mating systems, including the gene transmission advantage of mixed mating individuals relative to obligate outcrossers (Fisher 1941) and the inbreeding depression that is expected with increasing homozygosity (Lande and Schemske 1985; Charlesworth and Charlesworth 1987; Uyenoyama et al. 1993). Ecological arguments regarding the evolution of selfing are more recent. Costs include pollen (or seed) discounting, in which the production of selfed offspring reduces the number of gametes available for outcrossing (Holsinger et al. 1984; Lloyd and Schoen 1992). The discount is primarily important when inbreeding depression occurs, such that selfed offspring provide a reduced fitness benefit relative to outcrossed offspring. Possible ecological benefits of self pollination include an increase in reproductive assurance (autonomous seed production) when pollinators are rare or absent (especially important in marginal habitats, see Stebbins 1957), and an increase in colonization ability (because neither pollinators nor mating partners are required for seed production by autonomous selfers, see Baker 1955). Because these ecological benefits are tied to the (biotic) pollination environment, the importance of autonomy (and specifically reproductive assurance) should be habitat-specific (Schoen et al. 1996).

Both theory and data support the idea that selfing species are good colonizers (Baker 1955; Lloyd 1979; Price and Jain 1981; Barrett 1996). Island populations tend to have reduced allozyme diversity relative to mainland populations (Glover and Barrett 1987; Inoue et al. 1996; Frankham 1997) and selfing phenotypes (small flowers, less herkogamy) often occur on islands or at species range margins (Stebbins 1957; Schoen 1982; Barrett et al. 1989; Inoue et al. 1996). In *Clarkia xantiana*, a small-flowered, selfing subspecies is found in marginal areas where specialist *Clarkia* bees are uncommon (Runions and Geber 2000; Fausto et al. 2001). Other studies suggest that pollinator rarity is important for selfing but have limited data on the pollinator community (Schoen and Brown 1991; Ramsey and Vaughton 1996), and few studies have done the relevant manipulative experiments (i.e., emasculation) to clearly demonstrate the link between pollinator scarcity and increased selfing ability. In three manipulative studies of reproductive assurance, one (Eckert and Schaefer 1998) found no reproductive assurance benefit for *Aquilegia canadensis*, and another (Culley 2002) documented a reproductive assurance benefit in chasmogamous flowers of *Viola pubescens*. In both cases flowers are morphologically outcrossing but pollinator service is apparently unreliable. The third study, described below, also documented a reproductive assurance benefit, and was the first to show this benefit varied with floral phenotype (corolla size; Elle and Carney 2003). In general, the question of whether "marginality" is important through effects on pollinator service, as suggested here, or through effects of a more extreme or limiting abiotic environment, is not always clear (see next section).

I have used *Collinsia parviflora* (Scrophulariaceae) as a model system to investigate the evolution of selfing. This winter annual is common on vernally moist to dry grassy slopes, mossy rock outcrops, and beaches in western North America (Douglas et al. 1998). In my study area on Vancouver Island, British Columbia, flowering is initiated in late March through early May, depending on location. In BC, there is significant among-population variation in flower size in natural populations (Ganders and Krause 1986; Elle and Carney 2003; and see Table 1), which makes this species ideal for a study of how flower size variation is related to autonomous selfing ability and the pollinator community. Inter-population variation also begs the question of the taxonomy of the BC material; USA populations are categorized into two diploid species (*C. parviflora* and *C. grandiflora*) distinguished by flower size, but in BC all populations are

Table 1. Means ± SE of phenotypic traits measured for up to three offspring from up to 24 maternal families from seven populations of *C. parviflora* from British Columbia (sample size for variables below ranges from 49 to 70 plants within populations). Days to germination are the number of days between planting and emergence. Days to first flower are the number of days between germination and the first open flower on the plant. Autonomous selfing is (fruit set/flower production)*100. Means with the same letter are not significantly different as determined with Ryan's Q post-hoc test. Annual precipitation data are from Environment Canada, and are Climate Normals from the years 1971 to 2000 (JP and RB are 1981–2000; NH 1971–1992) from the climate station nearest to the study population.

Population and abbreviation		Corolla width (mm)	Corolla length (mm)	Days to germination	Days to first flower	Total flower production	Autonomous selfing (%)	Final plant height (cm)	Final plant dry weight (mg)	Annual precip. (mm)
Cowichan River	CR	8.15 ± 0.16 b	7.29 ± 0.11 b	11.66 ± 0.46 c	101.96 ± 1.69 a	24.25 ± 2.03 d	55.25 ± 3.56 c	12.86 ± 0.83 ab	407.3 ± 17.07 a	2022.2
Elk Falls	EF	9.47 ± 0.12 a	7.67 ± 0.10 a	12.01 ± 0.39 c	92.74 ± 1.05 b	29.87 ± 1.75 d	63.97 ± 2.42 c	14.27 ± 0.56 a	440.71 ± 13.21 a	1451.5
Jack Point	JP	7.08 ± 0.14 c	5.20 ± 0.08 c	15.08 ± 1.07 b	64.90 ± 1.34 d	47.02 ± 2.96 ab	74.50 ± 2.81 b	10.68 ± 0.70 b	336.02 ± 14.26 b	1140.9
Kin Beach	KB	6.61 ± 0.14 d	4.70 ± 0.09 d	14.24 ± 0.53 b	71.72 ± 1.72 c	38.22 ± 2.44 c	64.04 ± 3.26 c	12.18 ± 0.91 b	324.95 ± 16.02 b	1179.0
Nanoose Hill	NH	3.80 ± 0.06 f	3.62 ± 0.07 f	14.76 ± 1.17 b	57.88 ± 1.06 ef	42.73 ± 2.17 bc	83.18 ± 1.56 a	5.97 ± 0.31 c	200.23 ± 9.41 d	937.8
Rathtrevor Beach	RB	4.90 ± 0.08 e	3.84 ± 0.07 e	22.11 ± 0.90 a	61.43 ± 1.47 de	26.11 ± 2.33 d	86.64 ± 1.16 a	4.94 ± 0.34 c	175.77 ± 10.66 d	1098.5
Thetis Lake	TL	3.64 ± 0.04 f	3.32 ± 0.05 g	14.33 ± 1.24 b	56.84 ± 0.69 f	50.59 ± 3.04 a	79.94 ± 1.51 ab	10.46 ± 0.43 b	232.92 ± 9.15 c	911.5

tetraploid and "intermediate" -sized populations are common (Ganders and Krause 1986; Elle and Carney 2003). Among-population variation in flower size in B.C. may have arisen via hybridization between the two diploid species in the USA, followed by either differential colonization of sites or selection *in situ*. An analysis of the phylogeography of the species is planned; here I describe experiments aimed at understanding the importance of biotic (below) and abiotic (next section) selection for the observed population differentiation (Table 1).

In spring 2001, we performed a suite of experiments to assess the pollination environment and quantify reproductive assurance in *C. parviflora* (Elle and Carney 2003). When given a choice between large-flowered and small-flowered plants in experimental arrays, pollinators at all study sites (including sites where the local population of *C. parviflora* was small-flowered, and a site where it was large-flowered) over-visited large-flowers, suggesting that pollinator-mediated selection should lead to larger flower size in all populations. However, pollinator visitation rate is lower in small-flowered *C. parviflora* populations, consistent with the idea that small-flowered, putatively selfing populations evolved in areas where pollinators are limited. To determine whether selfing provides reproductive assurance in this species, and whether reproductive assurance varied with flower size, we manipulated the ability of plants to self autonomously. Using experimental arrays of potted plants, we compared seed production from plants from the largest- (EF and CR) and smallest- (TL and NH) flowered natural populations known (Elle and Carney 2003; and see Table 1). We emasculated two flowers on each of 47 large-flowered and 47 small-flowered plants in experimental arrays, and compared seed set on emasculated flowers to seed set on two control flowers on each plant. The experiment was conducted within both a large-flowered and a small-flowered *C. parviflora* population. Relative to control flowers, emasculated flowers of small-flowered plants produced very few seeds, but emasculated and intact (control) flowers on large-flowered plants produced similar numbers of seeds (Fig. 1). That is, large flowers were capable of full seed set (via outcrossing) when emasculated, but small flowers were not, and relied on autonomous selfing for full seed set. Thus, selfing does provide reproductive assurance for *C. parviflora*, and the reproductive assurance benefit varies, as predicted, with flower size (Elle and Carney 2003). There was also a trend for greater seed production in emasculated flowers of the size that "matched" the background population's flower size; that is, reproductive assurance via selfing was reduced under conditions when pollinators were potentially more efficient (due to some combination of visitation rates and pollen placement, Elle and Carney 2003). The relationship between flower size and autonomous selfing ability under natural pollination conditions is evidence of the importance of biotic selection for floral traits in *C. parviflora*. Despite the preponderance of hypotheses that explicitly link floral adaptations to autonomy, evidence such as that presented here is rare.

Abiotic selection

Although pollinator limitation is perhaps the most obvious place to look when studying floral adaptations and the evo-

Fig. 1. Seed production of emasculated flowers relative to intact flowers on large- and small-flowered *C. parviflora* under natural pollination conditions. $N = 47$ plants for each flower size category, two flowers per treatment per plant.

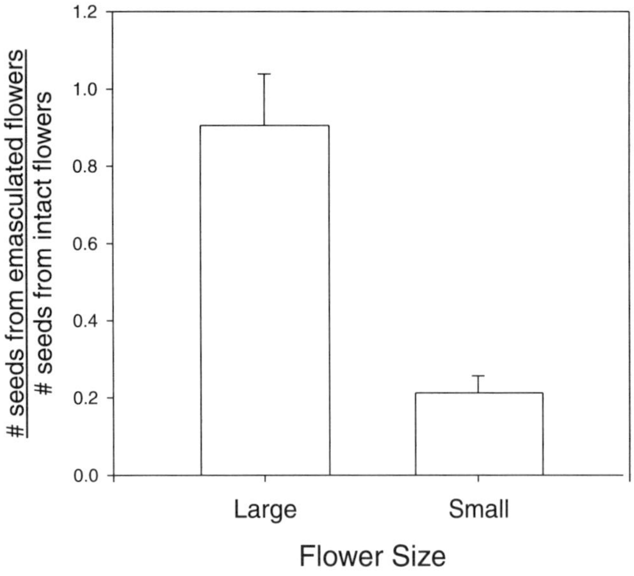

lution of selfing, it is possible that selfing is instead an evolutionary response to abiotic selection pressures. An understudied idea is that there may be selection for rapid plant development in ephemeral habitats, such as deserts that dry out rapidly after a short window of opportunity for plant growth and reproduction (Guerrant 1989; Diggle 1992; Aarssen 2000). Plants with rapid development often express paedomorphosis, a 'juvenilization' of floral traits such that plants that rapidly attain sexual maturity produce flowers similar in appearance to buds of more slowly developing relatives (Diggle 1992; Li and Johnston 2000). Rather than selection acting directly to decrease flower size, selection for rapidity may therefore indirectly result in the production of small flowers with sexual parts in close proximity, which subsequently are capable of autonomous selfing. That is, selfing may have evolved due to time limitation (Aarssen 2000).

An association between time to flowering and flower size has been shown in several plant species. The evolution of cleistogamous flowers from chasmogamous flowers is one example; cleistogamous flowers often resemble chasmogamous buds and have earlier sexual maturity (Mayers and Lord 1983, reviewed in Li and Johnston 2000; Porras and Munoz 2000). Several comparisons within and among species suggest that selfers have more rapid development than outcrossers, and are associated with more arid or more ephemeral habitats (Vasek 1964; Moore and Lewis 1965; Solbrig and Rollins 1977; Guerrant 1989, reviewed in Diggle 1992; Hill et al. 1992). Recently, Runions and Geber (2000) found accelerated sexual development in small-flowered, selfing *Clarkia xantiana* ssp. *parviflora*, relative to outcrossing *C. xantiana* ssp. *xantiana*, confirming earlier results of Moore and Lewis (1965). *Clarkia xantiana* ssp. *parviflora* is known to occur in drier, more marginal desert environments than ssp. *xantiana*, as noted above (Eckhart and Geber 1999), where specialist pollinators are rare (Fausto et al. 2001). The relative importance of drought and low pollinator service is difficult to ascertain, and indeed the two may be linked in *Clarkia* and other taxa. In winter annuals that flower very early in the spring, there is the potential for a trade-off between flowering early, when pollinators are less reliably present but there is time for seed maturation (after autogamy), or flowering later, when it is warm enough for increased pollinator activity (and thus outcrossing is more likely), but time during which the environment will remain tenable is limited. It is probable that the amount of inbreeding depression in the taxon, and the predictability of time limitation, will contribute to developmental solutions exhibited by different plant species.

In *C. parviflora*, I examined the hypothesis that small-flowered plants should reach reproductive maturity more quickly and self autonomously at a higher rate than large-flowered plants. Seeds were collected from seven natural populations that differed significantly in the width of the two banner petals in spring of 2000, and raised in a growth chamber to control for environmental influences on flower size and development time. Three seeds from each of 24 field-collected sibships per population were germinated under short days in a Conviron growth chamber (20°C/10 h day, 10°C/14 h night), and switched to long days (20°C/16 h day, 10°C/8 h night) eight weeks later. I scored the following traits on all plants: days to germination, days to first flower, corolla width and length, total flowers produced, total fruits produced (via autogamy), final plant height, and final aboveground dry weight. I derived the following variables for analysis: development time (days to first flower — days to germination) and autonomous selfing (total fruits produced/total flowers produced).

I used analysis of variance (ANOVA) to determine whether development time, autonomous selfing, vegetative size, and flower size differed among populations and families within populations. All variables were square root transformed to improve homogeneity of variances, except autonomous selfing, which was arcsine square root transformed (Sokal and Rohlf 1995). Population differences were evaluated by testing the population mean square against the family within population mean square. Differences among families within populations were evaluated by testing against the error, and are interpreted as evidence of a genetic component of phenotypic variation. Pearson correlation coefficients among all traits for all plants (N = 400+ plants, see Table 2) were calculated to estimate phenotypic correlations; sequential Bonferroni adjustments to an alpha significance value of 0.05 were subsequently performed (Rice 1989). Correlations among flower size, development time, and autonomous selfing are of primary interest in evaluating my hypothesis. To estimate whether there were genetic correlations between flower size and development time, I calculated trait means across the three plants per family within populations, and subsequently performed linear regression with corolla width as the independent variable and development time as the dependent variable for each population separately. SAS was used for all analyses.

There were significant differences among populations (all $P < 0.0001$) and among families within populations (all $P < 0.0008$) for all measured traits, indicating that phenotypic variation in the traits measured here has a genetic basis in *C. parviflora*. As expected, the highest level of autonomous

Table 2. Phenotypic correlations among all plant traits measured. See Table 1 and text for explanation of variables. Sample size is 431 plants for germination and total flower production, 415 plants for final plant weight, 414 plants for autonomous selfing rate, 413 plants for development time, 412 plants for corolla width and length, and 410 plants for height. An asterisk (*) denotes significance at $P <$ 0.05 after sequential Bonferroni adjustment of significance.

	Corolla length	Days to germination	Days to first flower	Total flower production	Autonomous selfing (%)	Final plant height	Final plant dry weight
Corolla width (mm)	0.911*	–0.177*	0.736*	–0.299*	–0.382*	0.470*	0.619*
Corolla length (mm)		–0.218*	0.785*	–0.343*	–0.388*	0.475*	0.584*
Days to germination			–0.274*	–0.204*	0.205*	–0.210*	–0.416*
Days to first flower				–0.410*	–0.390*	0.416*	0.583*
Total flower production					–0.032	0.128*	0.159*
Autonomous selfing (%)						–0.335*	–0.405*
Final plant height (cm)							0.452*

selfing was found in the smallest-flowered populations. Populations with smaller flowers germinated somewhat later, but flowered 1–2 months sooner than the largest-flowered populations in the common environment (Table 1).

Phenotypic correlations among all traits indicated that relatively small-flowered plants germinated later, flowered sooner, produced more flowers, had a higher level of autonomous selfing, and had a smaller final vegetative size than relatively large-flowered plants (Table 2). When data from all populations were combined, there was a positive relationship between flower size and development time (Fig. 2). There was little within-population variation for most traits, however, and when family means were calculated and the relationship between flower size and development time was assessed using regression, significance was only attained for two populations, and marginal significance for a third. For KB and JB, there was a significant positive relationship between mean flower size and mean development time (JP: $R^2 = 0.21$, $P =$ 0.046, N = 19 families; KB: $R^2 = 0.26$, $P = 0.015$, N = 22 families). For CR, there was a marginally significant negative relationship between these variables ($R^2 = 0.12$, $P =$ 0.065, N = 22; Fig. 3). The largest-flowered population (EF) and the three smallest-flowered populations (TL, NH, RB) exhibited no relationship between flower size and development time, suggesting that correlated evolution between these traits is unlikely in the future, though I cannot evaluate whether it occurred in the past. Finally, there was a positive relationship between mean annual precipitation at nearby climate stations (data from Environment Canada: climate normals 1971–2000; URL: http://www.climate.weatheroffice.ec.gc.ca /) and flower size in my study populations (Table 1 and Fig. 4). Precipitation levels fall below an arbitrary cutoff of 50 mm per month in April for TL, May for RB, NH, KB, and JP, and July for EF and CR.

My results suggest that an abiotic factor, precipitation, may be important for local differentiation in flower size and development time. Large-flowered plants with long development times are found in wetter areas in British Columbia,

Fig. 2. *Collinsia parviflora* with larger flowers have more days between germination and sexual maturity in the growth chamber than plants with smaller flowers. $N = 412$ plants across seven populations, $R^2 = 0.53$, $P < 0.0001$.

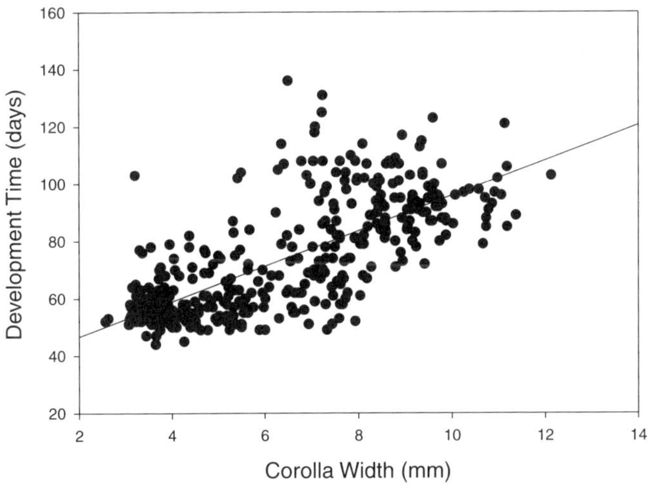

Fig. 3. The relationship between mean flower size and mean development time for growth-chamber-grown maternal families within seven populations of *C. parviflora*. The regression was only significant (or marginally so) for three of the seven populations studied, CR (solid line, $P = 0.07$), JP (dotted line, $P =$ 0.03) and KB (dashed line, $P = 0.02$; see text for statistics). Regression lines are not indicated for populations where the results were not significant (EF, NH, RB, TL).

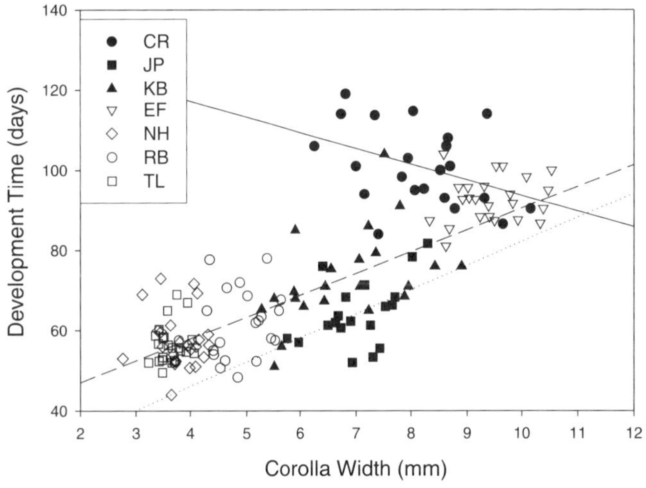

Fig. 4. Large-flowered populations occur in wetter habitats on Vancouver Island. Precipitation data is from Environment Canada, and corolla width was measured on plants raised in a growth chamber (see Table 1). $N = 7$ populations, $R^2 = 0.58$, $P = 0.048$.

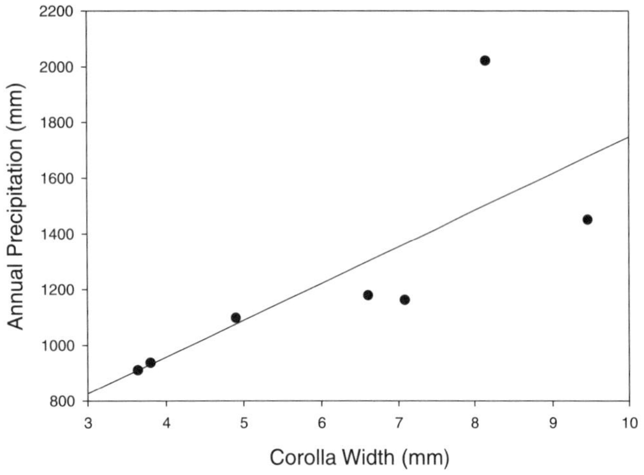

where there is more time for reproduction prior to the environment becoming untenable, a major premise of the time limitation hypothesis. The lack of a relationship between flower size and development time within the largest- and smallest-flowered populations could indicate that these populations are genetically fixed, in contrast to populations with more intermediate flower size, where there is the potential for further correlated evolution of size and timing of flowers. Future work will address whether differences among populations in the strength and direction of the relationship between flower size and development time are robust, and if so, will explore possible explanations for this pattern. One hypothesis for the observed reduction in flower size with development time in CR is that this population occurs in a very small fragment of appropriate habitat, and so plants may be experiencing inbreeding depression.

In *C. parviflora*, both biotic and abiotic selection may be important determinants of significant among-population variation in flower size and the autonomous selfing rate. The data presented here do not explicitly address whether local phenotypes are adaptive; a common-garden field experiment in different habitats is planned to address that point. However, the data do illustrate the importance of broadening one's scope to consider abiotic factors, a point suggested by Rajakaruna and Whitton (this volume, Chapter 13), as well as aspects of plant development, when studying the evolution of flower form.

Concluding comments

Interest in heterochrony (an evolutionary change in the phenotype due to a change in the relative timing of developmental processes) has increased in recent years, as we have begun to acknowledge its importance for the production of evolutionary novelties in plants (Li and Johnston 2000). At the same time, the pace of acquisition of new information about plant developmental processes has become almost exponential. Primarily through developmental genetics research on *Arabidopsis thaliana*, we know more than ever about when plants switch to flowering, and the molecular signals involved in the switch (Simpson et al. 1999; Araki 2001; Simpson and Dean 2002). Interestingly, there appears to be a relationship between development time and plant size for at least some of the genes studied (e.g., Johanson et al. 2000; Wang et al. 2000). Research in natural plant populations likewise suggests that the development/phenotype connection may be an important aspect of ecological adaptation.

We have made great strides towards understanding the genetic architecture of floral traits that are important in pollinator-mediated adaptation and speciation, but our understanding of the underlying developmental pathways for most of the floral traits in question is limited. In contrast, we know quite a bit about the molecular genetics of numerous developmental pathways, but these have primarily been studied in an ecological vacuum. A bridge between single-gene research and the continuous variation that is common in natural populations of species like *C. parviflora* has yet to be made, and may hold the key to the integration of molecular and ecological methodologies for the study of plant adaptation.

Acknowledgements

R. Carney's contribution to the reproductive assurance experiment is gratefully acknowledged. D. Bohn, J. Kumagai, and especially R. Woychuk provided invaluable assistance with the growth chamber experiment, and J. Biernaskie, A. Parachnowitsch, J. Whitton, and two anonymous reviewers provided constructive feedback on earlier versions of this manuscript. This research was supported by a grant from the Natural Sciences and Engineering Research Council of Canada.

References

Aarssen, L.W. 2000. Why are most selfers annuals? A new hypothesis for the fitness benefit of selfing. Oikos **89**: 606–612.

Antonovics, J. 1967. Evolution in closely adjacent plant populations. V. Evolution of self-fertility. Heredity **23**: 219–238.

Araki, T. 2001. Transition from vegetative to reproductive phase. Curr. Opin. Plant Biol. **4**: 63–68.

Baker, H.G. 1955. Self-compatibility and establishment after "long-distance" dispersal. Evolution **9**: 347–348.

Barrett, S.C.H. 1996. The reproductive biology and genetics of island plants. Phil. Trans.Roy. Soc. Lond. B Biol. Sci. **351**: 725–733.

Barrett, S.C.H., Morgan, M.T., and Husband, B.C. 1989. The dissolution of complex genetic polymorphism: the evolution of self-fertilization in tristylous *Eichhornia paniculata* (Pontederiaceae). Evolution **43**: 1398–1416.

Bradshaw, H.D.J., Wilbert, S.M., Otto, K.G., and Schemske, D.W. 1995. Genetic mapping of floral traits associated with reproductive isolation in monkeyflowers (*Mimulus*). Nature **376**: 762–765.

Case, A.L., and Barrett, S.C.H. 2001. Ecological differentiation of combined and separate sexes of *Wurmbea dioica* (Colchicaceae) in sympatry. Ecology **82**: 2601–2616.

Charlesworth, D., and Charlesworth, B. 1987. Inbreeding depression and its evolutionary consequences. Annu. Rev. Ecol. Syst. **18**: 237–268.

Chase, V.C., and Raven, P.H. 1975. Evolutionary and ecological relationships between *Aquilegia formosa* and *A. pubescens* (Ranunculaceae), two perennial plants. Evolution **29**: 474–486.

Crepet, W.L. 1984. Advanced (constant) insect pollination mechanisms: pattern of evolution and implications vis-a-vis Angiosperm diversity. Ann. Mo. Bot. Gard. **71**: 607–630.

Culley, T.M. 2002. Reproductive biology and delayed selfing in *Viola pubescens* (Violaceae), an understory herb with chasmogamous and cleistogamous flowers. Int. J. Plant Sci. **163**: 113-122.

Diggle, P.K. 1992. Development and the evolution of plant reproductive characters. *In* Ecology and Evolution of Plant Reproduction. *Edited by* R. Wyatt. Chapman & Hall, New York. pp. 326–355.

Douglas, G.W., Meidinger, D.V., and Pojar, J. 1998. Illustrated flora of British Columbia. Ministry of Environment, Lands, and Parks, Victoria, BC.

Eckert, C.G., and Schaefer, A. 1998. Does self–pollination provide reproductive assurance in *Aquilegia canadensis* (Ranunculaceae)? Am. J. Bot. **85**: 919-924.

Eckhart, V.M., and Geber, M.A. 1999. Character variation and geographic distribution of *Clarkia xantiana* A. Gray (Onagraceae): flowers and phenology distinguish two subspecies. Madrono **46**: 117–125.

Elle, E., and Carney, R. 2003. Reproductive assurance varies with flower size in *Collinsia parviflora* (Scrophulariaceae). Am. J. Bot. **90**: 888–896.

Elle, E., and Hare, J.D. 2002. Environmentally induced variation in floral traits affects the mating system in *Datura wrightii*. Funct. Ecol. **16**: 79–88.

Fausto, J.A.J., Eckhart, V.M., and Geber, M.A. 2001. Reproductive assurance and the evolutionary ecology of self-pollination in *Clarkia xantiana* (Onagraceae). Am. J. Bot. **88**: 1794–1800.

Fisher, R.A. 1941. Average excess and average effect of a gene substitution. Ann. Eugen. **11**: 53–63.

Fishman, L., Kelly, A.J., and Willis, J.H. 2002. Minor quantitative trait loci underlie floral traits associated with mating system divergence in *Mimulus*. Evolution **56**: 2138–2155.

Frankham, R. 1997. Do island populations have less genetic variation than mainland populations? Heredity **78**: 311–327.

Galen, C. 1989. Measuring pollinator-mediated selection on morphometric floral traits: bumble bees and the alpine skypilot, *Polemonium viscosum*. Evolution **43**: 882–890.

Galen, C., and Stanton, M.L. 1989. Bumble bee pollination and floral morphology: factors influencing pollen dispersal in the alpine skypilot, *Polemonium viscosum* (Polemoniaceae). Am. J. Bot. **76**: 419–426.

Galen, C., Zimmer, K.A., and Newport, M.E. 1987. Pollination in floral scent morphs of *Polemonium viscosum*: a mechanism for disruptive selection on flower size. Evolution **41**: 599–606.

Ganders, F.R., and Krause, G.R. 1986. Systematics of *Collinsia parviflora* and *C. grandiflora* (Scrophulariaceae). Madrono **33**: 63–70.

Glover, D.E., and Barrett, S.C.H. 1987. Genetic variation in continental and island populations of *Eichhornia paniculata* (Pontederiaceae). Heredity **59**: 7–17.

Grant, V. 1952. Isolation and hybridization between *Aquilegia formosa* and *A. pubescens*. El Aliso **2**: 341–360.

Grant, V. 1971. Plant Speciation. Columbia University Press, New York.

Guerrant, E.O.J. 1989. Early maturity, small flowers, and autogamy: A developmental connection? *In* The Evolutionary Ecology of Plants. *Edited by* J.H. Bock and Y.B. Linhart. Westview Press, Boulder, CO. pp. 61–84.

Hill, J.P., Lord, E.M., and Shaw, R.G. 1992. Morphological and growth rate differences among outcrossing and self-pollinating races of *Arenaria uniflora* (Caryophyllaceae). J. Evol. Biol. **5**: 559–573.

Hodges, S.A., and Arnold, M.L. 1994. Floral and ecological isolation between *Aquilegia formosa* and *Aquilegia pubescens*. Proc. Natl. Acad. Sci. U.S.A. **91**: 2493–2496.

Holsinger, K.E., Feldman, M.W., and Christiansen, F.B. 1984. The evolution of self fertilization in plants: a population genetic model. Am. Nat. **124**: 446–453.

Holtsford, T.P., and Ellstrand, N.C. 1992. Genetic and environmental variation in floral traits affecting outcrossing rate in *Clarkia tembloriensis* (Onagraceae). Evolution **46**: 216–225.

Inoue, K., Maki, M., and Masuda, M. 1996. Evolution of *Campanula* flowers in relation to insect pollinators on islands. *In* Floral Biology: studies on floral evolution in animal-pollinated plants. *Edited by* D.G. Lloyd and S.C.H. Barrett. Chapman and Hall, New York. pp. 377–400.

Johanson, U., West, J., Lister, C., Michaels, S., Amasino, R., and Dean, C. 2000. Molecular analysis of *FRIGIDA*, a major determinant of natural variation in *Arabidopsis* flowering time. Science (Washington, D.C.) **290**: 344–347.

Jonas, C.S., and Geber, M.A. 1999. Variation among populations of *Clarkia unguiculata* (Onagraceae) along altitudinal and latitudinal gradients. Am. J. Bot. **86**: 333–343.

Lande, R., and Schemske, D.W. 1985. The evolution of self-fertilization and inbreeding depression in plants. I. Genetic models. Evolution **39**: 24–40.

Li, P., and Johnston, M.O. 2000. Heterochrony in plant evolutionary studies through the twentieth century. Bot. Rev. **66**: 57–88.

Lloyd, D.G. 1979. Some reproductive factors affecting the selection of self-fertilization in plants. Am. Nat. **113**: 67–79.

Lloyd, D.G., and Schoen, D.J. 1992. Self- and cross-fertilization in plants. I. Functional dimensions. Int. J. Plant Sci. **153**: 358–369.

Malo, J.E., and Baonza, J. 2002. Are there predictable clines in plant-pollinator interactions along altitudinal gradients? The example of *Cytisus scoparius* (L.) Link in the Sierra de Guadarrama (Central Spain). Diversity Distrib. **8**: 365–371.

Mayers, A.M., and Lord, E.M. 1983. Comparative flower development in the cleistogamous species *Viola odorata*. I. A growth rate study. Am. J. Bot. **70**: 1548–1555.

Moore, D.M., and Lewis, H. 1965. The evolution of self-pollination in *Clarkia xantiana*. Evolution **19**: 104–114.

Porras, R., and Munoz, J.M. 2000. Cleistogamous capitulum in *Centaurea melitensis* (Asteraceae): heterochronic origin. Am. J. Bot. **87**: 925–933.

Price, S.C., and Jain, S.K. 1981. Are inbreeders better colonizers? Oecologia **49**: 283–286.

Primack, R.B. 1985. Longevity of individual flowers. Annu. Rev. Ecol. Syst. **16**: 15–37.

Ramsey, M., and Vaughton, G. 1996. Inbreeding depression and pollinator availability in a partially self-fertle perennial herb *Blandfordia grandiflora* (Liliaceae). Oikos **76**: 465–474.

Rice, W.R. 1989. Analyzing tables of statistical tests. Evolution **43**: 223–225.

Rick, C.M., Fobes, J.F., and Holle, M. 1977. Genetic variation in *Lycopersicon pimpinellifolium*: Evidence of evolutionary change in mating systems. Plant Syst. Evol. **127**: 139–170.

Runions, C.J., and Geber, M.A. 2000. Evolution of the self-pollinating flower in *Clarkia xantiana* (Onagraceae). I. Size and development of floral organs. Am. J. Bot. **87**: 1439–1451.

Sakai, A.K., and Weller, S.G. 1999. Gender and sexual dimorphism in flowering plants: a review of terminology, biogeographic patterns, ecological correlates, and phylogenetic approaches. *In*

Gender and sexual dimorphism in flowering plants. *Edited by* M.A. Geber, T.E. Dawson, and L.F. Delph. Springer-Verlag, Berlin. pp. 1–33.

Schemske, D.W., and Bradshaw, H.D.J. 1999. Pollinator preference and the evolution of floral traits in monkeyflowers (*Mimulus*). Proc. Natl. Acad. Sci. U.S.A. **96**: 11910–11915.

Schemske, D.W., and Horvitz, C.C. 1989. Temporal variation in selection on a floral character. Evolution **43**: 461–465.

Schoen, D.J. 1982. The breeding system of *Gilia achilleifolia*: variation in floral characteristics and outcrossing rate. Evolution **36**: 352–360.

Schoen, D.J., and Brown, A.H.D. 1991. Whole- and part-flower self-pollination in *Glycine clandestina* and *G. argyrea* and the evolution of autogamy. Evolution **45**: 1651–1664.

Schoen, D.J., Morgan, M.T., and Bataillon, T.M. 1996. How does self-pollination evolve? Inferences from floral ecology and molecular genetic variation. Phil. Trans. Roy. Soc. Lond. B. Biol. Sci. **351**: 1281–1290.

Scobell, S.A., and Scott, P.E. 2002. Visitors and floral traits of a hummingbird-adapted cactus (*Echinocereus coccineus*) show only minor variation along an elevational gradient. Am. Midl. Nat. **147**: 1–15.

Simpson, G.G., and Dean, C. 2002. *Arabidopsis*, the Rosetta Stone of flowering time? Science (Washington, D.C.) **296**: 285–289.

Simpson, G.G., Gendall, A.R., and Dean, C. 1999. When to switch to flowering. Annu. Rev. Cell Dev. Biol. **99**: 519–550.

Sokal, R.R., and Rohlf, F.J. 1995. Biometry: the principles and practice of statistics in biological research, 3rd ed. W. H. Freeman, New York.

Solbrig, O.T., and Rollins, R.C. 1977. The evolution of autogamy in species of the mustard genus *Leavenworthia*. Evolution **31**: 265–281.

Stebbins, G.L. 1957. Self fertilization and population variability in the higher plants. Am. Nat. **91**: 337–354.

Stebbins, G.L. 1970. Adaptive radiation of reproductive characteristics in angiosperms. I. Pollination mechanisms. Annu. Rev. Ecol. Syst. **1**: 307–326.

Thomson, J.D., and Wilson, P. 1996. How do flowers diverge? *In* Floral Biology: studies on floral evolution in animal-pollinated plants. *Edited by* D.G. Lloyd and S.C.H. Barrett. Chapman & Hall, New York. pp. 88–111.

Totland, O. 2001. Environment-dependent pollen limitation and selection on floral traits in an alpine species. Ecology **82**: 2233–2244.

Uyenoyama, M.K., Holsinger, K.E., and Waller, D.M. 1993. Ecological and genetic factors directing the evolution of self-fertilization. Oxford Surv. Evol. Biol. **9**: 327–381.

Van Dijk, H., Boudry, P., McCombie, H., and Vernet, P. 1997. Flowering time in wild beet (*Beta vulgaris* ssp. *maritima*) along a latitudinal cline. Acta Oecol. **18**: 47–60.

Vasek, F.C. 1964. The evolution of *Clarkia unguiculata* derivatives adapted to relatively xeric environments. Evolution **18**: 26–42.

Vogler, D., and Kalisz, S. 2001. Sex among the flowers: the distribution of plant mating systems. Evolution **55**: 202–204.

Wang, H., Zhou, Y., Gilmer, S., Whitwill, S., and Fowke, L.C. 2000. Expression of the plant cyclin-dependent kinase inhibitor ICK1 affects cell division, plant growth and morphology. Plant J. **24**: 613–623.

Winn, A.A., and Gross, K.L. 1993. Latitudinal variation in seed weight and flower number in *Prunella vulgaris*. Oecologia **93**: 55–62.

15
Polyploidy and plant adaptation: a framework for future research

Brian C. Husband

Abstract: Polyploidy, the multiplication of whole chromosome sets beyond the diploid condition, is a prevalent feature of plants, yet its role in adaptation is poorly explored. Here I consider three aspects of the process of adaptation that can be addressed through research on genome multiplication. First, polyploidy may be a useful tool for understanding the factors governing the rates of and limits to adaptation along ecological gradients. Through its increased capacity for diversity, autopolyploids can provide a useful test of the role of genetic variability in regulating species range limits. Second, with potentially large phenotypic effects, polyploidy offers an opportunity to examine what chromosomal variation contributes, beyond the effects of genic variation, to the response to selection. Until now, chromosomal change has been debated in the context of species divergence, but diploid-polyploid clines may enable researchers to study its role in the adaptive process. Third, polyploidy may provide insights into the role of stochastic forces in adaptive evolution. The evolutionary dynamics of polyploids and diploids are governed by positive frequency-dependent selection and consequently, rare polyploids will not establish unless they can exceed a threshold frequency. Near this threshold point, relatively small changes in cytotype frequency caused by sampling error associated with founder events and small population size can have significant evolutionary consequences. Each of these problems are of general significance to plant adaptation and will benefit from the integration of ecological and genomic approaches.

Introduction

Polyploidy, the multiplication of whole chromosome sets above the diploid condition, is a dominant feature of flowering plants. Up to 70% of all angiosperm species have experienced a genome multiplication at sometime in their evolutionary past, and the number of chromosome copies ranges from three ($2n = 3\times$) to eighty ($2n = 80\times$) (Masterson 1994; Otto and Whitton 2000). In addition, many species exhibit variation in ploidy within and among populations (Lewis 1980). Given its prevalence in the plant kingdom and its strong association with plant physiological, morphological and ecological characteristics (Stebbins 1950; Levin 1983, 2002), polyploidy is central to any discussion on the genetics and ecology of adaptation.

Polyploid ecology has historically been an active area of research. Past interest in the biosystematics of polyploid complexes led many researchers to describe the geographic ranges of polyploids and their diploid progenitors (Stebbins 1950; Ehrendorfer 1980). This work focused attention towards the kinds of polyploids (auto vs. allo) that exist and the ecological, physiological and morphological correlates of such genome multiplications. Starting in the 1980's, the research focus shifted to the population biology of polyploidy, which has concentrated on their pathways of formation (Thompson and Lumaret 1992; Soltis and Soltis 1993, 1999; Ramsey and Schemske 1998) and consequences for establishment and reproductive isolation (Lumaret and Barrientos 1990; Petit et al. 1999; Husband and Schemske 2000; Husband et al. 2002). My own work has explored these issues using species that are variable in ploidy, and where the polyploids are derived from genomes of a single diploid species (i.e., autopolyploid) (Husband and Schemske 1998; Burton and Husband 1999; Husband 2000; Husband and Schemske 2000; Husband et al. 2002). Surprisingly, however, little is known of the extent to which polyploidy can contribute to adaptation along ecological gradients or the population processes by which this occurs.

My approach here is to develop a conceptual framework for future research, rather than to review existing data. My primary questions, then, are, 'How can research on polyploidy provide insights into the process of adaptation along ecological gradients', and 'How will research in molecular biology and genomics facilitate this?' I will argue that research into genome multiplication can be useful for addressing three basic questions in plant adaptation: (1) What are the genetic and ecological limits to adaptive divergence?, (2) Can large major shifts in phenotype due to chromosomal variation contribute to adaptation, beyond typical genic responses to selection?, and (3) When will stochastic processes play a role in adaptive evolution? I will elaborate on each of these points and identify some of the challenges that lie ahead in resolving them. Where possible, I will include selected empirical research from the literature and from my own research on fireweed, *Chamerion angustifolium* (Onagraceae), to illustrate relevant points. This perennial herbaceous plant has a broad geographic distribution and inhabits open and disturbed environments across a wide range of lati-

Brian C. Husband. Department of Botany, University of Guelph, Guelph, ON N1G 2W1, Canada.
e-mail address: bhusband@uoguelph.ca
Correct citation: Husband, B.C. 2004. Polyploidy and plant adaptation: a framework for future research. *In* Plant Adaptation: Molecular Genetics and Ecology. Proceedings of an International Workshop held December 11–13, 2002, in Vancouver, British Columbia, Canada. *Edited by* Q.C.B. Cronk, J. Whitton, R.H. Ree, and I.E.P. Taylor. NRC Research Press, Ottawa, Ontario. pp. 119–126.

tudes and altitudes in the northern hemisphere. Most pertinent, *C. angustifolium* is variable in chromosome number. In North America, it consists of diploid ($2n = 2x = 36$) and autopolyploid ($2n = 3x$, $4x$, $5x$) individuals (Mosquin 1967; Mosquin and Small 1971; Husband and Schemske 1998; H.A. Sabara and B.C. Husband, unpublished) that can coexist in sympatry.

Limits to adaptation

Evolutionary biologists have long been interested in what dictates the rate and extent of adaptation along ecological gradients? In particular, the debate has focused on the role of two broad forces: (1) selection pressures set by the biotic or abiotic environment that drive divergence, or (2) constraints to evolutionary response imposed by the content of the genome (i.e., genetic variability, heritability). This dichotomy is an oversimplification in that they can operate in concert and vary in relative importance, depending on time and location, in regulating a population's response to selection. In their overview of the evolutionary significance of polyploidy, Otto and Whitton (2000) suggest that chromosome doubling may provide insights into the extent to which adaptive divergence is limited by genetic variability. Specifically, by increasing the number of gene copies per genome, polyploidization automatically increases the capacity for genetic diversity within an individual and population. If genetic variability is the primary constraint to adaptation, then increasing gene copy number should elevate the probability of beneficial mutations arising, thus enabling polyploids to expand their ecological ranges beyond the limits of their diploid progenitors. Autopolyploids will be particularly valuable in testing this idea since genome multiplication results in increased copy number without the confounding effects of hybridization.

While numerous data exist regarding the genetic attributes and geographic ranges of diploids and polyploids, rarely has this information been synthesized to address this question. Population genetic surveys of species composed of diploid and tetraploid populations confirm that, indeed, tetraploids are usually genetically more variable than their diploid progenitors (summarized in Table 8.3 in Levin 2002). Based on five mixed-ploidy species (Lumaret 1985, 1988; Ness et al. 1989; Soltis and Soltis 1989; Wolf et al. 1990; Shore 1991), tetraploids had a higher frequency of polymorphic isozyme loci, higher heterozygosity and more alleles per polymorphic locus than their diploid counterparts (Fig. 1). On average, 46% of all loci were polymorphic in tetraploids compared to 34% in diploids, and in tetraploid *Tolmiea menziesii*, 39% of all plants surveyed had three or four alleles in at least one of the eight loci studied (Soltis and Soltis 1989). These genetic patterns suggest that polyploids should be better able than diploids to respond to selection when genetic variability is limiting.

Surprisingly, the difference in genetic diversity observed does not translate into a greater ecological range in polyploids (Fig. 1). In a recent survey, Petit and Thompson (1999) found that, among 50 genera from the Pyrenees, diploids occurred in a larger range of habitats than did polyploids. Stebbins and Dawe (1987) reported no difference in range size between diploid and polyploidy species in the European flora. It is important to note that none of these ecological comparisons have controlled for phylogeny (although Stebbins and Dawe is improved) and thus any similarities in genetic and ecological characteristics among polyploid species or among diploids may be due strictly to common descent. This is important given that Warwick and Gottlieb (1985) found that closely related species often have similar patterns of genetic variability (although see Olmstead 1990). Clearly, there is a need for more quantitative data on the size of ecological ranges in diploids and their polyploid relatives, and the use of phylogenetic information to remove the confounding effects of common ancestry on range size. At this point, however, there is no evidence that genome duplication leads to a general increase in response to selection, leading to range expansion.

Fig. 1. Differences in (top) genetic diversity and (bottom) ecological range for diploid and tetraploid plant species. Genetic diversity values based on Soltis and Soltis 1989, Ness et al. 1989, Wolf et al. 1990, Lumaret 1985,1988, and Shore 1991; ecological range estimates were first reported by Petit and Thompson 1999.

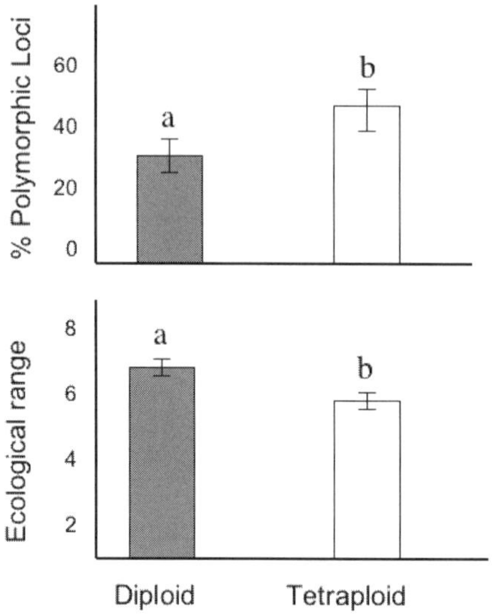

Although their range sizes may not differ, typically, the geographic ranges of polyploids are offset from the ranges of their diploid ancestors. For example, in *Chamerion angustifolium* (Onagraceae), there is little geographic overlap between diploids and autotetraploids within North America. Diploids occur at high latitudes but are replaced by tetraploids toward the southern edge of the Boreal Forest (Fig. 2, Mosquin and Small 1971). Within the overlap zone, in the Rocky Mountains, there is much variation in the frequency of cytotypes, and diploids give way to tetraploids as elevation decreases (Fig. 3, H.A. Sabara and B.C. Husband, unpublished). Thus, the geographic ranges of tetraploids and diploids are of similar breadth but only partially overlapping. As with many species, tetraploids have expanded in only one direction beyond the current boundaries of the diploid range. It is important to acknowledge that we cannot exclude the possibility that diploids once inhabited the current tetraploid range and were excluded after the genome duplication event(s) occurred. In either scenario, the tetraploid is able to compete and persist in a portion of the range that the diploid cannot, which I would argue represents a shift in effective ecological range.

Fig. 2. Geographic range of *Chamerion angustifolium* in North America. Light area is occupied by tetraploids, grey by diploids, dark hatching by both diploids and tetraploids. Modified from Mosquin and Small 1971.

Fig 3. Frequency of diploids in relation to elevation (m) in 46 populations of *Chamerion angustifolium* from the Canadian Rocky Mountains (H.A. Sabara and B.C. Husband unpublished).

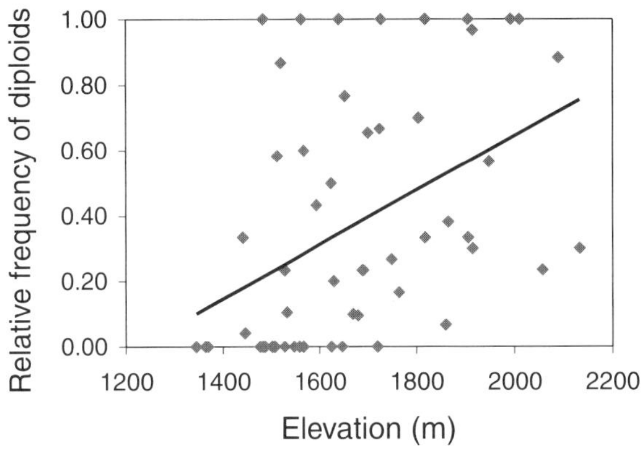

The unidirectional shift in geographic (and ecological) range in *C. angustifolium* reinforces the point that limits to adaptation are more complex than the simple ecology-versus-genetics dichotomy implies. Rather, the pattern suggests that the geographic limits of diploid ranges are genetically constrained at least in a portion of the range. Because range expansion only occurs in one direction, it seems likely that the genetic constraint on diploids is based more on the kinds of phenotypes available for adaptation than on the total magnitude of genetic diversity itself. Furthermore, the range of phenotypes created by genome multiplication may reflect the effects of increased genome size itself, and its indirect effects on cell size, development etc., rather than its effect on gene copy number and evolutionary potential.

Several studies have summarized the phenotypic consequences of chromosome doubling in extant (Levin 1983; Otto and Whitton 2000; Levin 2002) and newly synthesized (Ramsey and Schemske 2002) polyploids. Although there are some clear tendencies, we are still unable to predict for any given case what the direct phenotypic effects of increased genome size will be. For example, early plant surveys led biologists to conclude that polyploidy was more common at high latitudes and altitudes (Tischler 1935; Love and Love 1943; Muntzing 1936). However, Gustafsson (1948) and Clausen et al. (1940) argued that, in fact, when the ranges of polyploids are compared to their diploid progenitors in a paired manner, the trend is less consistent. A contrary case is found in fireweed and its sister species, *Chamerion latifolium* (Mosquin and Small 1971), where polyploids prevail at lower latitudes and altitudes. Similarly, flowering time is highly variable, with some polyploids flowering earlier and others later than their diploid progenitors (Husband and Schemske 2000). As a final example, polyploidy in mosses (as in other groups) has often been associated with increased cell size (described by Stebbins, 1950); however the cell size of any particular diploid line could not be predicted from the cell size of its haploid progenitor. These studies and many others highlight the difficulty in making generalizations about the effects of polyploidy on phenotypes (Stebbins 1950), other than that they will be affected.

Research on the genetic basis of ecologically important traits and the impact of dosage on gene expression can contribute greatly to our understanding of the effects of polyploidy on phenotypes. However, until this field develops fully, it will be increasingly important to design experiments that separate the effects of increased genome diversity from the effects of larger genome size per se. Synthesized polyploids (neopolyploids) will be particularly important in this regard, as they will retain the effects of genome size but not yet have the increased genetic diversity of established polyploids that can accrue from mutation, multiple origins and gene exchange (Bretagnolle and Thompson 1995, Ramsey and Schemske 2002). Transplant studies with synthesized and extant polyploids will help to clarify the extent to which species ranges are limited by insufficient genetic variability.

Saltatory responses to selection

As we have seen, genome multiplication can facilitate adaptation and substantial range shifts of a species along an ecological gradient. But what is the process by which this occurs? In a general sense, the answer rests with an understanding of the conditions under which chromosomal variation of large phenotypic effect can facilitates a response to selection, above and beyond what is possible through genic variation.

Since Darwin, biologists have debated the importance of phenotypic jumps versus small increments in the process of adaptive radiation (Mayr 1982; Futuyma 1998). The question was initially stimulated by Darwin's insistence that only continuous variants provide the fuel for natural selection and, after the Evolutionary Synthesis, by the observation of discontinuous variation among species and higher taxa observed in the paleontological record. While few practicing biologists would refute that chromosomal variants with large phenotypic effects can arise, a significant role for such saltatory shifts in phenotypic evolution have generally been refuted on two grounds: (1) genotypes with large phenotypic effects are relatively rare, and (2) such macro-mutations frequently have pleiotropic effects with strong negative effects on fitness. Here I examine the role of polyploid variation in the evolution of flowering time along an elevation gradient in *C. angustfolium*. First, I describe the contributions of polyploidy to variation in flowering time and then consider their frequency and correlated effects on fitness.

For a species that grows across such a wide range of latitudes and altitudes, and hence growing season durations, variation in flowering time will likely have large fitness consequences. This has been observed anecdotally across the altitudinal range for tetraploid *C. angustifolium*. At low elevations, tetraploids complete fruit development well within the growing season and produce a large number of seeds. In contrast, at high altitudes, plants are less likely to flower and those that do rarely develop beyond the flower bud stage by the time the freezing temperatures arrive in fall. As a result seed set is relatively low because it is too late in the year for regular pollinator visitation and seed development (Husband and Schemske 2000). Therefore, it is likely that flowering time will be under strong selection and variability in this trait may have originally contributed to expansion of the geographic range along the elevational gradient in which it is currently found.

Several lines of evidence suggest that variation in ploidy contributes significantly to flowering time variation in *C. angustifolium*. On average, autotetraploids begin to flower approximately 10 days after diploids. Such a delay has been observed consistently in field populations (Husband and Schemske 2000), common garden comparisons (Husband 2000) and in the greenhouse (Husband and Schemske 1997). Postponed flowering has also been observed, qualitatively, in a limited number of neopolyploids (n = 4) synthesized in the lab (B.C. Husband, unpublished). The relative contribution of polyploidy to flowering time variation was estimated from an analysis of phenotypic variation within and among nine *C. angustifolium* populations, three of which were diploid, three were tetraploid, and three contained both diploids and tetraploids (henceforth, n = 12). Plants were grown in a greenhouse and the time to flowering estimated for each plant (Fig 4). Mean time to flowering ranged from 8.5 to 38.1 days (after transplanting) among populations. Interestingly, 41% of this variance was attributable to ploidy. On average, diploids flowered nine days earlier than tetraploids (20 vs. 29 days after transplanting; $F = 5.2$, $P < 0.05$), a difference that was correlated with altitude. Significant but smaller portions of the variance were attributable to populations within ploidies (37%) and families within populations (2.5%). These results suggest that, indeed, polyploidy provides a large and consistent influence on flowering time. As with most polyploidy systems, it is unclear whether this delayed flowering is a product of chromosome doubling per se or selection afterward. However, the consistency of the pattern, preliminary observations of a small number of synthesized polyploid *C. angustifolium* and a general survey of neopolyploids (Ramsey and Schemske 2002) suggest that polyploids can contribute directly to the expansion of species ranges beyond the current boundaries of diploids.

The general significance of polyploidy to the evolution of flowering time depends in part on the availability of such chromosomal variation within populations. Historically, biologists have viewed polyploidization as a relatively rare and ancient event in the history of any polyploid lineage (Soltis and Soltis 1999). However, more recent molecular studies suggest that polyploidy is often polyphyletic within species and thus may arise recurrently (Soltis and Soltis 1999). For example, based on morphological, cpDNA, rDNA, and RAPDs, researchers have concluded that polyploid *Tragapogon mirus* may have arisen as many as nine times within the Palouse region of western United States (Cook et al. 1998). I would like to argue further that polyploid mutations may be a recurring source of phenotypic variation within populations, and that such variation is not readily detected because population cytogenetic surveys are often limited in sample size and recently produced polyploids can not be easily distinguished from diploids using conventional molecular markers.

Fig. 4. Mean time to flowering (days since transplanting) for 12 populations of diploid (dark diamonds) or tetraploid (light diamonds) *Chamerion angustifolium* grown in a greenhouse environment.

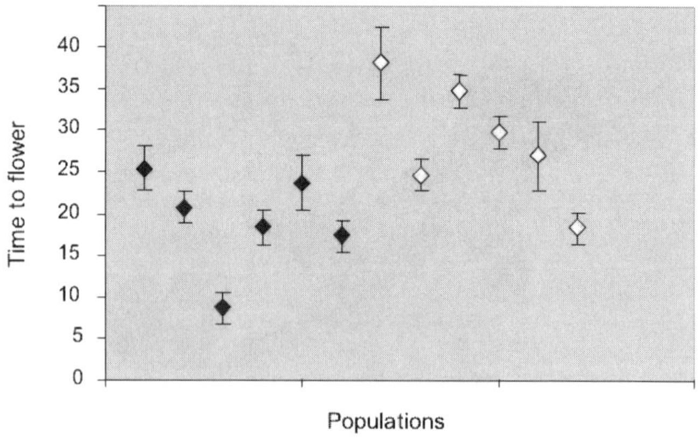

Several lines of evidence from studies of *C. angustifolium* indicate that polyploid mutation can be quite high, through the fusion of unreduced gametes (gametes with the somatic chromosome number) (Husband 2004). In a diallel crossing design among diploid, triploid and tetraploid plants, Burton and Husband (2001) inferred from the ploidy distribution of the offspring that approximately 3% of all gametes produced by diploids are unreduced ($n = 2x$). Moreover, triploids were capable of producing gametes with one, two or three chromosome sets, and thus could contribute significantly to the production of tetraploids. Taking this information as well as the relative fitness of triploids, diploids and tetraploids into account, Husband estimated that 2.3 of every 1000 zygotes produced per generation will be tetraploid (Husband 2004). In a separate study, in which we measured the ploidy of progeny from 30 families of a pure diploid population, we found that 4 of 750 offspring were tetraploid (B.C. Husband and P. Kron, unpublished), roughly confirming the estimate based on unreduced gamete production. Furthermore, we have recently been able to screen the DNA content of pollen directly, using flow cytometry. Using this approach, we have found that within pure diploid populations, 2.6% of all gametes are 2n (B.C. Husband and P. Kron, unpublished). Collectively, these results suggest that tetraploid offspring can be produced at a substantial rate each generation, which was, in fact, 10 times higher than estimates based on crop plants and 100 times higher than the average genic mutation rate (Ramsey and Schemske 1998). While more estimates of unreduced gametes and polyploid formation are necessary to understand the ecological and genetic factors regulating this process, our studies suggest that polyploid mutations can be sufficiently high and may serve as an important source of variation for phenotypic traits such as flowering time.

Another argument against the importance of discontinuous variation in adaptation is that macromutations frequently have correlated effects on phenotype that may offset any fitness benefits. Polyploids may be no exception, although few studies have examined the fitness effects of genome multiplication in natural populations and then related fitness to specific traits in a multivariate fashion. Nevertheless, the research is clear; shifts in ploidy are often associated with dramatic changes in flowering time, but simultaneous changes in other morphological and physiological characters may impose additional costs. In *Chamerion*, the negative pleiotropic effects associated with polyploidy include reduced seed set (Burton and Husband 2000), fewer flowers per inflorescence (Husband 2000) and slower development (Husband and Schemske 1997). Collectively, these correlated effects may explain the 40% reduction in fitness in tetraploids relative to diploids that we observed in a greenhouse environment (where, incidentally, flowering time doesn't effect the fitness measure). Such fitness costs may preclude chromosomal variants from contributing to a population's response to selection pressures.

In some cases, pleiotropic effects associated with polyploidy may enhance the response to selection. Recent theoretical models of the evolution of species ranges indicate that gene flow from high density, central populations to marginal populations of low density, may continually introduce maladapted genotypes into marginal populations and thus prevent any further expansion of the ecological and geographical range (Kirkpatrick and Barton 1997; Garcia-Ramos and Kirkpatrick 1997). By extension, polyploidy may facilitate ecological range expansion by minimizing the effects of gene exchange with diploids in the 'central' part of the range. Indeed, crosses between extant diploid and polyploid individuals demonstrate that gene flow between ploidies is diminished by reduced viability and fertility of triploid hybrids (Ramsey and Schemske 1998; Burton and Husband 2000). However, few estimates of gene flow between diploids and polyploids exist (Lumaret and Barrientos 1990; Brochmann et al. 1992; Husband and Schemske 2000; Husband and Sabara 2004) and our knowledge of reproductive barriers is far from comprehensive, particularly with respect to prezygotic barriers (Petit et al. 1999; Segraves and Thompson 1999; Husband and Schemske 2000). Our understanding of the magnitude of reproductive isolation between diploids and newly formed polyploids in natural populations is even smaller (Ramsey and Schemske 2002).

In *C. angustifolium* evidence gathered thus far suggests that gene exchange between extant diploids and tetraploids is highly reduced (Husband and Sabara 2004). In a greenhouse comparison, Burton and Husband (2000) found that synthesized triploids had only 9% of the fitness of diploids, a result of reduced fitness at several life stages including: seed maturation, germination and pollen viability. Also, diploids and tetraploids appear to be partially isolated due to pre-zygotic barriers such as pollinator fidelity (Husband and Schemske 2001), pollen precedence (Husband et al. 2002) and self-fertilization (Husband and Schemske 1995, 1997). In total, gene exchange was restricted to less than one percent of all matings, and most importantly, pre-zygotic barriers were particularly strong in preventing diploids from pollinating tetraploids (Husband and Schemske 2001; Husband and Sabara 2004). These results indicate that, in *C. angustifolium*, adaptation via genome duplication may be accompanied by a reduction in gene exchange with diploids, which in fact may facilitate ecological expansion beyond the diploid range.

Population thresholds and stochastic forces

Beginning with Fisher and Wright (Provine 1986), biologists have long debated the evolutionary significance of stochastic processes. Population genetic theory indicates that drift is likely when selection is weak and effective size of populations is small. The question remains, however, whether these circumstances are ever met and whether they can lead to evolutionary transitions that would not occur otherwise. At this time it appears that, in theory, drift in conjunction with selection can cause adaptive peak shifts, however, empirical evidence supporting the significance of this process is sparse (Coyne et al 1997). I suggest that polyploid evolution may represent a logical place to look for stochastic effects for two reasons. First, as already discussed, polyploids may be partially reproductively isolated from diploids and thus are released from the constraints of gene flow, which counteract genetic drift (Coyne et al. 1997). Second, the evolutionary dynamics of polyploids are particularly vulnerable to stochastic events in populations.

Levin (1975) showed that the interaction between diploid and polyploid cytotypes is defined by positive frequency-dependent selection. All else being equal, the minority cytotype will experience a mating disadvantage owing to the low probability of mating with a plant of the same cytotype and the strong infertility of triploid offspring derived from between-cytotype matings. As a result, polyploid evolution is defined by two stable equilibria; predominantly diploid and predominantly tetraploid. To increase in frequency, a rare cytotype requires a mechanism for crossing the frequency threshold (50%, when both cytotypes have equal fitness) (Felber 1991). One such mechanism is stochastic events due to small population size or founder events (Fowler and Levin 1984; Thompson and Lumaret 1992).

There are no direct tests of the role of stochastic processes in natural polyploid populations. However, many of the conditions necessary for stochastic forces to operate in *C. angustifolium* are currently met. First of all, a study of *C. angustifolium* populations (Husband 2000), whose cytotype frequencies were experimentally manipulated (100, 66, 50, 33, 0% diploids), confirmed that fitness is frequency-dependent, as predicted by Levin (1975). Specifically, diploid seed set declined with increasing tetraploid frequency, whereas tetraploid fitness was independent of frequency (Husband 2000). When combined into a relative fitness measure, tetraploid fitness increased linearly with frequency in the population. The threshold frequency, at which diploids and tetraploids have equal fitness and where drift may be most effective, was estimated as 42% tetraploids. In addition, *C. angustifolium* is a colonizer of open and disturbed habitats; thus, founder events are a common and ongoing process in this species. These separate pieces of evidence suggest that stochastic processes, in small populations and in association with colonization, may play a role in the transition of populations from diploidy to polyploidy. This process may, in part, explain the prevalence of diploid- and tetraploid-dominated populations, and the rarity of mixed populations, in many species (Burton and Husband 1999; Van Dijk et al. 1992), It may also have contributed to the success of tetraploid-dominated populations in the diploid-tetraploid contact zone in *C. angustifolium* throughout Wyoming and Montana (Fig. 5).

Fig. 5. Frequency distribution of diploids (vs. polyploids) in 23 populations of *Chamerion angustifolium* sampled in Wyoming and Montana.

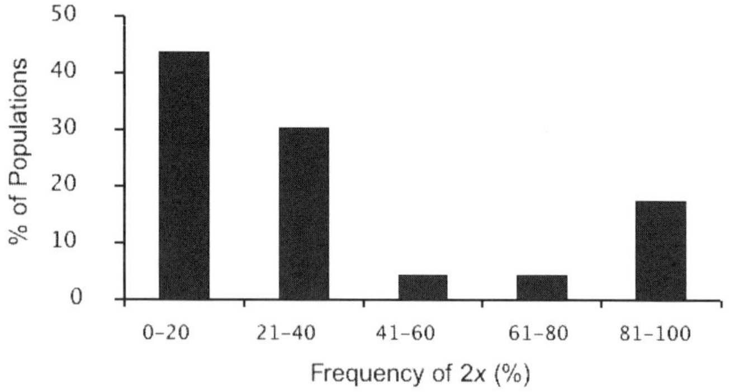

A role for ecological and genetic approaches

I have argued that polyploidy, or genome multiplication, can provide opportunities to further our knowledge of the rates, modes and constraints on adaptation in plants. Of course, what I have presented are mostly ideas. Clearly, much theoretical and empirical research is still required, particularly at the interface of ecology and genetics. Progress on the role of chromosomal variation in adaptation will benefit from advances in four areas, listed below.

Additional ecological research is required to understand the patterns of selection operating on polyploids along ecological gradients. In particular, more thorough surveys are needed to describe patterns of variation in ecologically important characters such as flowering time in species that vary in ploidy. Greenhouse or common garden comparisons should also be used to quantify the genetically-based variation in these traits, to relate phenotype to the local environment along an ecological gradient, and to evaluate the correlated phenotypic responses to genome multiplication that may affect the response to selection. In addition, field studies are necessary to assess the frequency of diploid and polyploid variants available within populations, to estimate the magnitude and direction of selection acting on ecological characters in mixed cytotype populations, and to evaluate the direct and indirect effects of polyploidy on fitness. To date, I am unaware of any research of this kind.

Molecular approaches will play an especially important role in future polyploidy research. First, additional genetic markers are needed to examine evolutionary relationships among polyploids and their diploid progenitors. This has and will be especially challenging for species that are polymorphic for chromosome number since (1) markers must be variable enough to resolve the phylogenies of populations within mixed-ploidy species (Soltis and Soltis 1999), and (2) complications can arise when interpreting these genetic data because reproductive isolation is often not complete between diploids and polyploids (Petit et al. 1999). Soltis and Soltis (1999) reviewed past research on the genetic relationships between diploid and polyploid species/populations. While several studies have now addressed this problem using molecular tools, many are restricted in geographic scale and genetic resolution, and do not apply rigorous phylogenetic analyses. Second, as with any research on character evolution, studies of adaptation through genome multiplication will benefit from comparative analyses of ploidy and ecologically important characters at a broad taxonomic scale (Silvertown et al. 1997). This approach will require better phylogenetic information for taxonomic groups that are variable in ploidy, combined with more complete information on ecological ranges and characters of adaptive significance. Together, these data can be used for testing the generality of hypotheses regarding correlated evolution of ecological ranges, morphological characters and polyploidy. To date, these analyses have not been conducted.

Our understanding of the process of adaptation can benefit from a better understanding of the genetic basis of ecologically important characters. In particular, knowledge of the effect of genome multiplication on the phenotype is central to building a predictive science, but until now this area has been treated as a black box. Quantitative trait loci and associated approaches can provide a first glimpse into the number of loci controlling ecological characters, such as flowering time, as well as their distribution throughout the genome (Ungerer et al. 2002). In addition, molecular and genomic approaches will be essential for identifying specific genes controlling such traits and their interactions with each other and their environment (Chou and Yang 1999; Johanson et al. 2000; Cronk 2001; Ratcliffe and Riechmann 2002). Perhaps the most critical and biggest challenge, however, is increasing our ability to predict the effects of genome multiplication on these ecologically important characters. This involves a better understanding of the function of genes controlling characters such as flowering time and the impact of dosage on their regulation and expression (Osborn et al. 2003). Such questions represent perhaps the most valuable contribution genomic research can make to the study of adaptation and polyploidy at this time.

Finally, one of the longest standing obstacles to polyploid research is the difficulty disentangling the effects of gene copy number from those of increased genome size. As mentioned earlier, many extant polyploids are genetically diverse relative to their diploid ancestors. This fact makes it difficult to attribute any one aspect of polyploidy to the shifts in geographic range. One approach that may help resolve this is to use newly synthesized polyploids (=neopolyploids). Neopolyploids can be produced by exposing them to mitosis disrupters (colchicine) or by manipulating the production and union of unreduced gametes (Bretagnolle and Thompson 1995). Unreduced gametes may be the ideal method as they mimic the most common process operating in natural populations and are least like to have additional genetic side effects (Ramsey and Schemske 2002). In the first few generations, neopolyploids will exhibit the direct effects of genome size (larger cell size, etc.) and any associated changes in gene expression (Osborn et al. 2003), but will not have had sufficient time to build up additional genic diversity through mutation and gene flow. By transplanting these plants into and beyond the diploid species range and evaluating their performance, we may finally begin to understand not only the reason for range shifts between diploids and polyploids but also the underlying constraints on adaptation in diploid species in general.

Acknowledgements

I thank Holly Sabara and Paul Kron for permission to include unpublished data, two anonymous reviewers and the Natural Sciences and Engineering Research Council of Canada, Premier's Research Excellence Award program and the Canada Research Chair program for financial support.

References

Bretagnolle, F., and Thompson, J.D. 1995. Gametes with the somatic chromosome number: mechanisms of their formation and role in the evolution of polyploid plants. New Phytol. **129**: 1–22.

Brochmann, C., Stedje, B., and Borgen, L. 1992. Gene flow across ploidal levels in *Draba* (Brassicaceae). Evol. Trends Plants **6**: 125–134.

Burton, T.L., and Husband, B.C. 1999. Population cytotype structure in the polyploid *Galax urceolata* (Diapensiaceae). Heredity **82**: 381–390.

Burton, T.L., and Husband, B.C. 2000. Fitness differences among diploids, tetraploids and their triploid hybrids in *Chamerion angustifolium* (Onagraceae): mechanisms of inviability and implications for polyploid evolution. Evolution **54**: 1182–1191.

Burton, T.L., and Husband, B.C. 2001. Fecundity and offspring ploidy in matings among diploid, triploid and tetraploid *Chamerion angustifolium* (Onagraceae): consequences for tetraploid establishment. Heredity **87**: 573–582.

Clausen, J., Keck, D.D., and Hiesey, W.M. 1940. Experimental studies on the nature of species. I. The effect of varied environments on western North American plants. Carnegie Inst. Washington, Publ. No. 520, 452 pp.

Cook, L.M., Soltis, P.S., Brunsfeld, S.J., and Soltis, D.E. 1998. Multiple independent formations of *Tragapogon* tetraploids (Asteraceae): evidence from RAPD markers. Mol. Ecol. **7**: 1293–3102.

Chou, M.-L., and Yang, C.-H. 1999. Late-flowering genes interact with early-flowering genes to regulate flowering time in *Arabidopsis thaliana*. Plant Cell Physiol. **40**: 702–708.

Coyne, J.A., Barton, N.H., and Turelli, M. 1997. A critique of Sewall Wright's shifting balance theory of evolution. Evolution **51**: 643–671.

Cronk, Q.C.B. 2001. Plant evolution and development in post–genomic context. Nat. Rev. Genet. **2**: 607–619.

Ehrendorfer, F. 1980. Polyploidy and distribution. *In* Polyploidy – Biological Relevance. Edited by W.H. Lewis. Plenum, New York. pp. 45–60.

Felber, F. 1991. Establishment of a tetraploid cytotype in a diploid population: effect of the relative fitness of cytotypes. J. Evol. Biol. **4**: 195–207.

Fowler, N.L., and Levin, D.A. 1984. Ecological constraints on the establishment of a novel polyploidy in competition with its diploid progenitor. Am. Nat. **124**: 703–711.

Futuyma, D.J. 1998. Evolutionary Biology, 3rd Edition. Sinauer Associates Inc. Sunderland, MA.

Garcia-Ramos, G., and Kirkpatrick, M. 1997. Genetic models of rapid evolutionary divergence in peripheral populations. Evolution **51**: 1–23.

Gustafsson, A. 1948. Polyploidy, life-form, and vegetative reproduction. Hereditas **34**: 1–22.

Husband, B.C. 2000. Constraints on polyploid evolution: a test of the minority cytotype exclusion principle. Proc. R. Soc. Lond. B Biol. Sci. **267**: 217–223.

Husband, B.C. 2004. The role of triploids in the evolutionary dynamics of mixed-ploidy populations. Biol. J. Linn. Soc. **82**: 537–546.

Husband, B.C., and Sabara, H.A. 2004. Reproductive isolation between autotetraploids and their diploid progenitors in fireweed, *Chamerion angustifolium* (Onagraceae). New Phytol. **161**: 703–713.

Husband, B.C., and Schemske, D.W. 1997. The effect of inbreeding in diploid and tetraploid populations of *Epilobium angustifolium* (Onagraceae): implications for the genetic basis of inbreeding depression. Evolution **51**: 737–746.

Husband, B.C., and Schemske, D.W. 1998. Cytotype distribution at a diploid-tetraploid contact zone in *Chamerion* (*Epilobium*) *angustifolium* (Onagraceae). Am. J. Bot. **85**: 1688–1694.

Husband, B.C., and Schemske, D.W. 2000. Ecological mechanisms of reproductive isolation and coexistence of diploid and tetraploid *Chamerion angustifolium*. J. Ecol. **88**: 1–14.

Husband, B.C., Schemske, D.W., Goodwillie, C., and Burton, T.L. 2002. Pollen competition as a unilateral mechanism of reproductive isolation between diploid and tetraploid *Chamerion angustifolium*. Proc. R. Soc. Lond. B Biol. Sci. **269**: 2565–2571.

Johanson, U., West, J., Lister, C., Michaels, S., Amasoni, R., and Dean, C. 2000. Molecular analysis of FRIGIDA, a major determinant of natural variation in *Arabidopsis* flowering time. Science (Washington, D.C.) **290**: 344–347.

Kirkpatrick, M., and Barton, N.H. 1997. Evolution of a species range. Am. Nat. **150**: 1–23.

Levin, D.A. 1975. Minority cytotype exclusion in local plant populations. Taxon **24**: 35–43.

Levin, D.A. 1983. Polyploidy and novelty in flowering plants. Am. Nat. **135**: 563–572.

Levin, D.A. 2002. The role of chromosomal change in plant evolution. Oxford Series in Ecology and Evolution. Oxford University Press, Oxford.

Lumaret, R., and Barrientos, E. 1990. Phylogenetic relationships and gene flow between sympatric diploid and tetraploid plants of *Dactylis glomerata* (Gramineae). Plant Syst. Evol. **169**: 81–96.

Masterson, J. 1994. Stomatal size in fossil plants: evidence for polyploidy in majority of flowering plants. Science (Washington, D.C.) **264**: 421–424.

Mayr, E. 1982. The growth of biological thought: diversity, evolution and inheritance. Belknap Press of Harvard University Press, Cambridge, MA.

Mosquin, T. 1967. Evidence for autopolyploidy in *Epilobium angustifolium* (Onagraceae). Evolution **21**: 713–719.

Mosquin, T., and Small, E. 1971. An example of parallel evolution in *Epilobium* (Onagraceae). Evolution **25**: 678–682.

Olmstead, R.G. 1990. Biological and historical factors influencing genetic diversity in the *Scutellaria angustifolia* complex (Labiatae). Evolution **44**: 54–70.

Osborn, T.C., Pires, J.C., Birchler, J.A., Auger, D.L., Chen, Z.J., Lee, H-S., Comai, L., Madlung, A., Doerge, R.W., Colot, V., and Martienssen, R.W. 2003. Understanding mechanisms of novel gene expression in polyploids. Trends Genet. **19**: 141–147.

Otto, S.P., and Whitton, J. 2000. Polyploid incidence and evolution. Annu. Rev. Genet. **34**: 401–437.

Petit, C., Bretagnolle, F., and Felber, F. 1999. Evolutionary consequences of diploid-polyploid hybrid zones in wild species. Trends Ecol. Evol. **14**: 306–311.

Petit, C., and Thompson, J.D. 1999. Species diversity and ecological range in relation to ploidy level in the flora of the Pyrenees. Evol. Ecol. **13**: 45–66.

Provine, W.B. 1986. Sewall Wright and evolutionary biology. Chicago University Press, Chicago, IL.

Ramsey, J., and Schemske, D.W. 1998. Pathways, mechanisms, and rates of polyploid formation in flowering plants. Annu. Rev. Ecol. Syst. **29**: 467–501.

Ramsey, J., and Schemske, D.W. 2002. Neopolyploidy in flowering plants. Annu. Rev. Ecol. Syst. **33**: 589–639.

Ratcliffe, O.J., and Riechmann, J.L. 2002. *Arabidopsis* transcription factors and the regulation of flowering time: a genomic perspective. Curr. Iss. Mol. Biol. **4**: 77–91.

Segraves, K.A., and Thompson, J.N. 1999. Plant polyploidy and pollination: floral traits and insect visits to diploid and tetraploid *Heuchera grossulariifolia*. Evolution **53**: 1114–1127.

Silvertown, J., Franco, M., and Harper, J.L. 1997. Plant life histories: ecology, phylogeny and evolution. Cambridge University Press. Cambridge.

Soltis, D.E., and Soltis, P.S. 1989. Genetic consequences of autopolyploidy in *Tolmeia* (Saxifragaceae). Evolution **43**: 586–594.

Soltis, D.E., and Soltis, P.S. 1993. Molecular data and the dynamic nature of polyploidy. Crit. Rev. Plant Sci. **12**: 243–273.

Soltis, D.E., and Soltis, P.S. 1999. Polyploidy: recurrent formation and genome evolution. Trends Ecol. Evol. **14**: 348–352.

Stebbins, G.L., Jr. 1950. Variation and evolution in plants. Columbia University Press, New York.

Stebbins, G.L., and Dawe, J.C. 1987. Polyploidy and distribution in the European flora: a reappraisal. Bot. Jahrb. Syst. **108**: 343–354.

Thompson, J.D., and Lumaret, R. 1992. The evolutionary dynamics of polyploid plants: origins, establishment and persistence. Trends Ecol. Evol. **7**: 302–307.

Ungerer, M.C., Halldorsdottir, S.S. Modliszewski, J.L., Mackay, T.F.C., and Purugganan, M.D. 2002. Quantitative trait loci for inflorescence development in *Arabidopsis thaliana*. Genetics **160**: 1133–1151.

Van Dijk, P., Hartog, M., and Van Delden, W. 1992. Single cytotype areas in autotetraploid *Plantago media* L. Biol. J. Linn. Soc. **46**: 315–331.

Warwick, S.I., and Gottlieb, L.D. 1985. Genetic divergence and geographic speciation in *Layia* (Compositae). Evolution **39**: 1236–1241.

16 Evolutionary genetics of self-incompatibility in a new "model" plant: *Arabidopsis lyrata*

Barbara K. Mable

Abstract: For practical purposes, initial eukaryotic genome sequencing projects have concentrated on species that have been used as models for previous genetic studies. However, examination of polymorphisms within and between species in relation to environmental adaptation is where the real power of genomics will lie. Recently, *Arabidopsis lyrata* has been proposed as a new model plant "species" because it is an outcrossing relative of the selfing model species *Arabidopsis thaliana*. The taxonomy of *A. lyrata* is confused at best, but former classifications have been based predominantly on geographic distribution. The current suggestion to synonymize populations from Europe, North America, and Asia into a single species runs the risk of overlooking potentially interesting variation in morphology, development, mating systems, and genomic restructuring that could be related to habitat types or environmental differences among localities. We have been studying the strength of sporophytic self-incompatibility (SSI) and ploidy variation in populations from across the species range. Contrary to previous reports, not all populations are diploid nor are they all self-incompatible. This variation increases the potential utility of this species for studying environmental adaptation while emphasizing the importance of considering the natural history and biology of a species when promoting use of new models.

Introduction

Model organisms have long been used in fields of biology directed towards finely dissecting the genetic mechanisms underlying processes important to all organisms, such as development and reproduction. In this age of comparative genomics, the necessity to concentrate (at least initially) on a few key organisms is unquestionable. With the completion of a number of eukaryotic genomes and, especially, completion of those that allow comparative analyses of closely related species and more complicated interactions such as host-vector-parasite systems (Table 1), powerful tools for determining the genetic basis for a wide number of processes are becoming available. In plants, most efforts are currently being concentrated on economically important groups (e.g., rice, tomatoes, soybeans, cotton, barley, coffee, alfalfa, rice, bean, sugar cane, potato, wheat, sorghum, wheat, broccoli, rapeseed, canola) as well as several tree species (eucalyptus, pines, poplar), but there is increasing interest in looking at naturally occurring plants to relate genomic differences to ecological adaptation.

The second step in genome analysis, relating sequences to functions, so far has been mainly concentrated in cellular and molecular biology laboratories (e.g., in plants, work reviewed by Kersten et al. 2002). The risk of using model species is that the biology of the organism may become reduced to a set of molecular processes, without relation to the whole organism or ecological consequences of the attributes under study. While evolutionary biologists and ecologists are beginning to exploit functional genomics databases (e.g., see "preview" by Gibson 2002), there is enormous potential for using these data to elucidate factors that affect environmental adaptation. Advances in this area will be most pronounced if molecular biologists and ecologists /evolutionary biologists join forces from the outset of new projects.

Variation in *Arabidopsis thaliana* ecotypes

As an example, *Arabidopsis thaliana* (Brassicaceae) has long been used as a genetic model because of its unusually fast generation time, its relative ease of transformation using *Agrobacterium*, and the ease of tissue culture in the laboratory. With the completion of its genome sequencing project in 2000 and with concentrated efforts in the area of proteomics (reviewed by Kersten et al. 2002), the already vast amount of genetic information on this species has increased dramatically. There are a growing number of excellent websites providing a resource for accumulating data on *A. thaliana* — Munich Information Center for Protein Sequences, (MIPS Schoof et al. 2002: http://mips.gsf.de/proj/thal/db); The Initiative for Genome Research (TIGR: http://www.tigr.org/tdb/e2k1/ath1/); and The Arabidopsis Information Resource (TAIR Rhee et al. 2003: http://www.arabidopsis.org/home.html). Although it is widely recognized that this model species can be divided into a large number of "ecotypes" (populations) with different responses to various treatments, it has been relatively rare for this variation to be related directly back to the habitats from which the "ecotypes" were originally sampled. To demonstrate, a PubMed

Barbara K. Mable.[1] Department of Botany, University of Guelph, Guelph, ON N1G 2W1, Canada. e-mail address: bmable@uoguelph.ca
Correct citation: Mable, B.K. 2004. Evolutionary genetics of self-incompatibility in a new "model" plant: *Arabidopsis lyrata*. *In* Plant Adaptation: Molecular Genetics and Ecology. Proceedings of an International Workshop held December 11–13, 2002, in Vancouver, British Columbia, Canada. *Edited by* Q.C.B. Cronk, J. Whitton, R.H. Ree, and I.E.P. Taylor. NRC Research Press, Ottawa, Ontario. pp. 127–134.

[1]Current address: Division of Environmental and Evolutionary Biology, University of Glasgow, Glasgow, Scotland, U.K. G12 8QQ. e-mail address: b.mable@bio.gla.ac.uk

Table 1. A partial list of eukaryotic genomes sequenced (for a complete list see http://www.genomesonline.org).

Species	Reference
Saccharomyces cerevisiae	Yeast Genome Directory 1997
Caenorhabditis elegans	*C. elegans* Sequencing Consortium 1998
Drosophila melanogaster	Adams et al. 2000
Homo sapiens	Lander et al. 2001
Arabidopsis thaliana	Arabidopsis Genome Initiative 2001
Schizosaccharomyces pombei	Wood et al. 2002
Oryza sativa japonica	Goff et al. 2002
Oryza sativa indica	Yu et al. 2002
Anopheles gambiae (mosquito vector)	Holt et al. 2002
Plasmodium falciparum (human malaria)	M.J. Gardner et al. 2002
P. yoelli yoelli (rodent malaria)	Carlton et al. 2002

(NCBI: http://www.ncbi.nlm.nih.gov:80/entrez) search using the phrase "Arabidopsis thaliana ecotypes" yielded 162 references from 1991 to 2002 with these words in the title or abstract. Of these abstracts, one mentioned the words "habitat and geography" (Miyashita et al. 1999), one mentioned "ecology" (Ungerer et al. 2002), 11 mentioned "environment" (Clarke et al. 1995; Mantyla et al. 1995; Innan et al. 1996; Rutherford and Masson 1996; van Der Schaar et al. 1997; Alonso-Blanco et al. 1998; Dennis et al. 1998; Orbovich and Taras'ev 1999; Cooley et al. 2001; Kliebenstein et al. 2001*a*, 2001*b*), 7 mentioned the adjective "geographic" (Innan et al. 1996, 1997; Ullrich et al. 1997; Yanovsky et al. 1997; Bergelson et al. 1998; Miyashita et al. 1999; Torabinejad and Caldwell 2000), but none included the word "adaptation". A large number of the remaining studies discussed variation in ecotypes but, at least in the abstracts, this was not related back to why the differences might exist based on differences in ecological pressures. This type of survey is limited but demonstrates that there has not been a strong emphasis on adaptation in the recent molecular genetics literature.

One reason that there has not been more interest in ecotypic variation in relation to naturally occurring habitats could be due to the assumption that selfing species (like *A. thaliana*) will be characterized by very low levels of genetic variability, which would limit the ability to detect geographic variation. Indeed, results from population surveys conducted to look at genetic divergence among ecotypes in relationship to geographic distance have mainly shown very low levels of genetic diversity and little evidence of isolation by distance. For example, Bergelson et al. (1998) compared 11 populations worldwide (including introduced populations) and 7 commonly used laboratory ecotypes using one mitochondrial (*Nad*5) and three nuclear (*Adh*, *Dhs*1, and *Gpa*1) genes and found no evidence of population structuring related to geographic location. However, they also found very low levels of divergence and it is possible that there was an insufficient level of variability in the gene regions used to have any statistical power. A study by Sharbel et al. (2000), on the other hand, sampled *A. thaliana* only from its original range (in Europe and Asia) and used amplified fragment length polymorphisms (AFLPs) to compare ecotypes. This study found significant isolation by distance but no "ecotype" phylogeny, which the authors concluded to be due to a past history of recombination. Such a star-like phylogeny was also found by Miyashita et al. (1999).

Nevertheless, there are a large number of interesting results from both molecular biology and ecology/evolution groups on variation among ecotypes in *A. thaliana* that could be used to design experiments to relate this variation back to habitat differences. For example, a number of recent studies have found ecotypic differences in susceptibility to various pathogens (e.g., Aguilar et al. 2002; Cui et al. 2002; Park et al. 2002; Wong et al. 2002); flowering time (Le Corre et al. 2002), inflorescence development (Suzuki et al. 2002; Ungerer et al. 2002), growth rate (Beemster et al. 2002), cell wall composition (Gardner et al. 2002), and metabolites (Reichelt et al. 2002). While this represents only a small sampling of the range of studies, it is now feasible to relate the genetics of variability for a wide array of processes to variation in micro or macro-environments experienced by particular ecotypes.

To avoid the historical disjunction between molecular geneticists and ecologists/evolutionary biologists in the future, one approach may be to *first* determine the extent and nature of variation (both phenotypic and genetic) in new model organisms, *before* they become widely used for molecular genetic studies. As an example, I will discuss such variation in a close relative of *A. thaliana*, *A. lyrata*, whose popularity as a model species is currently increasing (e.g., Nasrallah 2000; Mitchell-Olds 2001).

Arabidopsis lyrata as a model for the study of self-incompatibility

One area of research that has attracted the attention simultaneously of both molecular biologists/biochemists and population/evolutionary geneticists is the genetic control of self-incompatibility (SI) that occurs in a large number of plant species. This is because it is one of the best known examples of cell-cell interactions as well as being one of the best examples for balancing selection (e.g., Charlesworth et al. 2000). Although there are a number of ways that self-incompatibility is accomplished, the basic principle is that fertilization does not occur when the SI alleles expressed on the surface of the pollen match those expressed in the stigma. One type of SI that has been widely studied is "sporophytic" SI (SSI), in which the SI phenotype of the pollen is determined by the diploid genotype of its parent plant (i.e., protein products are deposited on the surface of the pollen by the pollen mother cells). Most of the original work on SSI was carried out on cultivated plants in the genus *Brassica,* but extending these studies to naturally occurring plants is currently a growing focus. Because *A. thaliana* reproduces by selfing, it does not provide a useful model in this case, but one of its closest relatives, *A. lyrata* is known to be self-incompatible and is becoming more widely used for studies of SI (Kusaba et al. 2001; Schierup et al. 2001; Kusaba et al. 2002; Charlesworth et al. 2003; Mable et al. 2003). *A. lyrata* was chosen as a model for these studies because it has been described as an obligately outcrossing, perennial, diploid species whose close evolutionary relationship to *A. thaliana* will allow exploitation of the vast amount of genomic information available on this more widely studied model. The validity of this statement will be evaluated below by discussing mating system and genome size variation that we have found among populations of *A. lyrata* sampled from five geographic regions (Table 2).

Taxonomy

The decision to "split" or "lump" species is often controversial from a taxonomic standpoint but also has implications for studies of genetic and ecological variation. Populations now included under the name "*Arabidopsis lyrata*" formerly belonged to four species in two different genera (*Arabis* and *Cardaminopsis*), predominantly based on geographic distribution and currently considered as subspecies of *A. lyrata* by some researchers: *A. lyrata lyrata* from North America (formerly *Arabis lyrata*), *A. lyrata petraea* from Europe (formerly *Cardaminopsis petraea*), *A. lyrata kamchatica* from northwestern North America and Asia (formerly *Arabis kamchatica*), and *A. lyrata kawasakiana* from Asia (formerly *Arabis kawasakiana*). There is some discrepancy about whether the last two (*kamchatica* and *kawasakiana*) should be considered one or two subspecies but, as there is currently insufficient data, they will be referred to separately here. While the recommendation for synonymy of these species by Al-Shehbaz et al. (1999) may be warranted based on both morphological and molecular data (Price et al. 1994), it is important to consider the variation that originally resulted in separate species distinctions when using these populations for such studies. Just as *A. thaliana* ecotypes may not be completely interchangeable, populations from across the range of this species group are likely to have experienced very different selective pressures that could result in real and potentially interesting variation.

In addition, even within geographic ranges, *A. lyrata* populations are found in very different types of habitats. Although each of the habitats in which it is found (rocky outcrops, mountain ridges, rocky creek beds, limestone pavements, and sand dunes) share the attributes of being low competition environments with limited nutrients and well drained, basic substrates, they do provide potentially distinct selective environments. For example, populations along ridges and cliff edges tend to exist in small very scattered patches with potentially limited migration between patches, whereas sand dune populations have more continuous habitats and can exist at very high densities in large patches (Mable, unpublished). It might be particularly interesting to look at populations in Alaska, where *A. lyrata lyrata*, *A. lyrata kamchatica*, and *A. lyrata petraea* are all thought to occur. There is also a great deal of morphological variation in leaf shape and flower size among populations of *A. lyrata* and sometimes even within greenhouse-raised families (Mable, unpublished). This variation has not yet been related to habitat type and relationships to geographic distributions have not been documented beyond keys to identify the original species.

The major reason for the increasing use of this group as a model is its phylogenetic relationship to *Arabidopsis thaliana*. Molecular systematic studies over the past 10–15 years have resulted in drastic taxonomic revisions, and the Brassicaceae is no exception. Recent efforts to resolve relationships in the Brassicaceae based on molecular analyses have suggested a revision of the genus *Arabidopsis* to include species formerly in *Arabis*, *Cardaminopsis* and *Cardamine* (Price et al. 1994; O'Kane and Al-Shehbaz 1997; Al-Shehbaz et al. 1999). Although *Arabidopsis* as currently defined is paraphyletic and requires further definition, in studies based on nuclear genes (Miyashita et al. 1998; Koch et al. 1999; Koch et al. 2000; Savolainen et al. 2000; Koch et al. 2001), the four subspecies of *A. lyrata,* as well as *Arabidospis halleri* (formerly *Arabis halleri*) and *Arabidopsis suecica* (formerly *Arabis suecica*), consistently form a monophyletic group that is more closely related to ecotypes of *A. thaliana* than to other species in the *Arabidopsis* sensu

Table 2. Populations of *A. lyrata* used in this study, indicating ploidy level (2x = diploid, 4x = tetraploid), self-incompatibility

Population	Ploidy	Self-compatibility	Subspecies designation
Iceland	2x*	SI	*petraea*
Austria	4x	SI	*petraea*
Japan	4x	SC	*kawasakiana*
Indiana	2x	SI	*lyrata*
Ontario	2x	SC	*lyrata*

*A small number of triploid individuals also have been identified from this population.

Fig. 1. Relationships among species in the *Arabidopsis sensu stricto* clade, extracted from Fig. 4 in Koch et al. 2000. Based on *Adh* sequences compiled from Genbank and their own sequences. NA refers to North America, GER to Germany, SWE to Sweden and CO to the Columbia strain of *A. thaliana.* Numbers indicate percentage bootstrap support for individual nodes (based on 1,000 replicates).

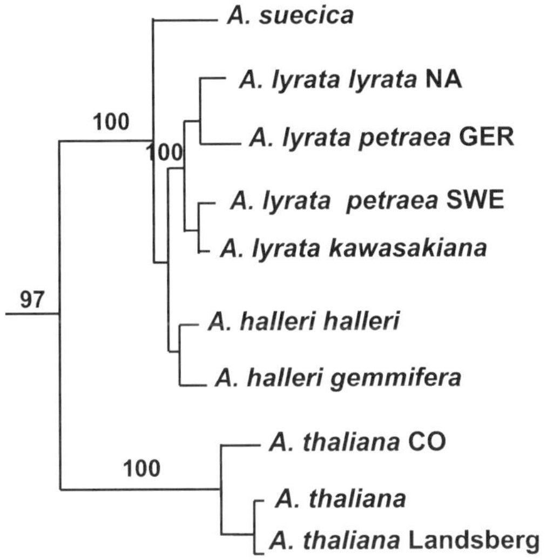

stricto clade (see Fig. 1, based on Koch et al. 2000). Thus, while some taxonomic confusion exists, the validity of using *A. lyrata* as a phylogenetically close relative of *A. thaliana* seems justified.

Early research on this species indicates that information based on the *A. thaliana* genome can easily be used to amplify homologous regions from *A. lyrata*. For example, microsatellite primers designed for *A. thaliana* genes often work well in *A. lyrata* (e.g., van Treuren et al. 1997; Clauss et al. 2002), and primers designed for *A. lyrata* to amplify the putative *SRK* locus, which controls the female side of the self-incompatibility recognition reaction, are able to amplify a pseudogene in *A. thaliana* (Kusaba et al. 2001). Thus, comparison of patterns and levels of genetic variation in traits related to reproductive mode or habitat within *A. lyrata* and between *A. lyrata* and *A. thaliana* are feasible.

Ploidy

Polyploidy is prevalent in plant species, with estimates ranging from 30 to more than 80% of angiosperms having polyploidy somewhere in their history (Levin 1983; Masterson 1994; Otto and Whitton 2000). Because polyploidy is often characterized by a return to diploid-like inheritance and chromosome pairing patterns, recognizing polyploidy is not always trivial. However, the clearest cases are when populations of a species occur as a polyploid series. North American and European populations of *A. lyrata* (i.e., subspecies *A. lyrata lyrata* and *A. lyrata petraea*) have been considered to be diploid with $2n = 2x = 16$ chromosomes (e.g., Nasrallah et al. 2000; Savolainen et al. 2000; Mitchell-Olds 2001), but tetraploid populations of both *A. lyrata kawasakiana* (Miyashita et al. 1998) and *A. lyrata kamchatica* (Mulligan 1964; Rollins 1966; Taylor and Mulligan 1968) have been described. In addition, a recent survey of populations of *A. lyrata petraea* from mainland Europe revealed a number of putatively tetraploid populations, based on allozyme data (Steve Ansell and Johannes Vogel, personal communication), which substantiate earlier reports of populations with $2n = 4x = 32$ chromosomes (Polatschek 1966).

We have been using one putatively tetraploid population of *A. lyrata kawasakiana* from Japan (collected by N. Miyashita) and one putatively tetraploid population of *A. lyrata petraea* from Austria (collected by S. Ansell) to study the strength and inheritance of self-incompatibility in polyploid *A. lyrata*. In the process of verifying the ploidy of these populations, we have discovered that not only are these populations tetraploid but they also differ in DNA content. Chromosome counts compared to diploids from Iceland ($2n = 2x = 16$) confirmed that both the Japanese and Austrian populations were tetraploid, each with $2n = 4x = 32$ chromosomes (Dart et al. 2004). However, analysis using flow cytometry indicated that, whereas the Japanese population was characterized by an average DNA content approximately twice that of the diploid Icelandic populations, that of the Austrian populations was only about 1.7 times (Fig. 2). Triploids resulting from crosses between the Icelandic (2x) and Japanese (4x) populations fell half-way between the two but overlapped the range of the Austrian tetraploids. This discrepancy may indicate that the two populations of tetraploids have arisen through different mechanisms (i.e., via hybridization between different pairs of species, or perhaps by allopolyploidy versus autopolyploidy) or that the Austrian populations have experienced extensive genomic rearrangements following their formation, as has been described for newly created *Brassica* (Song et al. 1995) and *Arabidopsis* (Comai et al. 2000) tetraploids. More detailed investigation of the geologic and genetic histories of these populations is required. Regardless of the reasons for differences in DNA content, if such differences can be found by looking at only two populations, it is likely that variation in chromosome numbers and/or DNA content may be widespread. This raises the point that, given the prevalence of polyploidy in plants in general, it is not justifiable to assume that DNA content and ploidy are constant in any angiosperm species.

Fig. 2. Plot of mean peak channel based on flow cytometric measurement of relative DNA content in populations of *A. lyrata* sampled from Iceland ($2n = 2x = 16$; *A. lyrata petraea*), Japan ($2n = 4x = 32$; *A. lyrata kawasakiana*) and Austria ($2n = 4x = 32$; *A. lyrata petraea*) as well as hybrids produced between the Icelandic and Japanese populations ($2n = 3x = 24$). Chromosome counts confirmed the ploidies of each population. Note that the Austrian tetraploids have about 1.7 times the DNA content of the diploids, whereas the Japanese tetraploids have nearly exactly twice the DNA content of the diploids.

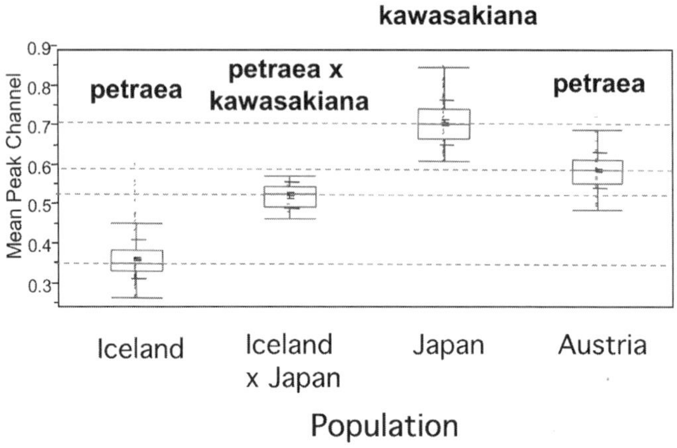

Growth habit

A. lyrata is thought to be a perennial plant that reproduces exclusively by outcrossing (e.g., Nasrallah 2000). At least in the greenhouse, we have noted that *A. lyrata* tends to be highly clonal under high nutrient conditions (at the expense of sexual reproduction) but that this varies by population. Diploid populations from Iceland, for example, produce numerous clones, whereas diploid populations from Indiana, U.S.A. tend to produce few, if any, clones and often not until they have flowered for an extensive period. Whereas plants from most populations will flower for 3–4 years in the greenhouse, some populations (most notably the Japanese population) appear to survive for only a single flowering season and thus display more of an annual growth pattern. We do not yet know whether this is an artifact of growth in the greenhouse environment or whether these differences would also be seen in field populations. This raises the point that when considering a new model species, observations about variability made in the greenhouse should be verified under field conditions.

Self-incompatibility

Our major interest in *A. lyrata* has been as a model to study self-incompatibility (SI), as a naturally occurring relative of the widely studied, cultivated genus, *Brassica*. The assumption that *A. lyrata* is obligately outcrossing is based on extensive observations that inbreeding is not tolerated in populations that exhibit strong evidence of self-incompatibility. In fact, in populations sampled from Indiana, U.S.A. (*A. lyrata lyrata*) and Iceland (*A. lyrata petraea*) we have been unable to raise individuals that are homozygous at the *S*-locus to sexual maturity, as they tend to become highly susceptible to pathogens and die at an early stage (Mable, unpublished). We do find evidence for partial SI in some combinations of genotypes (Mable et al. 2003), but overall, SI in these populations is very strong. In contrast, some populations of diploid *A. lyrata* from Ontario, Canada appear to be self-compatible. We first suspected a population sampled from Rondeau Provincial Park (Lake Erie) to be self-compatible based on observations of a few individuals that produced viable selfed progeny in the greenhouse; subsequent field and laboratory selfings confirmed that individuals in this population are predominantly self-compatible (Mable, Robertson, Dart, Di Berardo and Witham, unpublished). This is particularly unexpected considering the strong self-incompatibility found in the population from Indiana Dunes National Lakeshore (Lake Michigan) (Table 2). This has prompted a survey of several additional populations from the Ontario shores of Lakes Erie, Huron, and Georgian Bay, which has revealed substantial variation in strength of SI within and between populations.

Similar to the results from diploids, the two tetraploid populations examined also differ in strength of self-incompatibility. The Japanese population is highly self-compatible, with viable selfed seeds producing very vigorous progeny, whereas the Austrian population is highly self-incompatible based on greenhouse crosses within families.

The exciting possibility raised by these findings is that we may come closer to discovering the functionally important regions of the *S*-genes by comparing alleles in self-compatible and self-incompatible populations. Using the same PCR-based approach that we have used to identify the *S*-alleles in self-incompatible populations (Schierup et al. 2001; Charlesworth et al. 2003; Mable et al. 2003) we have thus far established that both the self-compatible diploids and the self-compatible tetraploids share alleles at the *S*-locus with self-incompatible populations (Mable, unpublished). We have also established that the Austrian tetraploids, while having a unique set of alleles, show similar patterns of dominance as diploid populations. This unexpected variation in self-incompatibility thus provides new avenues for research that would have been overlooked by more limited sampling of worldwide populations.

Discussion

If *Arabidopsis lyrata* is to be used as a model species, it is clear that substantial variation exists among populations; results based on one population may not directly apply to others. Much more work is required to establish whether variation in this species group is geographic, habitat related, or more randomly distributed. Rather than detracting from its use as a model, the level of variability in morphology, reproductive biology, strength of SI, ploidy, DNA content, and habitat types apparent in this species group makes it a much more interesting system to use as a model than more homogeneous species. Even in self-compatible *A. thaliana* it is apparent that important differences exist among ecotypes that could influence interpretation of results, but the relatively low levels of variation in this species make it difficult to relate such differences to geographic or habitat differences (e.g., Bergelson et al. 1998). Therefore, using a model that is known from the outset to show high levels of variability could provide a valuable tool, as long as each population is recognized as a potentially distinct entity.

While the taxonomy of *A. lyrata* may still require revision, it is the variation among populations sampled from Europe, Asia, and North America that is interesting, regardless of what names are used to describe them. Similarly, the ability to transfer genetic information between *A. lyrata* and *A. thaliana* remains an important tool whether or not *A. lyrata* is considered one species or several.

The occurrence of tetraploidy in a model species allows the application of research questions relating to the evolution of polyploidy that remain controversial and largely untested. For example, a common suggestion about polyploids is that they should exhibit higher levels of selfing than their diploid relatives. This is both because they should be able to tolerate higher levels of inbreeding due to the genetic buffering provided by "extra" gene copies and because chances for perpetuation of rare tetraploid individuals would be increased if they can reproduce without having to find another rare individual of the same ploidy level (e.g., Miller and Venable 2000; Husband and Schemske 1997). A related suggestion is that polyploidy will result in a breakdown in the SI system (Lewis 1943; 1947; Miller and Venable 2000), although this is not universally true (e.g., see Chawla et al. 1997 compared to Young et al. 2000). In fact, even in the gametophytic SI systems for which this phenomenon was originally described, breakdown of SI tends to be genotype specific and is not ubiquitous (Lewis 1943). Study of SI in tetraploid *A. lyrata* emphasizes that even in a single "species", breakdown of SI may not be related to ploidy *per se* but to other factors, which may be revealed by comparison of the SI and SC tetraploid populations described here. As well, extension of studies of the genetics of SI in diploid *A. lyrata* (Schierup et al. 2001; Mable et al. 2003; Charlesworth et al. 2003) provides the potential to examine in detail the inheritance of SI in a self-incompatible tetraploid for the first time.

The finding of diploid self-compatible populations is even more exciting because it provides a tool for discovering more about how SI works in sporophytic systems and allows extension to *A. lyrata* of many of the procedures developed for *A. thaliana* that rely on generating selfed lines. For comparison with similar surveys conducted in European populations (Schierup 1998; Clauss et al. 2002), we are currently conducting an extensive survey of strength of SI, effective selfing rates, ploidy, genetic diversity at allozyme and microsatellite loci, and number and types of *S*-alleles in populations of *A. lyrata* sampled from different habitats around the Great Lakes region.

The example provided by *A. lyrata* emphasizes the importance of conducting extensive ecological and genetic surveys to establish variation in natural populations in model species. By incorporating such variation into the study of genetic control of adaptive traits, advances in genome research might be used more efficiently to understand an important role of genetic change: allowing organisms to survive and reproduce in the face of changing environmental conditions.

Acknowledgements

I would like to thank P. Kron, S. Dart, and J. Beland for plant rearing and technical assistance, D. Charlesworth for continuing discussions, and N. Miyashita, T. Thorhallsdottir, and S. Ansell for providing seeds. Comments by S. Otto and an anonymous reviewer significantly improved the manuscript. This research was supported by the Natural Sciences and Engineering Research Council of Canada.

References

Adams, M.D., Celniker, S.E., Holt, R.A., Evans, C.A., Gocayne, J.D., Amanatides, P.G., Scherer, S.E., Li, P.W., Hoskins, R.A., Galle, R.F., and 185 others. 2000. The genome sequence of *Drosophila melanogaster*. Science (Washington, D.C.) **287**: 2185–2195.

Aguilar, I., Alamillo, J.M., Garcia-Olmedo, F., and Rodriguez-Palenzuela, P. 2002. Natural variability in the *Arabidopsis* response to infection with *Erwinia carotovora subsp. carotovora*. Planta **215**: 205–209.

Al-Shehbaz, I.A., O'Kane, S., and Price, R.A. 1999. Generic placement of species excluded from *Arabidopsis* (Brassicaceae). Novon **9**: 297–307.

Alonso-Blanco, C., El-Assal, S.E., Coupland, G., and Koornneef, M. 1998. Analysis of natural allelic variation at flowering time loci in the Landsberg erecta and Cape Verde Islands ecotypes of *Arabidopsis thaliana*. Genetics **149**: 749–764.

Arabidopsis Genome Initiative. 2001. Analysis of the genome sequence of the flowering plant *Arabidopsis thaliana*. Nature **408**: 796–815.

Beemster, G.T., De Vusser, K., De Tavernier, E., De Bock, K., and Inze, D. 2002. Variation in growth rate between *Arabidopsis* ecotypes is correlated with cell division and A-type cyclin-dependent kinase activity. Plant Physiol. **129**: 854–864.

Bergelson, J., Stahl, E., Dudek, S., and Kreitman, M. 1998. Genetic variation within and among populations of *Arabidopsis thaliana*. Genetics **148**: 1311–1323.

C. elegans Sequencing Consortium. 1998. Genome sequence of the nematode *C. elegans*: a platform for investigating biology. Science (Washington, D.C.) **282**: 2012–2018.

Carlton, J.M., Angiuoli, S.V., Suh, B.B., Kooij, T.W., Pertea, M., Silva, J.C., Ermolaeva, M.D., Allen, J.E., Selengut, J.D., Koo, H.L., Peterson, J.D., and 33 others. 2002. Genome sequence and comparative analysis of the model rodent malaria parasite *Plasmodium yoelii yoelii*. Nature **419**: 512–519.

Charlesworth, D., Awadalla, P., Mable, B.K., and Schierup, M.H. 2000. Population-level studies of multiallelic self-incompatibility loci, with particular reference to Brassicaceae. Ann. Bot. **85** (Suppl. A): 227–239.

Charlesworth, D., Mable, B.K., Schierup, M.H., Bartolomé-Husson, C., and Awadalla, P. 2003. Diversity and linkage of genes in the self-incompatibility gene family in *Arabidopsis lyrata*. Genetics **164**: 1519–1535.

Chawla, B., Bernatzky, R., Liang, W., Marcotrigiano, M. 1997. Breakdown of self-incompatibility in tetraploid *Lycopersicon peruvianum*: inheritance and expression of S-related proteins. Theor. Appl. Genet. **95**: 992–996.

Clarke, J.H., Mithen, R., Brown, J.K., and Dean, C. 1995. QTL analysis of flowering time in *Arabidopsis thaliana*. Mol. Gen. Genet. **248**: 278–286.

Clauss, M.J., Cobban, H., and Mitchell-Olds, T. 2002. Cross-species microsatellite markers for elucidating population genetic structure of *Arabidopsis* and *Arabis* (Brassicaceae). Mol. Ecol. **11**: 591–601.

Comai, L., Tyagi, A.P., Winter, K., Holmes-Davis, R., Reynolds, S.H., Stevens, Y., and Byers, B. 2000. Phenotypic instability and rapid gene silencing in newly formed *Arabidopsis* allotetraploids. Plant Cell **12**: 1551–1567.

Cooley, N.M., Higgins, J.T., Holmes, M.G., and Attridge, T.H. 2001. Ecotypic differences in responses of *Arabidopsis thaliana* L. to elevated polychromatic UV-A and UV-B+A radiation in the natural environment: a positive correlation between UV-B+A inhibition and growth rate. J. Photochem. Photobiol. B **60**: 143–150.

Cui, J., Jander, G., Racki, L.R., Kim, P.D., Pierce, N.E., and Ausubel, F.M. 2000. Signals involved in *Arabidopsis* resistance to *Trichoplusia ni* caterpillars induced by virulent and avirulent strains of the phytopathogen *Pseudomonas syringae*. Plant Physiol. **129**: 551–564.

Dart, S., Kron, P., and Mable, B.K. 2004. Characterizing polyploidy in *Arabidopsis lyrata* using chromosome counts and flow cytometry. Can. J. Bot. **82**(2): 185–197.

Dennis, E.S., Bilodeau, P., Burn, J., Finnegan, E.J., Genger, R., Helliwell, C., Kang, B.J., Sheldon, C.C., and Peacock, W.J. 1998. Methylation controls the low temperature induction of flowering in *Arabidopsis*. Symp. Soc. Exp. Biol. **51**: 97–103.

Gardner, M.J., Hall, N., Fung, E., White, O., Berriman, M., Hyman, R.W., Carlton, J.M., Pain, A., Nelson, K.E., Bowman, S., and 35 others 2002. Genome sequence of the human malaria parasite *Plasmodium falciparum*. Nature **419**: 498–511.

Gardner, S.L., Burrell, M.M., and Fry, S.C. 2002. Screening of *Arabidopsis thaliana* stems for variation in cell wall polysaccharides. Phytochemistry **60**: 241–254.

Gibson, G. 2002. Microarrays in ecology and evolution: a preview. Mol. Ecol. **11**: 17–24.

Goff, S.A., Ricke, D., Lan, T.H., Presting, G., Wang, R., Dunn, M., Glazebrook, J., Sessions, A., Oeller, P., Varma, H., and 45 others. 2002. A draft sequence of the rice genome (*Oryza sativa* L. ssp. *japonica*). Science (Washington, D.C.) **296**: 92–100.

Holt, R.A., Subramanian, G.M., Halpern, A., Sutton, G.G., Charlab, R., Nusskern, D.R., Wincker, P., Clark, A.G., Ribeiro, J.M., Wides, R., and 113 others 2002. The genome sequence of the malaria mosquito *Anopheles gambiae*. Science (Washington, D.C.) **298**: 129–149.

Husband, B.C., and Schemske, D.W. 1997. The effect of inbreeding in diploid and tetraploid populations of *Epibolium angustifolium* (Onagraceae): implications for the genetic basis of inbreeding depression. Evolution **51**: 737–746.

Innan, H., Tajima, F., Terauchi, R., and Miyashita, N.T. 1996. Intragenic recombination in the *Adh* locus of the wild plant *Arabidopsis thaliana*. Genetics **143**: 1761–1770.

Innan, H., Terauchi, R., and Miyashita, N.T. 1997. Microsatellite polymorphism in natural populations of the wild plant *Arabidopsis thaliana*. Genetics **146**: 1441–1452.

Kersten, B., Burkle, L., Kuhn, E.J., Giavalisco, P., Konthur, Z., Lueking, A., Walter, G., Eickhoff, H., and Schneider, U. 2002. Large-scale plant proteomics. Plant Mol. Biol. **48**: 133–141.

Kliebenstein, D.J., Gershenzon, J., and Mitchell-Olds, T. 2001*a*. Comparative quantitative trait loci mapping of aliphatic, indolic and benzylic glucosinolate production in *Arabidopsis thaliana* leaves and seeds. Genetics **159**: 359–370.

Kliebenstein, D.J., Lambrix, V.M., Reichelt, M., Gershenzon, J., and Mitchell-Olds, T. 2001*b*. Gene duplication in the diversification of secondary metabolism: tandem 2-oxoglutarate-dependent dioxygenases control glucosinolate biosynthesis in *Arabidopsis*. Plant Cell **13**: 681–693.

Koch, M.A., Bishop J.G., and Mitchell-Olds, T. 1999. Molecular systematics and evolution of *Arabidopsis* and *Arabis*. Plant Biol. **1**: 529–537.

Koch, M.A., Haubold, B., and Mitchell-Olds, T. 2000. Comparative evolutionary analysis of the chalcone synthase and alcohol dehydrogenase loci among different lineages of *Arabidopsis*, *Arabis* and related genera (Brassicaceae). Mol. Biol. Evol. **17**: 1483–1498.

Koch, M.A., Haubold, B., and Mitchell-Olds, T. 2001. Molecular systematics of the Brassicaceae: evidence from coding plastidic *matK* and nuclear *Chs* sequences. Am. J. Bot. **88**: 534–544.

Kusaba, M., Dwyer, K.G., Hendershot, J., Vrebalov, J., Nasrallah, J.B., and Nasrallah, M.E. 2001. Self-incompatibility in the genus *Arabidopsis*: characterization of the S locus in the outcrossing *A. lyrata* and its autogamous relative *A. thaliana*. Plant Cell **13**: 627–643.

Kusaba, M., Tung, C.-W., Nasrallah, M.E., and Nasrallah, J.B. 2002. Monoallelic expression and dominance interactions in anthers of self-incompatible *Arabidopsis lyrata*. Plant Physiol. **128**: 17–20.

Lander, E.S, Linton, L.M., Birren, B., Nusbaum, C., Zody, M.C., Baldwin, J., Devon, K., Dewar, K., Doyle, M., FitzHugh, W., Funke, R., and 245 others. International Human Genome Sequencing Consortium. 2001. Initial sequencing and analysis of the human genome. Nature **409**: 860–921.

Le Corre, V., Roux, F., and Reboud, X. 2002. DNA polymorphism at the FRIGIDA gene in *Arabidopsis thaliana*: extensive nonsynonymous variation is consistent with local selection for flowering time. Mol. Biol. Evol. **19**: 1261–1271.

Levin, D.A. 1983. Polyploidy and novelty in flowering plants. Am. Nat. **122**: 1–25.

Lewis, D. 1943. Physiology of incompatibility in plants. III. Autopolyploids. J. Genet. **45**: 171–185.

Lewis, D. 1947. Competition and dominance of incompatibility alleles in diploid pollen. Heredity **1**: 85–108.

Mable, B.K., Schierup, M.H., and Charlesworth, D. 2003. Estimating the number, frequency, and dominance of *S*-alleles in a natural population of *Arabidopsis lyrata* (Brassicaceae) with sporophytic control of self-incompatibility. Heredity **90**: 422–431.

Mantyla, E., Lang, V., and Palva, E.T. 1995. Role of abscisic acid in drought-induced freezing tolerance, cold acclimation, and accumulation of LT178 and RAB18 proteins in *Arabidopsis thaliana*. Plant Physiol. **107**: 141–148.

Masterson, J. 1994. Stomatal size in fossil plants: evidence for polyploidy in majority of angiosperms. Science (Washington, D.C.) **264**: 421–423.

Miller, J.S., and Venable, D.L. 2000. Polyploidy and the evolution of gender dimorphism in plants. Science (Washington, D.C.) **289**: 2335–2338.

Mitchell-Olds, T. 2001. *Arabidopsis thaliana* and its wild relatives: a model system for ecology and evolution. Trends Ecol. Evol. **16**: 693–699.

Miyashita, N.T., Kawabe, A., and Innan, H. 1999. DNA variation in the wild plant *Arabidopsis thaliana* revealed by amplified fragment length polymorphism analysis. Genetics **152**: 1723–1731.

Miyashita, N.T., Kawabe, A., Innan H., and Terauchi, R. 1998. Intra- and interspecific DNA variation and codon bias of the alcohol dehydrogenase (*Adh*) locus in *Arabis* and *Arabidopsis* species. Mol. Biol. Evol. **15**: 1420–1429.

Mulligan, G.A. 1964. Chromosome numbers of the family Cruciferae I. Can. J. Bot. **42**: 1509–1519.

Nasrallah, M.E. 2000. *Arabidopsis* species hybrids—emerging model systems for the analysis of species differences. J. Plant Growth Reg. **19**: 326–333.

Nasrallah, M.E., Yogeeswaran, K., Snyder, S., and Nasrallah, J.B. 2000. *Arabidopsis* species hybrids in the study of species differences and evolution of amphiploidy in plants. Plant Physiol. **124**: 1605–1614.

O'Kane, S., and Al-Shehbaz, I.A. 1997. A synopsis of *Arabidopsis* (Brassicaceae). Novon **7**: 323–327.

Orbovich, V., and Taras'ev, A. 1999. Genetic differences in plastic responses to density between ecotypes of *Arabidopsis thaliana*. Genetika **35**: 631–640.

Otto, S.P., and Whitton, J. 2000. Polyploidy incidence and evolution. Annu. Rev. Genet. **34**: 401–437.

Park, S.H., Hur, J., Park, J., Lee, S., Lee, T.K., Chang, M., Davi, K.R., and Kim, J. 2002. Identification of a tolerant locus on *Arabidopsis thaliana* to hypervirulent beet curly top virus CFH strain. Mol. Cell. **13**: 252–258.

Polatschek, A. 1966. Cytotaxonomische Beitraege zur Flora der Ostalpenlaender. Österr. Bot. Zeitschritft **113**: 1–10.

Price, R.A., Palmer, J.D., and Al-Shehbaz, I.A. 1994. Systematic relationships of *Arabidopsis*: a molecular and morphological perspective. *In* Arabidopsis. *Edited by* E.M. Myerowitz and C.R. Sommerville. Cold Spring Harbor Laboratory Press, Cold Spring Harbor, NY. pp. 7–19.

Reichelt, M., Brown, P.D., Schneider, B., Oldham, N.J., Stauber, E., Tokuhisa, J., Kliebenstein, D.J., Mitchell-Olds, T., and Gershenzon, J. 2002. Benzoic acid glucosinolate esters and other glucosinolates from *Arabidopsis thaliana*. Phytochemistry **59**: 663–671.

Rhee, S.Y., Beavis, W., Berardini, T.Z., Chen, G., Dixon, D., Doyle, A., Garcia-Hernandez, M., Huala, E., Lander, G., Montoya, M., Miller, N., Mueller, L.A., Mundodi, S., Reiser, L., Tacklind, J., Weems, D.C., Wu, Y., Xu, I., Yoo, D., Yoon, J., and Zhang, P. 2003. The Arabidopsis Information Resources (TAIR): a model organism database providing a centralized, curated gateway to *Arabidopsis* biology, research, materials and community. Nucleic Acids Res. **31**: 224–228.

Rollins, R.C. 1966. Chromosome numbers of Cruciferae. Contrib. Gray Herbarium **197**: 43–65.

Rutherford, R., and Masson, P.H. 1996. *Arabidopsis thaliana* sku mutant seedlings show exaggerated surface-dependent alteration in root growth vector. Plant Physiol. **111**: 987–998.

Savolainen, O., Langley, C.H., Lazzaro, B.P., and Fréville, H. 2000. Contrasting patterns of nucleotide polymorphism at the alcohol dehydrogenase locus in the outcrossing *Arabidopsis lyrata* and the selfing *Arabidopsis thaliana*. Mol. Biol. Evol. **17**: 645–655.

Schierup, M.H. 1998. The effect of enzyme heterozygosity on growth in a strictly outcrossing species, the self-incompatible *Arabis petraea* (Brassicaceae). Hereditas **128**: 21–31.

Schierup, M.H., Mable, B.K., Awadalla, P., and Charlesworth, D. 2001. Identification and characterization of a polymorphic receptor kinase gene linked to the self-incompatibility locus of *Arabidopsis lyrata*. Genetics **157**: 387–399.

Schoof, H., Zaccaria, P., Gundlach, H., Lemcke, K., Rudd, S., Kolesov, G., Arnold, R., Mewes, H.W., and Mayer, K.F. 2002. MIPS Arabidopsis thaliana Database (MAtDB): an integrated biological knowledge resource based on the first complete plant genome. Nucleic Acids Res. **30**: 91–93.

Sharbel, T.F., Haubold, B., and Mitchell-Olds, T. 2000. Genetic isolation by distance in *Arabidopsis thaliana*: biogeography and postglacial colonization of Europe. Mol. Ecol. **9**: 2109–2118.

Song, K., Lu, P., Tang, K., and Osborn, T.C. 1995. Rapid genome change in synthetic polyploids of *Brassica* and its implications for polyploid evolution. Proc. Natl. Acad. Sci. U.S.A. **92**: 7719–7723.

Suzuki, M., Takahashi, T., and Komeda, Y. 2002. Formation of corymb-like inflorescences due to delay in bolting and flower development in the corymbosa2 mutant of *Arabidopsis*. Plant Cell Physiol. **43**: 298–306.

Taylor, R.L., and Mulligan, G.A. 1968. Chromosome numbers of vascular plants of British Columbia. Syesis **10**: 12, 5–138.

Torabinejad, J., and Caldwell, M.M. 2000. Inheritance of UV-B tolerance in seven ecotypes of *Arabidopsis thaliana* L. Heynh. and their F1 hybrids. J. Hered. **91**: 228–233.

Ullrich, H., Lattig, K., Brennicke, A., and Knoop, V. 1997. Mitochondrial DNA variations and nuclear RFLPs reflect different genetic similarities among 23 *Arabidopsis thaliana* ecotypes. Plant Mol. Biol. **33**: 37–45.

Ungerer, M.C., Halldorsdottir, S.S., Modliszewski, J.L., Mackay, T.F., and Purugganan, M.D. 2002. Quantitative trait loci for inflorescence development in *Arabidopsis thaliana*. Genetics **160**: 1133–1151.

van Der Schaar, W., Alonso-Blanco, C., Leon-Kloosterziel, K.M., Jansen, R.C., van Ooijen, J.W., and Koornneef, M. 1997. QTL analysis of seed dormancy in *Arabidopsis* using recombinant inbred lines and MQM mapping. Heredity **79**: 190–200.

van Treuren, R., Kuittinen, H., Kärkkäinen, K., Baena-Gonzalez, E., and Savolainen, O. 1997. Evolution of microsatellites in *Arabis petraea* and *Arabis lyrata*, outcrossing relatives of *Arabidopsis thaliana*. Mol. Biol. Evol. **14**: 220–229.

Wong, C.E., Carson, R.A., and Carr, J.P. 2002. Chemically induced virus resistance in Arabidopsis thaliana is independent of pathogenesis-related protein expression and the NPR1 gene. Mol. Plant-Microbe Interact. **15**: 75–81.

Wood, V., Gwilliam, R., Rajandream, M.A., Lyne, M., Lyne, R., Stewart, A., Sgouros, J., Peat, N., Hayles, J., Baker, S., and 125 others. 2002. The genome sequence of *Schizosaccharomyces pombe*. Nature **415**: 871–880.

Yanovsky, M.J., Casal, J.J., and Luppi, J.P. 1997. The VLF loci, polymorphic between ecotypes Landsberg erecta and Columbia, dissect two branches of phytochrome A signal transduction that correspond to very-low-fluence and high-irradiance responses. Plant J. **12**: 659–667.

Yeast Genome Initiative. 1997. The yeast genome directory. Nature **387** (6632) suppl.: 5.

Young, A.G., Millar, C., Gregory, E., and Langston, A. 2000. Sporophytic self-incompatibility in diploid and tetraploid races of *Rutidosis leptorrhynchoides* (Asteraceae). Aust. J. Bot. **48**: 667–672.

Yu, J., Hu, S., Wang, J., Wong, G.K., Li, S., Liu, B., Deng, Y., Dai, L., Zhou,Y., Zhang, X., and 91 others. 2002. A draft sequence of the rice genome (*Oryza sativa* L. ssp. *indica*). Science (Washington, D.C.) **296**: 79–92.

17
Natural variation among accessions of *Arabidopsis thaliana*: beyond the flowering date, what morphological traits are relevant to study adaptation?

X. Reboud, V. Le Corre, N. Scarcelli, F. Roux, J.L. David, T. Bataillon, C. Camilleri, D. Brunel, and H. McKhann

Abstract: A set of 240 wild type accessions of *Arabidopsis thaliana* were cultivated in two contrasting conditions, with or without application of a cold treatment prior to cotyledon emergence. In each condition, 20 primary and 3 derived variables were measured that describe the phenology and morphology of each plant. The cold pre-treatment greatly modified the time to flowering but diversely affected accessions (according to their need for vernalization).

Three criteria were then used to identify a minimum subset within the 23 variables to be measured under a time-saving constraint. We combined a criterion of highest genetic heritability so that environmental experimental variance is lowered, a criterion of highest Spearman coefficient so that the rank of any accessions remained stable across treatments and, last, a criterion of highest contribution of the variable to the seed production as estimated in a multilinear regression analysis. Applying this 'three criteria' procedure lead us finally to propose a minimum set of five variables: flowering precocity, maximum plant height, height to first flower, number of flowering heads and mean distance between siliques as best describing variation in life history traits expressed by the whole collection. We believe that these variables definitely affect the ability of *Arabidopsis* to adjust its life cycle to ecological conditions including the intensity of the interspecific competition prevailing in the environment. The core collection of 24 *Arabidopsis* accessions that was chosen for maximizing molecular diversity was confirmed here to also maximize most of each trait's variability. Some linkage disequilibrium expressed at the whole genome level when considering accessions from very diverse origin is probably responsible for the conservation of such a high diversity. These 24 accessions could thus provide an important resource for natural variation to be exploited in the identification of quantitative trait loci (QTL), in genotype/phenotype association studies or exploration of ecological and evolutionary relations.

Introduction

The collections maintained in stock centers often consist of a very large number of accessions making them hard to handle and sometimes even to exploit. Strategies to construct core collections have been designed to tackle this problem so that a restricted set of accessions is available that encompasses the range of diversity of the full collection. This strategy has been applied successfully to *Arabidopsis thaliana* by surveying genetic diversity using SNPs present at 10 fragments spread throughout the genome (two per chromosome) (McKhann et al. 2004). The resulting core collection consists of 24 accessions selected from a larger collection of 265 wild type *Arabidopsis* from diverse origins.

On the other hand, if we wish to characterize a large number of accessions in a comparative study, it would be advantageous to reduce the time spent on measuring each accession. This is clearly another optimization problem. Of all the morphological and phenological traits that could be measured, which ones best describe the variation between accessions? Of course, we do not expect a unique list of traits, as each specific study will require the precise analysis of a relevant set of characteristics. However, some traits are more sensitive to environmental noise than others and thus more difficult to analyze experimentally. Some traits are also highly correlated so that most of the information collected by measuring one trait can be used to infer the values of the

X. Reboud, V. Le Corre, and F. Roux. UMR 1210 'Biologie et Gestion des Adventices', INRA, B.P. 86510, 21065 Dijon cedex, France.
C. Camilleri, D. Brunel, and H. McKhann. Centre National de Recherche en Génomique, 2 rue Gaston Crémieux, 91057 Evry cedex, France et Station de Génétique et Amélioration des Plantes, Route de Saint-Cyr, INRA, 78026 Versailles cedex, France.
N. Scarcelli. IRD Montpellier, UMR 1097 — Equipe DYNADIV, 911, avenue Agropolis, BP 64501, 34394 Montpellier cedex 5, France.
J.L. David and T. Bataillon. UMR 1097 Diversité et Génomes des Plantes Cultivées, INRA, Domaine de Melgueil, 34130 Mauguio, France.
Correct citation: Reboud, X., LeCorre, V., Scarcelli, N., Roux, F., David, J.L., Bataillon, T., Camilleri, C., Brunel, D., and McKhann, H. 2004. Natural variation among accessions of *Arabidopsis thaliana*: beyond the flowering date, what morphological traits are relevant to study adaptation? *In* Plant Adaptation: Molecular Genetics and Ecology. Proceedings of an International Workshop held December 11–13, 2002, in Vancouver, British Columbia, Canada. *Edited by* Q.C.B. Cronk, J. Whitton, R.H. Ree, and I.E.P. Taylor. NRC Research Press, Ottawa, Ontario. pp. 135–142.

other. Moreover, some variables have more impact on the final morphology, phenology and fitness of accessions so that natural selection probably acted more intensely on these characters. For all these reasons, there will be sets of variables that are more informative than others. When the information on the precise ecological conditions prevailing in the area of origin of the samples is not available or when very diverse accessions are to be compared in a limited number of experiments under necessarily standardized conditions, what are the most appropriate set of variables to measure?

The sampling effort for *Arabidopsis* has been far from uniform over the species' geographic range, some areas providing many more accessions than others, and therefore, it does not reflect an underlying variation in species abundance (Alonso-Blanco and Koornneef 2000). The ecological characterization of the sampling origin also remains astonishingly scarce for most accessions and only few studies have provided some phenotypic characterization of *Arabidopsis* plants collected from different geographical origins. The traits most often analyzed are those affecting flowering time. The variation of this character is presumed to be of ecological significance for adaptation and highly variable in *Arabidopsis* (list of references at: http://www.dpw.wau.nl/natural/resources/literature.htm).

To respond or not to cold treatment: that is the question…

Variation in flowering responses after cold treatment has been the object of intense analysis. At the two extremes are plants that require a cold treatment to induce flowering and those that do not respond to cold. These two flowering "strategies" are thought to be adaptive. For the first group, cold sensitivity is considered as a means of synchronizing vegetative plant development during autumn and mild winter periods, with flowering occurring in early spring after a cold winter. With the ability to tolerate frost, these accessions remain at the rosette stage during the autumn when they set reserves to ensure the earliest possible resumption of growth before interspecific competition begins in the spring. Their ecological niche would thus be a very narrow window for rapid development and seed setting before taller and stronger plants species develop. Consequently, such accessions may suffer the fitness cost of being unable to complete a second generation within the same year even under favorable conditions because of lack of the required conditions to induce flowering.

The second group, being rather insensitive to cold treatment, is able to flower as soon as plants have reached a given biomass or accumulated day × degree threshold. In ecological niches where the favorable growing period is short, these cold-insensitive plants will also only reproduce once a year; whereas in optimal conditions they may complete more than one cycle per year and then outperform the genotypes requiring cold to induce flowering. There is however a second indirect expectation that these cold-insensitive genotypes could have only been selected under environmental conditions where virtually no local competition exists. Hence, flowering time, which has been shown to be under complex genetic determinism (Sheldon et al. 2000), has probably evolved under diversifying selection (Le Corre et al. 2002).

Other forces may act on the optimum time for flowering. Physiological constraints dictate that the smaller the plant, the smaller its pollen and seed output. In a rosette species such as *Arabidopsis thaliana* where bolting puts an end to further resource accumulation (there will be no more leaves), flowering is expected to be positively correlated with total seed production. Thus, there are probably opposing forces acting on optimal flowering period in any particular condition: completing more than one cycle and/or avoiding competition requires short generation times on one hand, while total yield requires bigger plants produced with long generation times on another hand. Indeed, the fact that the flowering trait is complex can be viewed as the consequence of the necessity to locally adjust the flowering time cycle in any possible way. Different *Arabidopsis* populations experiencing selection for early flowering would thus differ in their evolutionary response to the trade-off between total productivity and flowering time. This suggests that other genetically determined traits may be involved in ecological adaptation. The identification and study of these traits would allow us to better understand trade-offs between flowering time and other life history traits. Using both vernalized and non-vernalized treatments on a set of *Arabidopsis* accessions enables the identification of a set of traits having a high impact on fitness over a range of environmental conditions. We have therefore explored variation in morphological traits describing several aspects of inflorescence architecture among 240 accessions of *Arabidopsis thaliana* from stock center collections. Using the two contrasting conditions of three weeks vernalization at 4°C versus no cold treatment, we tested the stability of these traits in the two treatments, their genetic component (heritability), as well as their contribution to total seed production.

The work described here thus contributes to the selection of morphological and phenological traits that capture and describe the greatest proportion of variation in life history traits in *Arabidopsis* accessions. The identification of such traits will allow optimization of procedures that attempt to describe the variation in life histories between accessions. We then validate the core collection of 24 accessions by determining how well it captures the diversity in these chosen traits.

Material and methods

Choice of multiple criteria to optimize a set of informative variables

Contribution to seed production

Examining the most important components of the seed production such as vegetative biomass or number of flowering units is a common practice to estimate fitness values in selfing plant species in which male and female reproductive fitness are both reflected in seed set. Identification of factors contributing to seed productivity and quality is also a common exercise for plant breeders aiming to improve crop yield. However, identification of the most adaptive traits in wild plants like *Arabidopsis* is sometimes hindered by the plasticity of the response when measured in standard laboratory conditions. Under absence of cold treatment and infinite time to flower, the latest plants to flower would present rosettes with high number of leaves and astonishing large size, well outside the natural size range of rosettes observed under natural conditions. Having accumulated a lot of resources, these plants can be highly productive but only under those particular artificial conditions. Experimental measures may thus sometimes be only poorly informative on how accessions indeed vary under their natural ecological range. This fact is acknowledged in the "genotype × environment interaction, or reaction norms" literature (Pigliucci and Schlichting 1997; Sultan 2000 and references therein). It may therefore be misleading to rely on the single criterion of correlation with total seed production to decide what variables are more valuable when measured in a restricted subset. So another criterion also has to be applied when considering the most valuable traits to be measured.

High genetic heritability

To determine the adaptive significance of a trait, another key criterion is its potential to respond to selection in a given environment, measured by its genetic variance in this environment. The higher this variance compared to that induced by environmental heterogeneity, the higher the heritability of the trait. Unfortunately, heritability also strongly depends on the choice of accessions as well as on the experimental conditions under which a given trait is measured. Further, a trait may be highly heritable but stable across a range of environmental conditions or highly heritable but conversely strongly affected by an environmental gradient (the genetic component building the phenotype in each particular condition). Therefore, the degree of plasticity of a trait may be unrelated to its heritability value and another criterion could still valuably be added.

Conservation of accessions ranking among environments

Finally, to progress in the understanding of plant adaptation, we could compare the same traits among several conditions in search for the most stable patterns. Doing so will not necessarily focus our choice on the ecologically most relevant traits because the interesting traits may be those showing high plasticity or because stability could first of all reflect the evenness of an underlying selective force across the natural species range that has built up that trait. Still, the underlying assumption in the choice of a stability criterion would be that this strategy will necessarily maximize the probability that a laboratory-measured response will reflect most precisely the response expected under natural conditions. For example, trichome number at the surface of the leaf is considered as stable across environmental variations such as degree of humidity or day length. Therefore, the trichome number as observed in greenhouse conditions reflects the value expected under natural conditions. As few data are available on the ecological origin of the accessions, we have no way to either reproduce the natural range of conditions or to get *a priori* knowledge of the most interesting variables, and thus, in these particular conditions, the quality of the measure to reflect the 'natural' value has also to be taken into account.

Hence, highest impact on fitness, highest heritability, and highest stability across variable conditions (here cold treatment) will be the three criteria under which we will compare the beneficial output for each of the 23 variables that have been measured here for more than 1400 plants (240 plants × 2 conditions × 3 replicates).

Plant material

The *Arabidopsis thaliana* collection used for morphological data consisted of 240 accessions, primarily obtained from the Arabidopsis Biological Resource Center (ABRC) at the Ohio State University and Nottingham (NASC) stock centers or collected by French (Lavigne et al. 2001; Le Corre et al. 2002) and Japanese (Todokoro et al. 1995) groups. From this set of 240 accessions plus an additional 25, a core collection has been generated (McKhann et al. 2004) using the maximization strategy of SNP allelic richness and Nei's diversity index employing the MSTRAT software (Gouesnard et al. 2001). This strategy has been shown to perform particularly well when accessions are from populations with restricted gene flow or when accessions are primarily selfing such as is the case of *Arabidopsis* (Bataillon et al. 1996). Twenty-four accessions could capture the majority (384 out of 399) of all the SNPs discovered by analyzing 10 fragments (two per chromosome) of 517–660 bp. This set of 24 accessions belonging to the so-called "core collection" is described in Table 1. All accessions were submitted to one round of multiplication prior to this experiment to homogenize seed age and maternal effects.

Collecting morphological data

All plants were placed in the greenhouse on the same day during Spring 2001 under the same growing conditions, but they differed in their pre-planting treatment: 3 weeks at 4°C for vernalization versus direct seed sowing for the non-vernalized plants. Each accession was represented by three plants in each treatment. The two sets of plants (vernalized versus non-vernalized) were grown separately according to an incomplete balanced randomized block design (the replicates are randomized within each block; blocks are not mixed but instead are maintained as group units so that two replicates can never be in the same tray). Eight trays were used as block units for each of the two treatments. On each tray, plants were separated from each other by 3 cm so that rosettes experienced some competition by the end of their development. The outermost line was planted with *A. thaliana* accession Ler to limit possible border effects and was excluded from further analysis. During growth, several primary traits were measured: time to first flower (FLOR); time to first mature silique as monitored by silique yellowing (MATURE); rosette diameter after 21 days (DIAM21); height from soil to first flower (H1FL) and number of leaves on the rosette at bolting (LEAF). All other traits were measured at harvest. They were: the diameter of the first axis at rosette emergence (STEMDIA); the height from soil to first silique (H1SIL) as well as maximum plant height (HMAX); the cumulated length of all primary or secondary branches bearing siliques separately on the primary axis (LSILAP) and on the secondary axis (LSILAS); the numbers of cauline leaves (CAULIN), flowering axes (AXIS), primary and secondary branches on the first and secondary axes (NRAM1AP, NRAM1AS, NRAM2AP, NRAM2AS respectively), number of green and mature siliques on first and secondary axes (SILVAP, SILVAS, SILMAP, SILMAS respectively). Last, derived traits were the total number of

Table 1. Core collection of 24 accessions of *Arabidopsis thaliana*.

Stock center number	Versailles number	Name	Country	Latitude	Longitude
N1094	162AV	Ct-1	Italy	37.3	15.06
N1436	224AV	Oy-0	Norway	60.23	6.13
N929	236AV	Shahdara	Tadjikistan	37.29	71.3
N1030	180AV	Blh-1	Czech Republic	48.49	16.45
N1028	172AV	Bur-0	Eire	53.07	–9.04
	25AV	JEA	France	43.41	7.2
N902	166AV	Cvi-0	Cape Verde Islands	16	–24
N1244	157AV	Ita-0	Morocco	34.04	–4.12
N1186	101AV	Ge-0	Switzerland	46.12	6.1
N1656	178AV	Alc-0	Spain	40.29	–3.22
N1534	62AV	St-0	Sweden	59.19	18.03
N1064	163AV	Can-0	Canary Islands	28	–15.3
	8AV	PYL-1	France	44.39	–1.1
N22491	266AV	Konchezero	Russia	62.07	34.01
N1380	94AV	Mt-0	Libya	32.34	22.46
N1368	215AV	Mh-1	Poland	53.31	20.12
	257AV	Sakata	Japan	38.55	139.5
N1210	200AV	Gre-0	U.S.A	43.11	–85.15
N1122	83AV	Edi-0	United Kingdom*	50.57	–3.13
N1286	70AV	Kn-0	Lithuania	54.54	23.54
N1564	91AV	Tsu-0	Japan	34.19	129.19
	252AV	Akita	Japan	39.43	140.06
N968	42AV	Bl-1	Italy*	44.29	11.2
N1538	92AV	Stw-0	Russia	52.57	36.04

*Botanical garden.

flowering heads (HEADS) measured by summing the variable AXIS to the four variables NRAM1AP, NRAM1AS, NRAM2AP, NRAM2AS; the total silique production (TOT SIL), a sum of the four variables SILVAP, SILVAS, SIL MAP, SILMAS; and the mean distance between siliques (LEN), as total length of silique areas (LSILAS + LSILAP) divided by adjusted number of branches (TOTSIL – HEAD) to take into account that there is always one more silique than inter-silique number on each branch.

These variables thus describe the height and volume occupied by the inflorescence, its degree of ramification and compactness.

Heritability of traits

The heritability of each trait was computed independently for each treatment using decomposition into several variance components as estimated by the VARCOMP procedure in SAS®. Here, the block (tray) effect was considered as fixed while the accessions were defined as a random factor. Assuming that accessions were completely homozygous, their associated variance component was thus considered an estimation for the genetic variance V(G), while the within-accession-within-tray component (error factor) was taken as an estimation of the environmental error V(E). Assuming that maternal effect was negligible, since all seeds came from mother plants previously multiplied under the same green house conditions, heritability, h^2, was then computed as the proportion of the variance under the genetic component V(G)/ (V(G)+V(E)). As in this experiment we could not further decompose the variance to access the additive part of the genetic variance, h^2 as presented here is a *sensu lato* heritability.

Analyzing stability between treatments

The Spearman's rank correlation was taken as a simple measure of the maintenance of the same accession ranking when changing from the vernalized to non-vernalized treatment. A variable with a high Spearman coefficient would ensure that an accession behaves similarly relative to other accessions irrespective of the cold pre-treatment; i.e., a small value in one environment would be matched by a small value in the other. Thus, variables showing opposite reaction norms in different accessions according to the treatment would definitely not be favored by this criterion. We also plotted cluster trees of the 23 variables using the Pearson coefficient as a distance measure to illustrate the relationship among variables and the degree of pattern conservation among variables across treatments.

Extracting fitness components from the inflorescence architecture

With such a high number of accessions, nearly all traits were found to be significantly correlated to fitness. Thus, using a forward stepwise option, variables were ranked by the number of steps necessary to get them to enter into a multilinear model estimating the total silique production (TOT SIL). This was carried out for each treatment independently using stepwise probabilities $p_{\text{enter}} = 0.15$, $p_{\text{remove}} = 0.05$ to enter or remove variables from the model, respectively. Under both treatments, nine variables introduced in the model were sufficient for the linear regression coefficient R^2 to reach high values. As there was no clear further improvement of the model fitting when entering supplementary variables, we stopped the sorting at step nine. In Table 2, only those nine variables kept in the model have their rank at entering (all others being labeled "*ne*" for not entered in the model). Ranking orders were then compared between the two treatments and the most often encountered variables were kept in the list of primary components of fitness. Of course, as they were directly correlated to total seed production, variables giving number of siliques or length of silique areas (which poorly represent architecture) were excluded from this analysis. Our approach therefore optimizes a complementary set of variables that best models the seed production.

Results and discussion

Heritabilities and correlations among traits

The 20 primary and 3 indirect variables studied exhibited a wide range of heritability values from virtually zero for the number of first branches on the secondary axis (NRAM1AS) in the vernalized treatment to 0.87 for flowering time (FLOR) in the non-vernalized treatment. With about 240 degrees of freedom, all heritability values exceeding 0.04 were statistically highly significant ($p < 0.001$). In most cases, the cold treatment tended to homogenize phenotypic expression across accessions, leading to lower heritability values. It is easy to illustrate that choosing the most significant variables according to this single criterion would have been misleading. For example, the heritability index considered alone would lead to keeping both flowering date (FLOR) and number of rosette leaves at flowering (LEAF) because they had the highest h^2 values in both treatments. However, it has long been established that these two variables are so tightly correlated that many teams in fact use the number of leaves on the rosette as their index of flowering precocity. Part of the measuring effort would thus be lost by measuring both of these rather redundant variables.

With 23 variables measured under two conditions, we could compare the correlation between pairs of variables as well as the correlation for a single variable between treatments (paired by accession number). We found that, in almost all cases, the correlation between two traits measured in the same environment (treatment) was higher than the correlation for the same trait between the two treatments. This observation demonstrates that the necessary equilibrium between different plant compartments such as roots and leaf area or other physiological constraints as shaped by the growth environment prevails over the genetic background in determining the relationship among traits. The comparison between the two cluster trees presented in Fig. 1 illustrates the degree of conservation of the relationship between variables. Some clusters such as the one containing the flowering time FLOR, the time to silique maturation MATURE and the number of leaves on the rosette at bolting LEAF are indeed highly conserved. Note also that variables such as NRAM1AP and NRAM1AS, although referring to a same character (number of ramifications but on a different flowering axis) can belong to different clusters, suggesting that their timing, underlying physiological and/or genetic pathways are different.

Relationships between traits and fitness

Individual fitness, measured as the total number of siliques, was shown to result from somewhat different sets of explicative traits according to the treatment considered. Without vernalization, greater variation was observed among accessions in the length of their vegetative phase, mainly because more accessions were late flowering. In this case, more plants could accumulate enough resources to develop a complex architecture and the number of siliques produced then depended on the maximum plant height, the number of siliques on the main reproductive axis and the total number of reproductive axes. In contrast, under vernalization, the vegetative phase was reduced for most plants so that the production of siliques mainly depended on the number of ramifications on the first axis. Thus, the effective significance of the trade-off between precocity and seed production seems strongly shaped by environmental conditions. Whatever the treatment, all accessions flowering with less than 22 rosette leaves at bolting suffered a fitness cost that can only be compensated in their original environment by some other advantage conferred by earliness.

Stability of traits across vernalized versus non-vernalized conditions

As shown in Table 2, the most stable traits were also those having the highest heritability values, so that the two criteria of tththehigh heritability and high ranking stability of the accessions across treatments, were found here to be very consistent. This result was not obvious, since the variation at regulatory genes underlying traits with a strong genetic basis may result in highly variable norms of reaction to environmental conditions, and thus to a high "instability" for the trait. However, in the case of flowering time (and related traits), the stability in the ranking of accessions may be explained by the underlying regulation mechanism. Flowering time depends on the repressing action of the MADS-box regulatory gene FLC (Michaels and Amasino 1999; Sheldon et al. 2000), whose level of expression is constitutively upregulated by several genes (FRIGIDA and other genes in the "autonomous pathway") that determine flowering time in the absence of vernalization, and eventually, down-regulated by

Table 2. Ranking of 23 morphological variables under vernalized and non-vernalized treatments using their respective values calculated using three different criteria. See Materials and Methods for details on each criterion. Columns are coefficient and variable ranks under each treatment, respectively.

		Criterion 1				Criterion 2		Criterion 3	
		Heritability				Regression stepwise entering rank		Spearman correlation	
Nickname	Variable	Non-vernalized		Vernalized		Non-vernalized	Vernalized	Coefficient	Variable rank
LEAF	Number of leaves	0.82	**2**	0.73	**1**	6	*ne*	0.602	**1**
FLOR	Flowering time	0.86	**1**	0.69	**2**	7	*ne*	0.583	**2**
NRAM1AP	Number of primary branches on first axis	0.72	**4**	0.63	**3**	*ne*	ne	0.548	**4**
CAULIN	Number of cauline leaves	0.75	**3**	0.62	**4**	*ne*	*ne*	0.580	**3**
MATURE	Mature silique time	0.71	**5**	0.61	**5**	*ne*	7	0.503	**7**
H1SIL	Height of first silique	0.59	**8**	0.56	**6**	9	5	0.511	**5**
HMAX	Maximum height	0.53	**9**	0.51	**7**	1	2	0.507	**6**
STEMDIA	Main stem diameter	0.44	13	0.51	**8**	8	6	0.406	13
LEN	Mean length between siliques	0.41	15	0.47	**9**	3	3	0.397	14
LSILAP	Cumulated length of siliques along first axis	0.43	14	0.15	10	*ne*	*ne*	0.498	**8**
SILVAP	Number of green siliques on first axis	0.40	16	0.38	11	*ne*	*ne*	0.367	19
NRAM2AP	Number of secondary branches on first axis	0.39	17	0.38	12	*ne*	**4**	0.378	16
SILMAP	Number of siliques on first axis	0.49	11	0.35	13	*ne*	*ne*	0.200	22
H1FL	Height of first flower	0.53	10	0.34	14	*ne*	8	0.481	**9**
HEAD	Total heads	0.38	19	0.33	15	**4**	1	0.451	10
TOTSIL	Total siliques	0.39	18	0.32	16	*ne*	*ne*	0.367	19
DIAM21	Rosette diameter at day 21	0.37	20	0.27	17	*ne*	*ne*	0.371	17
AXIS	Number of axes	0.09	22	0.20	18	*ne*	**9**	0.360	20
SILVAS	Number of green siliques on second axis	0.62	7	0.15	19	*ne*	*ne*	0.319	21
LSILAS	Cumulated length of siliques along second axis	0.47	12	0.33	20	*ne*	*ne*	0.368	18
SILMAS	Number of mature siliques on second axis	0.15	21	0.11	21	*ne*	*ne*	0.383	15
NRAM2AS	Number of secondary branches on second axis	0.69	**6**	0.00	22	**5**	*ne*	0.451	10
NRAM1AS	Number of primary branches on second axis	0.04	23	0.00	23	**2**	*ne*	0.411	12

ne = *not entered* into the multiple regression model.

vernalization. As effect of a given vernalization treatment probably depends on the constitutive level of expression of FLC, the more late flowering accessions in the absence of vernalization are also the more late-flowering ones after vernalization.

Here we propose a set of the most relevant variables based on the quality of the data collected as estimated via our multiple criteria applied here. These variables are flowering time (or the number of rosette leaves), maximum plant height, mean distance between siliques, total number of flowering heads and height to first silique (as well as main stem diameter to a lesser degree). It is clear from Fig. 1 that this set of variables covers different parts of the cluster trees. It demonstrates the low degree of redundancy between these variables (that otherwise would fall on the same branch of the cluster). Retaining only these variables allows accurate mod-

Fig. 1. Cluster trees between the 23 variables.

Trees were constructed separately under each condition using complete (conservative) clustering option and Pearson coefficient as distance measure. 100 re-sampling bootstraps on accessions were performed on each tree to test mean distance between variables, cluster branches and robustness of nodes. Bootstrap values exceeding 60% are added on graphs.

Variables: The four variables are highlighted in grey (LEN: mean distance between siliques; H1SIL: height from soil to first silique; HEAD: total number of flowering heads; HMAX: maximum plant height) and flowering time (FLOR) in italic. Belonging to different cluster branches, these four variables have limited redundancy (especially under the non-vernalized treatment). Note that NRAM1AP and NRAM1AS do not cluster tightly although they both refer to ramification but on a different flowering axis.

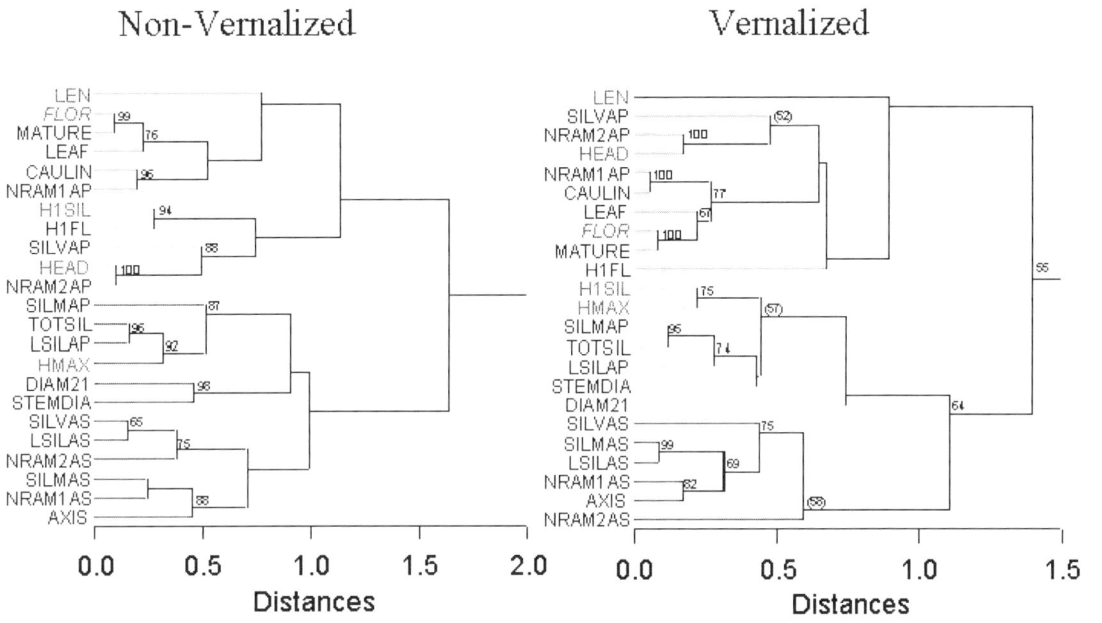

eling of total seed production in both treatments with R^2 = 0.948 and 0.952 for vernalized and non-vernalized treatments, respectively. With the exception of H1SIL under non-vernalized conditions (p = 0.005), all variables have high significant impact on the regression (p < 0.001) and thus all affect seed production.

Diversity in a core collection of 24 accessions

The core collection of 24 accessions selected for maximizing molecular diversity was then tested for its ability to cover the total range of morphological variation as well as its distribution evenness for each of the five more interesting variables. Because both vernalized and non-vernalized data sets are very similar, the illustration presented in Fig. 2 as a bar chart for each variable separately or Scatterplot matrix for pairs of variables is from the non-vernalized data set only.

We demonstrated that the core collection of 24 accessions contained significantly more variation for morphological traits than random samples of the same size (drawn from the full collection of 240 accessions) when analyzed globally on a set of morphological characters (data not shown). This is also illustrated here for individual trait distribution (bar charts in Fig. 2). For the core collection, the distribution for each trait is flattened compared to that of the whole sample and extreme values are often represented (as seen for FLOR, HMAX and HEAD). At the same time, the core collection is not only a subset of the most extreme accessions but rather includes accessions that regularly cover the whole range. In other words, the core collection contains accessions that occupy most of the morphological space and not a restricted set of accessions belonging only to the external envelope bordering that morphological space. Similarly, the areas covered by the core collection in the two by two plots are close to the maximum area of the whole sample. With 10% of the total number of the accessions (24 out of 240), these regions always exceed 120% of the mean areas covered by 24 accessions taken at random.

Many scientific teams in the field of population genetics and evolution have developed programs to sample, describe and relate the variation present in natural populations to adaptation (e.g., Van Tienderen et al. 2002). In the model plant *Arabidopsis thaliana*, such diversity is often found in accessions collected over a wide range of ecological and geographic conditions along the species distribution. An approach using quantitative trait loci (QTL) on a subset of crosses between some of these natural accessions can allow the identification and isolation of genes responsible for the control of a particular trait. This is not possible when using the mutant approach because of the narrow genetic background of these lines. Natural accessions submitted to contrasting environmental conditions over long time periods have often developed a unique original combination of genes or alleles to fit a local optimum. Analyzing the diversity of the genetic responses developed by different accessions may allow the identification of different (parts of) metabolic pathways affecting a trait. It is therefore a complementary approach to the mutant analysis strategy for establishing gene functions.

Fig. 2. Representativeness of the 24 accessions belonging to the core collection for the flowering time and each of the four added variables. Evenness of the distribution is depicted on the bar charts, and the surrounding shapes demonstrate the diversity covered by the core collection for each pair of variables as compared to a world-wide set of accessions.

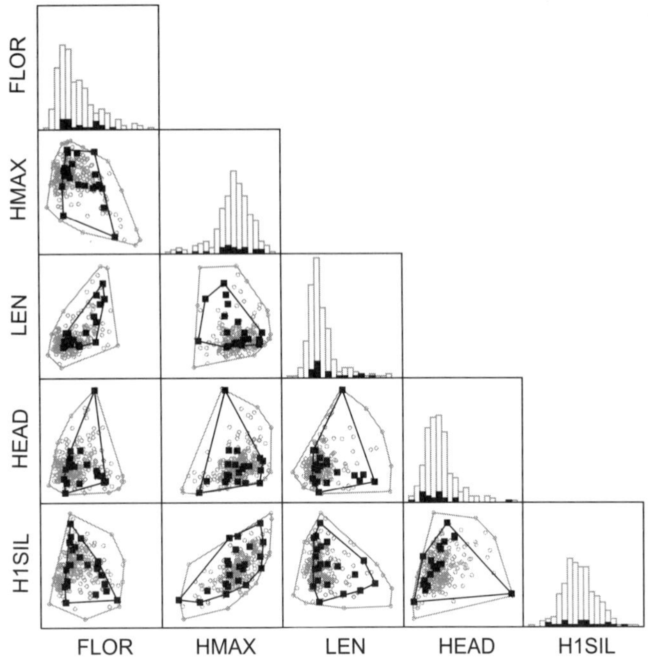

Conclusion

We have used multiple approaches to analyze 23 morphological characteristics from 240 accessions grown under two different cold treatments to search for environmentally stable and heritable fitness components in wild accessions of *Arabidopsis thaliana*. Four traits were selected in addition to the already well-established flowering precocity: maximum plant height, mean distance between siliques, number of flowering heads and height to first silique. All of these traits are easy to measure and could significantly reduce the effort needed to characterize natural diversity and to obtain a rough estimation of fitness. We believe that these traits reflect the phenotypic ability to thrive in a more or less competitive environment as determined by contrasting climatic conditions including the winter cold period.

This study also demonstrated the validity of the core collection strategy as a key method to seek natural diversity. Indeed, the core collection of 24 accessions maximizing the molecular diversity at the DNA level was confirmed here to cover most of each variable range. A reasonable amount of diversity was thus available in such a small sample and the core collection could be used to study the way genetic variation modifies the physiological relationships between traits in different contrasting environments, to search for allelic polymorphism in some candidate genes known to be involved in a trait and eventually to produce segregating populations for QTL mapping and cloning.

Acknowledgements

We are grateful to Michel Schoutith and Maurice Bourlier for their technical assistance, to Paul Neve who helped clarify the ideas presented in the introduction. We also thank Aurélie Bérard for help in generating the core collection. This study was supported by the 'Aracore' grant from French Research and Technology Minister.

References

Alonso-Blanco, C., and Koornneef, M. 2000. Naturally occurring variation in *Arabidopsis*: an underexploited resource for plant genetics. Trends Plant Sci. **5**: 22–29.

Bataillon, T.M., David, J.L., and Schoen, D.J. 1996. Neutral genetic markers and conservation genetics: simulated germplasm collections. Genetics **144**: 409–417.

Gouesnard, B., Bataillon, T.M., Decoux, G., Rozale, C., Schoen, D.J., and David, J.L. 2001. MSTRAT: An algorithm for building germ plasm core collections by maximizing allelic or phenotypic richness. J. Hered. **92**: 93–94.

Lavigne, C., Reboud, X., Lefranc, M., Porcher, E., Roux, F., Olivieri, I., and Godelle, B. 2001. Evolution of genetic diversity in metapopulations: *Arabidopsis thaliana* as an experimental model. Genet. Select. Evol. **33** (Suppl. 1): S399–S423.

Le Corre, V., Roux, F., and Reboud, X. 2002. DNA polymorphism at the FRIGIDA gene in *Arabidopsis thaliana*: extensive nonsynonymous variation is consistent with local selection for flowering time. Mol. Biol. Evol. **19**: 1261–1271.

McKhann, H.I., Camilleri, C., Berard, A., Bataillon, T., David, J.L., Reboud, X., Le Corre, V., Caloustian, C., Gut, I.G., and Brunel, D. 2004. Nested core collections maximizing genetic diversity in *Arabidopsis thaliana*. Plant J. **38**: 193–202.

Michaels, S.D., and Amasino, R.M. 1999. FLOWERING LOCUS C encodes a novel MADS domain protein that acts as a repressor of flowering. Plant Cell **11**: 949–956.

Pigliucci, M., and Schlichting, C.D. 1997. Reaction norms of *Arabidopsis* V. Flowering time controls phenotypic architecture in response to nutrient stress. J. Evol. Biol. **3**: 285–301.

Sheldon, C.C., Finnegan, E.J., Rouse, D.T., Tadege, M., Bagnall, D.J., Helliwell, C.A., Peacock, W.J., and Dennis, E.S. 2000. The control of flowering by vernalization. Curr. Opin. Plant Biol. **3**: 418–422.

Sultan, S.E. 2000. Phenotypic plasticity for plant development, function and life history. Trends Plant Sci. **5**: 537–542.

Todokoro, S.R., Terauchi, R., and Kawano, S. 1995. Microsatellite polymorphisms in natural populations of *Arabidopsis thaliana* in Japan. Jap. J. Genet. **70**: 543–554.

Van Tienderen, P.H., de Haan, A., van der Linden, C.G., and Vosman, B.L. 2002. Biodiversity assessment using markers for ecologically relevant traits. TREE **17**: 577–582.

18
The study of ancient adaptation: a case study of a phytochrome gene pair from early-diverging angiosperms

Sarah Mathews

Abstract: Recent phylogenetic, ecological, and physiological analyses of flowering plants suggest that members of the earliest diverging lines initiated seedling development in shaded environments or under water, processes that may have required innovation in light-sensing mechanisms. Strong evidence for episodic selection in the photoreceptor phytochrome A (phyA) early in the history of angiosperms suggests that structural changes in phyA were adaptive. The functions of phyA are unique among angiosperm phytochromes, allowing them to initiate development in deep shade and/or in response to brief pulses of light. The possibility that these functions originated in newly evolving angiosperms and were important to their initial establishment is being tested in two ways. The first involves surveying seed plants and early-diverging angiosperms to determine the distribution of phyA functions characterized in *Arabidopsis*. The second involves testing the functional and fitness effects of altering the positively selected sites by using sequence constructs to transform *phyA* null mutants of *Arabidopsis*. This case study highlights the utility of tests for selection to extend the study of ancient adaptation to cases in which detailed protein structural data are lacking and to provide functional insights missing from the study of model species.

Introduction

Evolutionary biologists seek to understand both past events and present day processes, requiring the synthesis of data from "two quite disparate disciplines" (Antonovics 1987). Studies of past events traditionally have been in the realms of systematics and the comparative method, focusing on the history of lineage splitting and on the origin of traits and their evolution. More recently, these have expanded to include investigations of adaptation at the molecular level from which past episodes of selection may be inferred, and comparative studies of development, from which the developmental changes leading to alternate phenotypes may be inferred. Studies of present day evolutionary processes mainly have been in the realms of population or ecological genetics, focusing on the forces by which traits originate and are maintained in populations or species. More recently, these have expanded to include identification of the genes underlying adaptive traits. Regardless of the starting point, it is an arduous task to (1) identify traits that are adaptive, (2) infer their past history, (3) characterize their underlying biological mechanisms, and (4) identify the ecological agents of selection. Here I describe the investigation of a photoreceptor gene pair in early-diverging flowering plants in order to highlight the strengths for studying adaptive evolution of studies that start with past events, in this case, with evidence of ancient episodic selection. This study began with a demonstration that positive selection occurred in the photosensory domain of phytochrome A (phyA) during its divergence from phytochrome C (phyC) and that this episode of selection occurred early in the history of flowering plants (Mathews et al. 2003). To link this episode of selection with a change in phenotype, or functional process, the phylogenetic distribution of physiological responses mediated by phyA in early-diverging species and of phyA-like responses in other seed plants is being investigated. To determine the roles that the positively selected sites have on phyA function, mutagenesis experiments are being used to test the physiological and fitness effects of changing these amino acids in *Arabidopsis*. Together these experiments will test the hypothesis that innovation in the function of phyA conferred an adaptive advantage to newly evolving angiosperms.

Sarah Mathews. Arnold Arboretum of Harvard University, 22 Divinity Avenue, Cambridge, MA 02138, U.S.A.
e-mail address: smathews@oeb.harvard.edu
Correct citation: Mathews, S. 2004. The study of ancient adaptation: a case study of a phytochrome gene pair from early-diverging angiosperms. *In* Plant Adaptation: Molecular Genetics and Ecology. Proceedings of an International Workshop held December 11–13, 2002, in Vancouver, British Columbia, Canada. *Edited by* Q.C.B. Cronk, J. Whitton, R.H. Ree, and I.E.P. Taylor. NRC Research Press, Ottawa, Ontario. pp. 143–152.

Finding molecular adaptation in phytochrome A

Phytochrome function and phylogeny

Through the process of photomorphogenesis, plants modify their development according to the cues they receive from shade, reflected light, direct light, and directional light. Phytochromes are among the photoreceptors that plants use to process these environmental light signals that are so crucial to their competition and survival. They are essential during all of the major developmental transitions, germination, de-etiolation, and flowering, and they play a principal role in shade avoidance (reviewed by Fankhauser and Chory 1997).

Phytochrome evolution in land plants is marked by a series of gene duplications that have led to independently evolving and functionally distinct lines (Mathews and Sharrock 1997). A duplication preceding the origin of seed plants resulted in two distinct lines that persist in all extant seed plants. Phylogenetic analyses suggest that subsequent duplications occurred in each of these lines, leading to the four major forms found in angiosperms. These are phytochromes A, B, C and E (phyA-C, E), encoded by *PHYA-C*, and *E* (Mathews et al. 1995; Mathews and Sharrock 1997). *PHYA* and *PHYC* form one related pair; *PHYB* and *PHYE* form a second related pair. The model plant species, *Arabidopsis thaliana*, has an additional phytochrome gene, *PHYD*, which results from a recent duplication of *PHYB* (Mathews et al. 1995) that occurred along the branch leading to Brassicaceae *sensu stricto* (K. McBreen and S. Mathews unpublished). In cycads, *Ginkgo*, and conifers, a duplication in the *PHYA/C*-related line led to *PHYN* and *PHYO*, but the *PHYB/E*-related line, *PHYP*, apparently did not diversify in other seed plants except for Pinaceae (Schneider-Poetsch et al. 1998; Clapham et al. 1999; Schmidt and Schneider-Poetsch 2002; S. Mathews unpublished). These phylogenetic analyses also suggest that the duplication leading to *PHYA* and *PHYC* occurred prior to the origin of angiosperms (Mathews et al. 1995; Mathews and Sharrock 1997; Mathews and Donoghue 1999), although its position relative to the origin of gymnosperm lines remains unclear and the question of whether separate duplications occurred in gymnosperms and on the line to angiosperms remains open (Fig. 1). *PHYA* and *PHYC* are about 50% identical (Sharrock and Quail 1989) and they have distinct functions (Casal et al. 1997; Halliday et al. 1997; Qin et al. 1997; Eichenberg et al. 2000; Franklin et al. 2003; Monte et al. 2003).

Altered selective constraints in the evolution of *PHYA* and *PHYC*

Relative rate tests provided evidence that functional constraints on *PHYA* and *PHYC* shifted during their divergence (Mathews and Sharrock 1997; Alba et al. 2000; Mathews et al. 2003). In relative rate tests, a comparison is made of the estimated number of substitutions that have accumulated in two related species or genes during their divergence from a third, more distantly related species or gene. If sequences remain under the same constraints during their divergence, they should accumulate a similar number of substitutions. However, these tests do not allow us to (1) distinguish whether shifts in rates resulted from relaxation of selective constraints, from episodic selection, or from a combination of these two processes, (2) identify the phylogenetic branch on which constraints changed, nor (3) identify individual amino acid mutations that might have been important in functional divergence. A test that addresses all three of these questions is the branch-sites test for selection developed by Yang and Nielsen (2002). This test can identify episodes of selection that occurred a long time ago and involved only a small number of sites. Whether this test identifies all the sites at which important changes might have occurred is not known. For example, Mathews et al. (2003) mapped amino acid changes onto the phylogeny of *PHYA* and *PHYC*, providing evidence that sites not identified in tests for selection might have contributed to functional divergence.

The branch-sites test has twice been applied to data sets of *PHYA* and *PHYC* sequences. Yang and Nielsen (2002) analyzed the data set of Alba et al. (2000), which included full-length coding sequences but did not include any species that diverged early in the history of angiosperms, nearer to the origin of the genes themselves. No outgroup sequence was

Fig. 1. Phytochrome phylogeny within land plant phylogeny. This summary phylogeny of *PHY1* and *2* in ferns, *PHYN-P* in gymnosperms, and *PHYA-C* and *E* in angiosperms is consistent with published and unpublished gene phylogenies (Mathews and Sharrock 1997; Schneider-Poetsch et al. 1998; Schmidt and Schneider-Poetsch 2002; S. Mathews, unpublished data). The branch separating the *PHYN/O* and *PHYA/C* pairs (gray line) is not unambiguously supported and the relationships among these genes remain under investigation. Taxon-specific phytochrome diversifications within conifers and angiosperms are not shown. *Amborella* and the water lilies diverge from the rest of the angiosperms before all other groups originate.

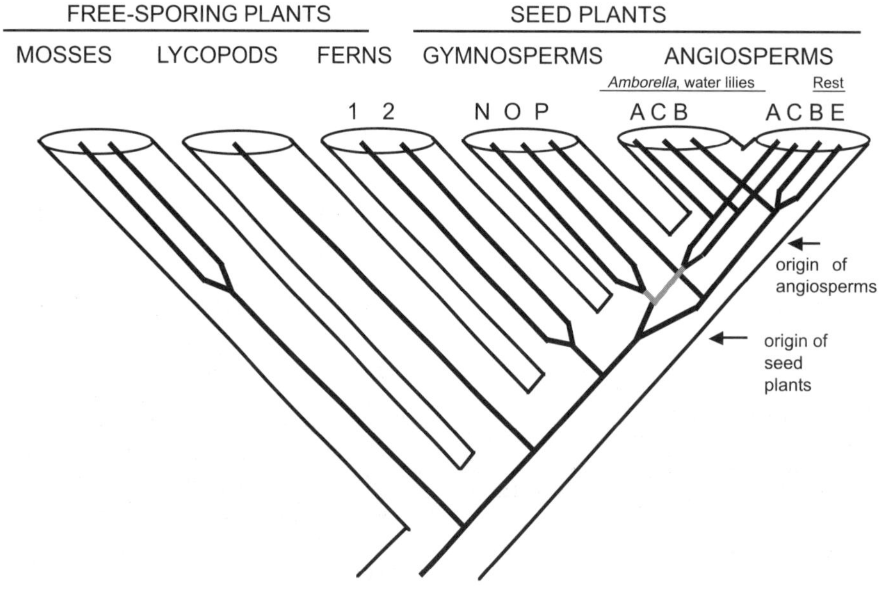

used in the analysis, and sites potentially under selection were identified with reference to the *PHYA* sequence from maize, a derived monocot. Thus, although they identified a number of positively selected sites, the branch on which selection occurred could not be determined. Mathews and colleagues (Mathews et al. 2003) analyzed the data set of Mathews and Donoghue (2000), which included photosensory domain sequences (~ 1.2 kb) from representatives of each of the early diverging angiosperm lines as well as representatives of eudicot and monocot clades nested within angiosperms. They included the published *PHYA/PHYC*-related sequence from *Picea PHYO* (U60264) in order to root the phylogeny. Fig. 2 shows a summary phylogeny of the *PHYA* (45) and *PHYC* (45) sequences included in their study, with branch lengths proportional to the number of amino acid changes that map unambiguously to the *PHYA* (32 changes) and *PHYC* (7 changes) branches. Mathews et al. (2003) showed that an elevation in non-synonymous rates resulted from an episode of selection involving eleven of the amino acid sites that change on the branch leading to all angiosperm *PHYA*. These sites lie within two subdomains of the phytochrome photosensory domain, one of which includes the site of chromophore attachment. Changes in some of the sites involve changes in charge that are conserved in all *PHYA* and that may have influenced interactions between the protein and chromophore and/or between the photosensory and regulatory domains. Conversely, they found no evidence of positive selection on *PHYC*. Moreover, they found no evidence of altered constraints within either the *PHYA* or *PHYC* clades. Instead, most amino acid sites (95%) were selectively constrained, and the ratio of non-synonymous to synonymous substitutions on branches within the *PHYA* clade did not differ from the ratio on the branches within the *PHYC* clade. Thus, positive selection at a handful of sites, rather than relaxation of selective constraints, apparently played a major role in the evolution of the photosensory domain of phytochrome A. This episode of selection occurred very early in the history of flowering plants, raising two questions. Does the signal of adaptive evolution in phyA at the base of the flowering plants reflect the origin of phyA function in angiosperms? Might any of the potentially unique functions of phyA have conferred an adaptive advantage to newly evolving angiosperms?

Linking phyA with a phenotype or functional process

Among angiosperm phytochromes, phyA has unique functions (summarized in Fig. 3) that enhance its capacity to serve a transient role under conditions where an extremely high sensitivity is required (Furuya and Schäfer 1996). This results from the fact that phyA accumulates in dark-grown seedlings but decays rapidly in the light, due both to degradation and downregulation (Somers et al. 1991), and from its role as mediator of very low fluence responses (VLFRs), which require exposure to only millisecond pulses of broad spectrum light (Botto et al. 1996). It also is unique in that the far-red high irradiance response (FR-HIR) depends on the conversion of far-red absorbing phytochrome (Pfr) to red absorbing phytochrome (Pr) induced by the absorption of continuous far-red light (FRc; Shinomura et al. 2000). In contrast, the other angiosperm phytochromes (including phyC) are light stable, do not accumulate in the dark to the same levels as phyA and their activities rely primarily on the conversion of Pr to Pfr in response to continous (Rc) or pulsed red light (R). An exception to this is FRc induction of seed germination in *Arabidopsis*, mediated by phyE (Hennig et al. 2002). However, FRc usually plays an inhibitory role in seed germination (Casal and Sanchez 1998) and, as with phyB and phyC, the primary activities of phyE are induced by absorption of either continuous (Rc) or pulsed red light (R).

Fig. 2. Summary of the phylogeny of *PHYA* and *PHYC* (45 sequences each) from early-diverging angiosperms presented by Mathews et al. (2003), rooted with the *PHYO* sequence from *Picea* (Pinaceae). Branch lengths are proportional to the total number of amino acid changes that map unambiguously to the *PHYA* (32) and *PHYC* (7) branches using parsimony. The same ancestral states for the *PHYA* branch are reconstructed at these sites using maximum likelihood. The ratios of nonsynonymous to synonymous substitutions (ω) were estimated under branch-sites model B of Yang and Nielsen (2002) for four site classes, with the branch to all *PHYA* (bold line) designated as the foreground branch. A proportion of the sites (p_0) are conserved on all branches of the tree; a proportion (p_1) are evolving under relaxed constraints; a proportion (p_3) are conserved throughout most of the tree but may be under selection along the foreground branch; a proportion (p_4) are evolving under relaxed constraint throughout most of the tree but may be under selection along the foreground branch. The significantly better fit of this model when compared in a likelihood ratio test with the site-specific model M3, which assumes just two ratios (ω_0 and ω_2) throughout the tree, together with finding that $\omega_2 > 1$, constitutes a test of selection along the branch to all *PHYA* and provides strong evidence that ~5% of the sites in the photosensory domain sequences of *PHYA* were under selection early in the history of angiosperms (Mathews et al. 2003).

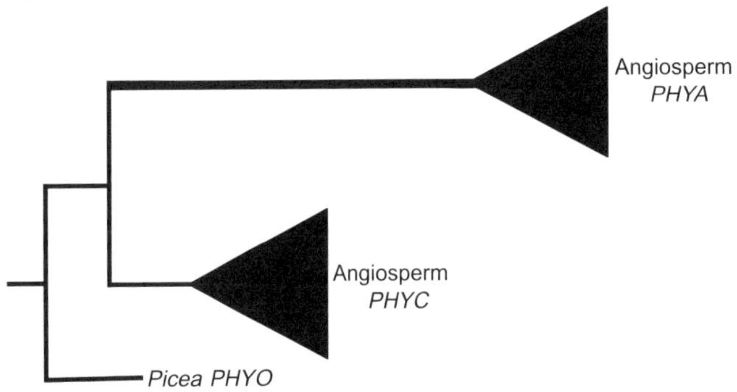

Parameters estimated for the phylogeny of *PHYA* and *PHYC* under site-specific model M3 and branch-sites model B

M3
$p_0 = 0.73$, $p_1 = 0.27$
$\omega_0 = 0.03$, $\omega_1 = 0.20$

Model B
$p_0 = 0.70$, $p_1 = 0.25$, $(p_2 + p_3) = 0.05$
$\omega_0 = 0.03$, $\omega_1 = 0.2$, $\omega_2 = 9.78$

Model M3 versus Model B
$2\Delta l = 11.22$; $P = 0.004$

Fig. 3. Phylogeny of *Arabidopsis* phytochromes. Columns to the right of each gene indicate the stability in light of their product, their photosensory specificity and physiological response mode, and the process that they induce in a particular response mode. phyA is the primary mediator of responses to FR and of very low fluence responses; phyB is the primary mediator of responses to R and the R/FR ratio. FR-HIR, far-red high irradiance response; VLFR, very low fluence response; R-LFR, red low fluence response; R/FR, red to far-red light ratio.

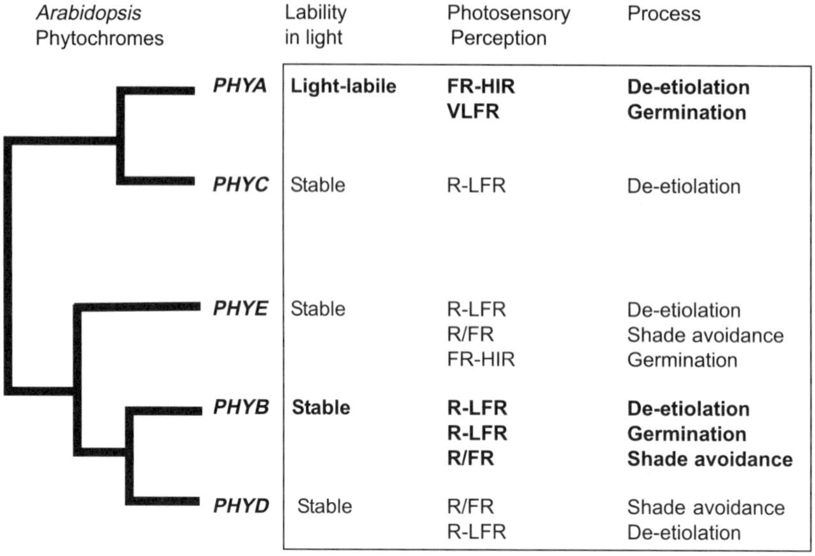

While phyA is known to have functions that are unique among angiosperm phytochromes, it is important to establish when these functions arose. If they occur in other seed plants and are encoded by an orthologous gymnosperm gene, or if functional innovation in phyA occurred later in angiosperm history, it is less likely that the origin of phyA function/s conferred an adaptive advantage to early angiosperms. To infer the ancestral functions of phyA in angiosperms requires characterization of phytochrome functions in basal angiosperms and in extant gymnosperms (conifers, cycads, *Ginkgo*, and Gnetales) and identification of the genes that encode them. It also requires better resolution of the gene phylogeny. Gymnosperms have two candidate orthologs of *PHYA*, *PHYN* and *PHYO*. Previous phylogenetic analyses have suggested that separate duplications led to the *PHYA/C* and *PHYN/O* pairs as shown in Fig. 1. However, this is best viewed as an open question. Current analyses with additional data cannot rule out the possibility that a single gene duplication early in the history of seed plants led to *PHYA* and *PHYN* on one side of the split and *PHYC* and *PHYO* on the other side.

The FR-HIR and its distribution in seed plants

Phytochrome null mutants of *Arabidopsis*, tomato, and rice demonstrate that phyA is the primary photoreceptor for the FR-HIR (Nagatani et al. 1993; Parks and Quail 1993; Whitelam et al. 1993; van Tuinen et al. 1995; Takano et al. 2001). The far-red high irradiance response mediates de-etiolation in response to FRc. Etiolation is viewed as a specialized developmental pathway that is used to delay development in the dark or under low light conditions (e.g., McNellis and Deng 1995). Etiolated angiosperm seedlings require a light signal to de-etiolate, or to develop into a compact, green seedling with fully expanded seed leaves. Some members of the conifer family Pinaceae have a rudimentary FR-HIR (Burgin et al. 1999). However, the degree of etiolation, and thus the need to de-etiolate, varies among conifers; many are green in the dark (Burgerstein 1900; Bogorad 1950; Mukai et al. 1992). Fernbach and Mohr (1990) also demonstrated that in the conifer *Pinus sylvestris*, FRc fails to inhibit hypocotyl elongation, an important component of de-etiolation. *Ginkgo* etiolates (Burgerstein 1900; Chinn and Silverthorne 1993) and also shows elements of the FR-HIR de-etiolation response (Christensen et al. 2002). Some cycads (*Cycas* and *Zamia*), and some Gnetales (*Ephedra* and *Gnetum*) also etiolate (Bower 1882; Burgerstein 1900), but their ability to respond to FRc needs to be investigated.

The VLFR and its distribution in seed plants

Phytochrome A also is the primary photoreceptor for very low fluence responses, allowing dark-imbibed seeds to germinate in response to millisecond pulses of broad spectrum light (Botto et al. 1996). The VLFR also contributes to de-etiolation under low light conditions (Smith et al. 1997). No evidence has been presented that plants other than angiosperms have VLFRs and the condition in early-diverging angiosperms remains unknown. The VLFR operates via a downstream pathway different than that of the FR-HIR (Yanovsky et al. 1997; Cerdán et al. 1999; Casal 2000; Yanovsky et al. 2002); thus these two phyA functions might have originated at different times.

The distribution of a light-labile pool of phytochrome in seed plants

Contributing factors to the transient role and the sensitivity of phyA in angiosperms are its accumulation in the dark and rapid decay in the light. Burgin et al. (1999) detected a pool of phytochrome in dark-grown seedlings of two conifers (both Pinaceae). This pool did decay in the light, but at a significantly slower rate than did the phyA pool in cucumber and oats. A similar finding has been reported for *Ginkgo* (Christensen et al. 2002), but other seed plants and early-diverging angiosperms remain to be characterized.

Together, these observations suggest that other seed plants are unable to de-etiolate completely in response to FRc, that they lack very low fluence responses, and that the phytochrome pool present in dark-grown tissues is more stable in the light than is angiosperm phyA. Thus, several of the functions of phyA appear to be unique to angiosperms. Since the distribution of phyA responses in the earliest angiosperms remains unknown, the possibility remains that the unique functions of phyA arose after the angiosperms had already established. However, it is reasonable to hypothesize that the evidence of positive selection in phyA may reflect the origin and/or enhancement of its functions in these species. If this hypothesis were correct, then (1) we would expect extant descendants of the earliest-diverging lineages to show the full range of responses that are attributed to phyA in model species, and (2) we would expect them to be lacking or incompletely developed in other seed plants. Phytochrome-mediated responses in early-diverging flowering plants, as well as in additional seed plants, are currently being characterized in order to test this hypothesis. The utility of this approach requires the assumption that extant descendants of early diverging lineages are appropriate ecological analogs of the first angiosperms, an assumption which is addressed below.

Does phyA increase fitness?

The ecology of early angiosperms

The question here is not just whether phyA increases fitness of extant species, but whether it might have conferred an adaptive advantage to newly evolving angiosperms, which faced the challenge of establishing in a landscape dominated by ferns and gymnosperms. The failure of *phyA* null mutants of *Arabidopsis* to survive in the shade suggests that its functions do increase fitness (Yanovsky et al. 1995). Moreover, since phyB inhibits de-etiolation in the shade, it is reasonable to infer that evolution of a photoreceptor to counter this inhibition (assuming it was present in early-diverging angiosperms) would be advantageous in dimly lit environments. However, since the ability to initiate development in the shade would provide an adaptive advantage only if early angiosperms colonized shaded environments, it is important to understand their ecology.

The angiosperm crown group originated approximately 140 to 190 million years ago (Sanderson and Doyle 2001) and fundamentally changed the terrestrial landscape. With the rise of the angiosperms, several lineages of plants that had been dominant since the Triassic declined in dominance or became extinct (Hickey and Doyle 1977; Doyle 1978; Knoll 1984; Friis et al. 1987; Lidgard and Crane 1988; Crane et al. 1995; Lupia et al. 1999). Competing hypotheses have emphasized two divergent ecological settings and plant morphologies associated with the initial establishment of flowering plants (Doyle and Donoghue 1993). With respect to the light environment, the first angiosperms have been portrayed either as understory woody plants of wet tropical rain forests (Arber and Parkin 1907; Bessey 1915; Bews 1927; Axelrod 1952; Takhtajan 1969; Thorne 1974; Cronquist 1988), or as semi-woody to herbaceous colonizers of disturbed, open habitats (Stebbins 1974; Doyle and Hickey 1976; Hickey and Doyle 1977; Doyle and Donoghue 1986; Taylor and Hickey 1992; 1996; Wing and Boucher 1998).

Recent phylogenetic analyses have identified the extant descendants of the earliest-diverging angiosperm lineages, which include shrubs and woody vines that occur in shady habitats (Mathews and Donoghue 1999; 2000; Parkinson et al. 1999; Qiu et al. 1999; Soltis et al. 1999; 2000; Barkman et al. 2000; Doyle and Endress 2000; Graham et al. 2000; Savolainen et al. 2000; Zanis et al. 2002). These are *Amborella* and a clade including the families Austrobaileyaceae, Illiciaceae, Schisandraceae, and Trimeniaceae. The water lilies (Nymphaeales) also diverge early, but probably not before the lineage containing the woody understory species *Amborella trichopoda* of New Caledonia (Zanis et al. 2002). Implications for character evolution of the recently discovered fossil aquatic plant *Archaefructus liaoningensis* remain unclear in light of the low support for its placement in phylogenetic trees (Sun et al. 2002); however, similarities to *Cabomba* suggest that it may be a member of the water lily lineage (James Doyle, personal communication). The ecology and physiology of these basal angiosperm lineages have been investigated in a series of studies by Feild and colleagues (Feild et al. 2000, 2001, 2003*a*, *b*), and most recently, the ecophysiological data have been used to infer how early angiosperms functioned in their habitats (Feild et al. 2004). These results suggest that neither of the earlier hypotheses of angiosperm origins were completely correct but that the earliest angiosperms were woody plants that occupied shady, but disturbed, and/or streamside habitats. These results also suggest that extant members of early diverging lineages are appropriate ecological analogs of the first angiosperms.

Phytochrome-mediated responses to shade

Two important responses of seedlings to shade are mediated by angiosperm phytochromes. Because canopy shade is FR-enriched relative to R, a result of the strong absorbance of R by chlorophyll, the phyA-mediated FR-HIR is important for de-etiolation in dense shade. The importance of phyA in dimly lit environments is supported by the observation that *phyA* mutants of *Arabidopsis* die prematurely when grown in deep shade (Yanovsky et al. 1995). Furthermore, there is variation in light sensitivities for de-etiolation in natural populations (Maloof et al. 2000), and an *Arabidopsis* ecotype with an altered *PHYA* sequence shows reduced sensitivity to FR (Maloof et al. 2001). Both of these observations are consistent with the hypothesis that the de-etiolation response is fine-tuned to the light environment (Maloof et al. 2000). Equally critical to angiosperm seedlings is shade avoidance, especially for shade-intolerant species. Shade avoidance, which is mediated primarily by phyB, is one of the most important competitive strategies that flowering plants possess (Smith and Whitelam 1997), and it is known to confer an adaptive advantage (e.g., Schmitt et al. 1995; Dudley and Schmitt 1996; Schmitt et al. 1999; Maloof et al. 2000). Shade-avoiding angiosperms respond to crowding, perceived as a reduction in the ratio of R to FR, by enhancing elongation growth in order to project their leaves into unattenuated daylight (Smith 2000). Other aspects of shade avoidance come into play if this is not successful, including accelerated flowering and seed production (Smith 2000). Elements of shade avoidance have been noted in conifers (Morgan et al. 1983; Warrington et al. 1988; Fernbach and Mohr 1990) and may be mediated by phyP, the gymnosperm homolog of phyB (Fig. 1). In its promotion of extension growth, absorption of FRc by phyB inhibits de-etiolation in dense shade (McCormac et al. 1992). It also has been argued that shade avoidance may be too costly for seedlings, which have only limited resources (Smith et al. 1997). Thus, seedlings emerging in the shade may require a mechanism through which early shade avoidance reactions are suppressed and through which the inhibitory action of phyB on extension growth is overcome, a mechanism that may be provided by the FR-HIR mediated by phyA (Smith et al. 1997). After de-etiolation, phyA would not interfere with shade avoidance responses because it decays rapidly in the light (Smith et al. 1997).

The VLFR also contributes to de-etiolation under low light conditions, although perhaps not enough to confer an advantage (Smith et al. 1997). It has been suggested that the very low fluence germination response is most important in open environments (Smith 1995), where there would be adequate light for seedling development following germination in response to brief soil disturbances. Moreover, seeds of many angiosperms germinate well in darkness, especially large seeds with adequate resources that are produced by many shade tolerant species (Smith 1995). However, most descendants of early-diverging angiosperms have small seeds and this is inferred to be the ancestral state (Feild et al. 2004). Feild et al. (2003*a*) highlight recent findings suggesting that small seeds can be viable in shady but disturbed habitats (Metcalfe and Grubb 1997; Grubb 1998) and they argue that small seeds can be lodged more easily in the small microsites in which seedlings of most of these species establish. These include small understory gaps created by limbfalls, in landslips, and/or on unstable substrates and rotting logs. They note that traits that would enhance the colonization of such sites would have been important. Light intensities are likely to fluctuate in such habitats, and seeds and seedlings might experience brief exposure due to disturbance. In these instances, the VLF seed germination in response to brief pulses of light could promote seedling establishment.

Together the inferred phylogeny, ecology, and physiology of early angiosperms suggest that they occupied habitats where functions that increased their capacity to establish and survive in dimly lit habitats were important. Given the signal of episodic selection very early in the history of angiosperms, it is reasonable to speculate that phyA may have conferred an adaptive advantage to species colonizing these habitats, through increasing the ability of seedlings to de-etiolate in shade, through increasing the light-sensitivity of seed germination, or through innovation in both of these functions. There is precedence for the acquisition of single amino acid changes that allow a species to occupy a new environment (e.g., Perutz 1983). If changes at any of the positively selected sites in phyA enhanced the sensitivity of phyA to FRc or to millisecond pulses of light, they may have had a similar effect. This hypothesis can be tested using mutagenesis experiments to compare phytochrome-mediated responses of wild-type *Arabidopsis* with those of *Arabidopsis phyA* null mutants bearing sequence constructs with the inferred ancestral states at the positively selected amino acid sites. If there is a connection between these sites and the known functions of phyA, altering them should alter one or more of these functions. The fitness of *phyA* null mutants carrying these constructs can then be determined in order to test the hypothesis that one or more of these functions could have conferred an adaptive advantage. The fitness of shade avoidance phenotypes has been similarly tested using *phyB* mutants (Schmitt et al. 1995; Pigliucci and Schmitt 1999).

The origin of phyA and the radiation of angiosperms

If phyA provided a fitness advantage to early diverging angiosperms as it does for *Arabidopsis*, it is possible that innovation in phyA function might represent the acquisition of a trait that allowed angiosperms to better exploit sites in which competition with ferns and gymnosperms was reduced, i.e., that it might represent a "key innovation" that promoted the radiation of angiosperms. However, the earliest-diverging lineages are not species-rich (Mathews and Donoghue 1999), and the phylogenetic distribution of species-rich and species-poor clades within angiosperms suggests that at least some of the traits promoting speciation do not characterize angiosperms as a whole (Magallón and Sanderson 2001). Feild and colleagues speculate that the understory ecology of early species may have restricted the diversity of pre-Aptian angiosperms and of living basal lineages, and that evolution of greater sun-tolerance may have been required for the ecological expansion of angiosperms (Feild et al. 2004). They also suggest that vegetative flexibility evolved in the understory phase and that this may have been a factor promoting their diversification into other habitats. Other innovations that may have favored the overall success of the angiosperms include the closed carpel, self-incompatibility, the evolution of vessels, triploid endosperm, the herbaceous habit, rapid rates of reproduction and growth, shade avoidance, and refinement of the floral development program (Doyle and Donoghue 1986; 1993; Bond 1989; Smith 2000; Kramer and Irish 2000). Nonetheless, the evidence for molecular adaptation in phyA may reflect its importance to the initial establishment of the angiosperms in disturbed understory and/or streamside habitats. Furthermore, the phyA study provides a model for testing the importance of other traits that might have had a significant role in subsequent radiations.

Concluding remarks

The study of ancient adaptation

It has been argued that to achieve a full understanding of adaptive change, it is necessary to identify molecular changes, which are directly responsible, and that a combination of phenotypic and phylogenetic evidence is not sufficient (Golding and Dean 1998: 355). In a review of six studies from vertebrates and bacteria, Golding and Dean (1998) illustrated the role of protein structural information and site-directed mutagenesis in the endeavor to link a phenotype with evidence of molecular adaptation. The phytochrome study further demonstrates that sites for mutagenesis experiments can be identified without highly detailed knowledge of protein structure, which is lacking for phytochromes (Montgomery and Lagarias 2002). The availability of likelihood-based tests for selection that can detect ancient episodic selection involving a small number of sites (e.g., Yang and Nielsen 2002) extends the study of ancient adaptation to the many cases in which protein structural data are not available. Together these studies demonstrate that the study of past events need not stop with describing episodes of selection at the molecular level.

Once positively selected sites are identified, the synthesis of proteins with selected sites returned to their ancestral states provides a powerful test of the functional effects of amino acid replacements. This does not guarantee that all sites with adaptive significance will be identified. However, the utility of the approach has been well documented (for examples and citations see: Golding and Dean 1998; Chang and Donoghue 2000; Watt and Dean 2000). An important point to emerge from these studies is that major functional shifts usually require just a few amino acid substitutions. A limitation of these studies was that the phenotypes of altered proteins were investigated but their effects on fitness were not. However, in some cases, convergent functional shifts involved the same amino acid replacements in paralogous proteins (e.g., Chang et al. 1995; Yokoyama 1997), which lends support to the adaptive nature of these changes. In the phytochrome case, constructs bearing the ancestral residue at a selected site can be used to transform *phyA* null mutants of *Arabidopsis* for *in vivo* functional and fitness tests, as well as being expressed in yeast for *in vitro* studies of protein activity. An important complementary test remains to be devised. Although it is possible to place a modified *PHYA* sequence in *Arabidopsis thaliana*, its ecology differs greatly from species like *Amborella trichopoda*. One is a weedy herb of open habitats, the other is a woody shade-tolerant shrub. The two also differ with respect to genetic background; thus amino acid changes in the two systems might have different effects. It would be desirable to test the constructs in *Amborella*, which is not currently possible, or to find a way to test the fitness of phyA-mediated responses of woody shade-tolerant plants by somehow altering phyA expression. For example, it might be possible to test the far-red high irradiance de-etiolation response by using a brief light exposure to reduce the phyA pool before subjecting the emerging seedlings to continuous FR.

The phytochrome study also illustrates a challenge that will sometimes be associated with determining the origin of a phenotypic or functional variant in studies of ancient adaptation. Functional knowledge is often derived from model species that are quite distant from the branch on which selection occurred. In these cases it is necessary to determine the origin of the functional variant by characterizing the phenotypes of exemplars from additional clades. At the same time, the phytochrome study highlights how the approach can provide functional insights missing from the study of model species. Many of the phytochrome mutants so far isolated in genetic screens using *Arabidopsis* have identified amino acid residues that are conserved, and possibly important, in all phytochromes (e.g., Quail et al. 1995; Maloof et al. 2001; Yanovsky et al. 2002). For example, one natural accession of *Arabidopsis thaliana* shows decreased sensitivity to FR and its phyA differs by a single amino acid from that of the lab strain Col-O. When this site was altered in phyB, sensitivity to low-fluence R was also reduced, suggesting that the site is generally important for modulating light response in other phytochromes in addition to phyA (Maloof et al. 2001). Similarly, an amino acid residue that is necessary for the phyA-mediated FR-HIR (Yanovsky et al. 2002) is also necessary for phyB-mediated response to continuous red light (Wagner and Quail 1995), suggesting that this site also is important in more than one phytochrome. In contrast, the phylogenetic approach described here has identified amino acid residues that are uniquely altered in phyA. Manipulation of these sites has the potential to provide novel insight into the structural determinants of phyA function.

Acknowledgements

Three individuals have been critical to the work described here. The duplicate gene data set to address basal angiosperm phylogeny was gathered in collaboration with Michael Donoghue. The tests for selection and altered constraints were conducted in collaboration with Gordon Burleigh. Taylor Feild has shared invaluable insight into the ecology and physiology of early angiosperms in discussions and manuscripts. Justin Borevitz and Rex Cocroft provided helpful comments on the manuscript. This material is based upon work supported by the National Science Foundation under Grant No. 9806397. Any opinions, findings, and conclusions or recommendations expressed in this material are those of the author and do not necessarily reflect the views of the National Science Foundation.

References

Alba, R., Kelmenson, P.M., Cordonnier-Pratt, M.M., and Pratt, L.H. 2000. The phytochrome gene family in tomato and the rapid differential evolution of this family in angiosperms. Mol. Biol. Evol. **17**: 362–373.

Antonovics, J. 1987. The evolutionary dis-synthesis: which bottles for which wine? Am. Nat. **129**: 321–331.

Arber, E.A.N., and Parkin, J. 1907. On the origin of angiosperms. Bot. J. Linn. Soc. **38**: 29–80.

Axelrod, D.I. 1952. A theory of angiosperm evolution. Evolution **6**: 29–60.

Barkman, T.J., Chenery, G., McNeal, J.R., Lyons-Weiler, J., Elisens, W.J., Moore, G., Wolfe, A.D., and Depamphilis, C.W. 2000. Independent and combined analyses of sequences from all three genomic compartments converge on the root of flowering plant phylogeny. Proc. Natl. Acad. Sci. U.S.A. **97**: 13166–13171.

Bessey, C.E. 1915. The phylogenetic taxonomy of flowering plants. Ann. Mo. Bot. Gard. **2**: 109–164.

Bews, J.W. 1927. Studies in the ecological evolution of angiosperms. New Phytol. **26**: 1–21.

Bogorad, L. 1950. Factors associated with the synthesis of chlorophyll in the dark in seedlings of *Pinus jeffreyi*. Bot. Gaz. **111**: 221–241.

Bond, W.J. 1989. The tortoise and the hare — ecology of angiosperm dominance and gymnosperm persistence. Biol. J. Linn. Soc. **36**: 227–249.

Botto, J.F., Sanchez, R.A., Whitelam, G.C., and Casal, J.J. 1996. Phytochrome A mediates the promotion of seed germination by very low fluences of light and canopy shade light in *Arabidopsis*. Plant Physiol. **110**: 439–444.

Bower, F.O. 1882. The germination and embryogeny of *Gnetum gnemon*. Quart. J. Microsc. Sci. 22: 278–298.

Burgerstein, A. 1900. Ueber das Verhalten der Gymnospermen-Keimlinge im Licht und im Dunkeln. Ber. Deutsch. Bot. Ges. **18**: 168–184.

Burgin, M.J., Casal, J.J., Whitelam, G.C., and Sanchez, R.A. 1999. A light-regulated pool of phytochrome and rudimentary high-irradiance responses under far-red light in *Pinus elliotti* and *Pseudotsuga menziesii*. J. Exp. Bot. **50**: 831–836.

Casal, J.J. 2000. Phytochromes, cryptochromes, phototropin: photoreceptor interactions in plants. Photochem. Photobiol. **71**: 1–11.

Casal, J.J., and Sanchez, R.A. 1998. Phytochromes and seed germination. Seed Sci. Res. **8**: 317–329.

Casal, J.J., Sanchez, R.A., and Yanovsky, M.J. 1997. The function of phytochrome A. Plant Cell Environ. **20**: 813–819.

Cerdán, P.D., Yanovsky, M.J., Reymunds, F.C., Nagatani, A., Staneloni, R.J., Whitelam, G.C., and Casal, J.J. 1999. Regualation of phytochrome B signaling by phytochrome A and FHY1 in *Arabidopsis thaliana*. Plant J. **18**: 499–507.

Chang, B.S.W., Crandall, K.A., Carulli, J.P., and Hartl, D.L. 1995. Opsin phylogeny and evolution: A model for blue shifts in wavelength regulation. Mol. Phylogenet. Evol. **4**: 31–43.

Chang, B.S.W., and Donoghue, M.J. 2000. Recreating ancestral proteins. Trends Ecol. Evol. **15**: 109–114.

Chinn, E., and Silverthorne, J. 1993. Light-dependent chloroplast development and expression of a light-harvesting chlorophyll a/b-binding protein gene in the gymnosperm *Ginkgo biloba*. Plant Physiol. **103**: 727–732.

Christensen, S., LaVerne, E., Boyd, G., and Silverthorne, J. 2002. *Ginkgo biloba* retains functions of both type I and type II flowering plant phytochrome. Plant Cell Physiol. **43**: 768–777.

Clapham, D.H., Kolukisaoglu, H.Ü., Larsson, C.-T., Qamaruddin, M., Ekberg, I., Wiegmann-Eirund, C., Schneider-Poetsch, H.A.W., and Von Arnold, S. 1999. Phytochrome types in *Picea* and *Pinus*. Expression patterns of *PHYA*-related types. Plant Mol. Biol. **40**: 669–678.

Crane, P.R., Friis, E.M., and Pederson, K.R. 1995. The origin and early diversification of angiosperms. Nature **374**: 27–33.

Cronquist, A. 1988. The evolution and classification of flowering plants. The New York Botanical Garden, New York.

Doyle, J.A. 1978. Origin of angiosperms. Annu. Rev. Ecol. Syst. **9**: 365–392.

Doyle, J.A., and Donoghue, M.J. 1986. Seed plant phylogeny and the origin of angiosperms: an experimental cladistic approach. Bot. Rev. **52**: 321–431.

Doyle, J.A., and Donoghue, M.J. 1993. Phylogenies and angiosperm diversification. Paleobiology **19**: 141–167.

Doyle, J.A., and Endress, P.K. 2000. Morphological phylogenetic analysis of basal angiosperms: comparison and combination with molecular data. Int. J. Plant Sci. **161**: S121–S153.

Doyle, J.A., and Hickey, L.J. 1976. Pollen and leaves from the mid-Cretaceous Potomac Group and their bearing on early angiosperm evolution. *In* Origin and early evolution of angiosperms. *Edited by* C.B. Beck. Columbia University Press, New York. pp. 139–206.

Dudley, S.A., and Schmitt, A. 1996. Testing the adaptive plasticity hypothesis: Density-dependent selection on manipulated stem length in *Impatiens capensis*. Am. Nat. **147**: 445–465.

Eichenberg, K., Baurle, I., Paulo, N., Sharrock, R.A., Rudiger, W., and Schafer, E. 2000. *Arabidopsis* phytochromes C and E have different spectral characteristics from those of phytochromes A and B. FEBS Lett. **470**: 107–112.

Fankhauser, C., and Chory, J. 1997. Light control of development. Annu. Rev. Cell Dev. Biol. **13**: 203–229.

Feild T.S., Arens, N.C., and Dawson, T.E. 2003*a*. Emerging perspectives on the ancestral ecophysiology of angiosperms. Int. J. Plant Sci. **164:** S129–S142.

Feild T.S., Arens, N.C., Doyle, J.A., Dawson, T.E., and Donoghue, M.J. 2004. Dark and disturbed: a new image of early angiosperm ecology. Paleobiology **30**: 82–107.

Feild, T.S., Franks, P.J., and Sage. T.L. 2003*b*. The shade-adapted ecophysiology of *Austrobaileya scandens*. Int. J. Plant Sci. **164**: 313–334.

Feild, T.S., Zwieniecki, M.A., Brodribb, T., Jaffré, T., Donoghue, M.J., and Holbrook, N.M. 2000. Structure and function of tracheary elements in *Amborella trichopoda*. Int. J. Plant Sci. **161**: 705–712.

Feild, T.S., Brodribb, T., Jaffre, T., and Holbrook, N.M. 2001. Responses of leaf anatomy, photosynthetic light-use, and xylem hydraulics to light in the basal angiosperm *Amborella trichopoda* (Amborellaceae). Int. J. Plant Sci. **162**: 999–1008.

Fernbach, E., and Mohr, H. 1990. Coaction of blue/ultraviolet-A light and light absorbed by phytochrome in controlling growth of pine (*Pinus sylvestris* L.) seedlings. Planta **180**: 212–216.

Franklin, K.A., Davis, S.J., Stoddart, W.M., Vierstra, R.D., and Whitelam, G.C. 2003. Mutant analyses define multiple roles for phytochrome C in *Arabidopsis* morphogenesis. Plant Cell **15**: 1981–1989.

Friis, E.M., Chalconer, W.G., and Crane, P.R. 1987. The origins of angiosperms and their biological consequences. Cambridge University Press, Cambridge.

Furuya, M., and Schäfer, E. 1996. Photoperception and signaling of induction reactions by different phytochromes. Trends Plant Sci. **1**: 301–306.

Golding, G.B., and Dean, A.M. 1998. The structural basis of molecular adaptation. Mol. Biol. Evol. **15**: 355–369.

Graham, S.W., Reeves, P.A., Burns, A.C.E., and Olmstead, R.G. 2000. Microstructural characters in noncoding chloroplast DNA: interpretation, evolution, and utility of indels and inversions in basal angiosperm phylogenetic inference. Int. J. Plant Sci **161**: S83–S96.

Grubb, P.J. 1998. Seeds and fruits of tropical rainforest plants: interpretation of the range of seed size, degree of defense, and flesh/seed quotients. *In* Dynamics of tropical communities. *Edited by* D.M. Newberry, H.H.T. Prins, and N.D. Brown. Blackwell Scientific, Oxford. pp. 1–24.

Halliday, K.J., Thomas, B., and Whitelam, G.C. 1997. Expression of heterologous phytochromes A, B, or C in transgenic tobacco plants alters vegetative development and flowering time. Plant J. **12**: 1079–1090.

Hennig, L., Stoddart, W.M., Dieterle, M., Whitelam, G.C., and Schäfer, E. 2002. Phytochrome E controls light-induced germination in *Arabidopsis*. Plant Physiol. **128**: 194–200.

Hickey, L.J., and Doyle, J.A. 1977. Early Cretaceous fossil evidence for angiosperm evolution. Bot. Rev. **43**: 1–104.

Knoll, A.H. 1984. Patterns of extinction in the fossil record of vascular plants. *In* Extinctions. *Edited by* M.H. Nitecki. University of Chicago Press, Chicago. pp. 21–68.

Kramer, E.M., and Irish, V.F. 2000. Evolution of the petal and stamen developmental programs: evidence from comparative studies of the lower eudicots and basal angiosperms. Int. J. Plant Sci. **161**: S29–240.

Lidgard, S., and Crane P.R. 1988. Quantitative analyses of the early angiosperm radiation. Nature **331**: 344–346.

Lupia, R., Lidgard, S., and Crane, P.R. 1999. Comparing palynological abundance and diversity: implications for biotic replacement during the Cretaceous angiosperm radiation. Paleobiology **25**: 305–340.

Magallón, S., and Sanderson, M.J. 2001. Absolute diversification rates in angiosperm clades. Evolution **55**: 1762–1780.

Maloof, J., Borevitz, J.O., Dabi, T., Lutes, J., Nehring, R.B., Redfern, J.L., Trainer, G.T., Wilson, J.M., Asami, T., Berry, C.C., Weigel, D., and Chory, J. 2001. Natural variation in light sensitivity of *Arabidopsis*. Nature Genet. **29**: 441–446.

Maloof, J., Borevitz, J.O., Weigel, D., and Chory, J. 2000. Natural variation in phytochrome signaling. Semin. Cell Dev. Biol. **11**: 523–530.

Mathews, S., Burleigh, J.G., and Donoghue, M.J. 2003. Adaptive evolution in the photosensory domain of phytochrome A in early angiosperms. Mol. Biol. Evol. **20**: 1087–1097.

Mathews, S., and Donoghue, M.J. 1999. The root of angiosperm phylogeny inferred from duplicate phytochrome genes. Science (Washington, D.C.) **286**: 947–950.

Mathews, S., and Donoghue, M.J. 2000. Basal angiosperm phylogeny inferred from duplicate phytochromes A and C. Int. J. Plant Sci. **161**: S41–S55.

Mathews, S., Lavin, M., and Sharrock, R.A. 1995. Evolution of the phytochrome gene family and its utility for phylogenetic analyses of angiosperms. Ann. Mo. Bot. Gard. **82**: 296–321.

Mathews, S., and Sharrock, R.A. 1997. Phytochrome gene diversity. Plant Cell Environ. **20**: 666–671.

McCormac, A.C., Whitelam, G.C., Boylan, M.T., Quail, P.H., and Smith, H. 1992. Contrasting responses of etiolated and light-adapted seedlings to red:far-red ratio: a comparison of wild type, mutant and transgenic plants has revealed differential functions of members of the phytochrome family. J. Plant Physiol. **140**: 707–714.

McNellis, T.W., and Deng, X.-W. 1995. Light control of seedling morphogenetic pattern. Plant Cell **7**: 1749–1761.

Metcalfe, D.J., and Grubb, P.J. 1997. The responses to shade of seedlings of very small-seeded tree and shrub species from tropical rain forest in Singapore. Funct. Ecol. **11**: 215–221.

Monte, E., Alonso, J.M., Ecker, J.R., Zhang, Y., Li, X., Young, J., Austin-Phillips, S., and Quail, P.H. 2003. Isolation and characterization of *phyC* mutants in *Arabidopsis* reveals complex crosstalk between phytochrome signaling pathways. Plant Cell **15**: 1962–1980.

Montgomery, B.L., and Lagarias, J.C. 2002. Phytochrome ancestry: sensors of bilins and light. Trends Plant Sci. **7**: 357–366.

Morgan, D.C., Rook, D.A., Warrington, I.J., and Turnbull, H.L. 1983. Growth and development of *Pinus radiata* D. Don: the effect of light quality. Plant Cell Environ. **6**: 691–701.

Mukai, Y., Tazaki, K., Fujii, T., and Yamamoto, N. 1992. Light-independent expression of three photosynthetic genes, *cab, rbcS*, and *rbcL*, in coniferous plants. Plant Cell Physiol. **33**: 859–866.

Nagatani, A., Reed, J.W., and Chory, J. 1993. Isolation and initial characterization of *Arabidopsis* mutants that are deficient in phytochrome A. Plant Physiol. **102**: 269–277.

Parkinson, C.L, Adams, K.L., and Palmer, J.D. 1999. Multigene analyses identify the three earliest lineages of extant flowering plants. Curr. Biol. **9**: 1485–1488.

Parks, B.M., and Quail, P.H. 1993. hy8, a new class of *Arabidopsis* long hypocotyl mutants deficient in functional phytochrome A. Plant Cell **5**: 39–48.

Perutz, M.F. 1983. Species adaptation in a protein molecule. Mol. Biol. Evol. **1**: 1–28.

Pigliucci, M., and Schmitt, J. 1999. Genes affecting phenotypic plasticity in *Arabidopsis*: pleiotropic effects and reproductive fitness of photomorphogenic mutants. J. Evol. Biol. **12**: 551–562.

Qin, M., Kuhn, R., Moran, S., and Quail, P.H. 1997. Overexpressed phytochrome C has similar photosensory specificity to phytochrome B but a distinctive capacity to enhance primary leaf expansion. Plant J. **12**: 1163–1172.

Qiu, Y.-L., Lee, J., Bernasconi-Quadroni, F., Soltis, D.E., Soltis, P.S., Zanis, M., Zimmer, E.A., Chen, Z., Savolainen, V., and Chase, M.W. 1999 The earliest angiosperms: evidence from mitochondrial, plastid and nuclear genomes. Nature **402**: 404–407.

Quail, P.H., Boylan, M.T., Parks, B.M., Short, T.W., and Wagner, D. 1995. Phytochromes: photosensory perception and signal transduction. Science (Washington, D.C.) **268**: 675–680.

Sanderson, M.J., and Doyle, J.A. 2001. Sources of error and confidence intervals in estimating the age of angiosperms from *rbc*L and 18S rDNA data. Am. J. Bot. **88**: 1499–1516.

Savolainen, V., Chase, M.W., Hoot, S.B., Morton, C.M., Soltis, D.E., and Bayer, C. 2000. Phylogenetics of flowering plants based on combined analysis of plastid *atp*B and *rbc*L gene sequences. Syst. Biol. **49**: 306–362.

Schmidt, M., and Schneider-Poetsch, H.A.W. 2002. The evolution of gymnosperms redrawn by phytochrome genes: The Gnetatae appear at the base of the gymnosperms. J. Mol. Evol. **54**: 715–724.

Schmitt, A., Dudley, S.A., and Pigliucci, M. 1999. Manipulative approaches to testing adaptive plasticity: Phytochrome-mediated shade-avoidance in plants. Am. Nat. **154**: S43–54.

Schmitt, A., McCormac, A.C., and Smith, H. 1995. A test of the adaptive plasticity hypothesis using transgenic and mutant plants disabled in phytochrome-mediated elongation responses to neighbors. Am. Nat. **146**: 937–953.

Schneider-Poetsch, H.A.W., Kolukisaoglu, Ü., Clapham, D.H., Hughes, J., and Lamparter, T. 1998. Non-angiosperm phytochromes and evolution of vascular plants. Physiol. Plant. **102**: 612–622.

Sharrock, R.A., and Quail, P.H. 1989. Novel phytochrome sequences in *Arabidopsis thaliana*: Structure, evolution and differential expression of a plant regulatory photoreceptor family. Genes Dev. **3**: 1745–1757.

Shinomura, T, Uchida, K., and Furuya, M. 2000. Elementary processes of photoreception by phytochrome A for high-irradiance response of hypocotyl elongation in *Arabidopsis*. Plant Physiol. **122**: 147–156.

Smith, H. 1995. Physiological and ecological function within the phytochrome family. Annu. Rev. Plant Physiol. Plant Mol. Biol. **46**: 289–315.

Smith, H. 2000. Phytochromes and light signal perception by plants — an emerging synthesis. Nature **407**: 585–591.

Smith, H., and Whitelam, G.C. 1997. The shade avoidance syndrome: multiple responses mediated by multiple phytochromes. Plant Cell Environ. **20**: 840–844.

Smith, H., Xu, Y., and Quail, P.H. 1997. Antagonistic but complementary actions of phytochromes A and B allow optimum seedling de-etiolation. Plant Physiol. **114**: 637–641.

Soltis, P.S., Soltis, D.E., and Chase, M.W. 1999. Angiosperm phylogeny inferred from multiple genes as a tool for comparative biology. Nature **402**: 402–404.

Soltis, P.S., Soltis, D.E., Chase, M.W., Mort, M.E., Albach, D.C., Zanis, M.J., Savolainen, V., Hahn, W.H., Hoot, S.B., Fay, M.F., Axtell, M., Swenson, S.M., Prince, L.M., Kress, W.J., Nixon, K.C., and Farris, J.S. 2000. Angiosperm phylogeny inferred from a combined data set of 18s rDNA, *rbc*L, and *atp*B sequences. Bot. J. Linn. Soc. **133**: 381–461.

Somers, D.E., Sharrock, R.A., Tepperman, J.M., and Quail, P.H. 1991. The *hy3* long hypocotyls mutant of *Arabidopsis* is deficient in phytochrome B. Plant Cell **3**: 1263–1274.

Stebbins, G.L. 1974. Flowering plants, Evolution above the species level. Harvard University Press, Cambridge.

Sun, G., Ji, Q., Dilcher, D.L., Zheng, S., Nixon, K.C., and Wang, X. 2002. Archaefructaceae, a new basal angiosperm family. Science (Washington, D.C.) **296**: 899–904.

Takano, M., Kanegae, H., Shinomura, T., Miyao, A., Hirochika, H., and Furuya, M. 2001. Isolation and characterization of rice phytochrome A mutants. Plant Cell **13**: 521–534.

Takhtajan, A.L. 1969. Flowering plants: origin and dispersal. Smithsonian Institution, Washington, D.C.

Taylor, D.W., and Hickey, L.J. 1992. Phylogenetic evidence for the herbaceous origin of angiosperms. Plant Syst. Evol. **180**: 137–156.

Taylor, D.W., and Hickey, L.J. 1996. Evidence for and implications of an herbaceous origin for angiosperms. *In* Flowering plant origin, evolution, and phylogeny. *Edited by* D. Taylor and L. Hickey. Chapman and Hall, New York. pp. 232–266.

Thorne, R.F. 1974. A phylogenetic classification of the Annoniflorae. Aliso **8**: 147–209.

Van Tuinen, A., Kerckhoffs, L.H.J., Nagatani, A., Kendrick, R.E., and Koornneef, M. 1995. Far-red light insensitive, phytochrome A deficient mutants of tomato. Mol. Gen. Genet. **246**: 133–141.

Wagner, D., and Quail, P.H. 1995. Mutational analysis of phytochrome B identifies a small COOH-terminal-domain region critical for regulatory activity. Proc. Natl. Acad. Sci. U.S.A. **92**: 8596–8600.

Warrington, I.J., Rook, D.A., Morgan, D.C., and Turnbull, H.L. 1988. The influence of simulated shadelight and daylight on growth, development and photosynthesis of *Pinus radiata, Agathis australis,* and *Dacrydium cupressinum*. Plant Cell Environ. **11**: 343–356.

Watt, W.B., and Dean, A.M. 2000. Molecular-functional studies of adaptive genetic variation in prokaryotes and eukaryotes. Annu. Rev. Genet. **34**: 593–622.

Whitelam, G.C., Johnson, E., Peng, J., Carol, P., Anderson, M.L., Cowl, J.S., and Harberd, N.P. 1993. Phytochrome A null mutants of *Arabidopsis* display a wild-type phenotype in white light. Plant Cell **5**: 757–768.

Wing, S.L., and Boucher, L.D. 1998. Ecological aspects of the Cretaceous flowering plant radiation. Annu. Rev. Earth Planetary Sci. **26**: 379–421.

Yang, Z., and Nielsen, R. 2002. Codon-substitution models for detecting molecular adaptation at individual sites along specific lineages. Mol. Biol. Evol. **19**: 908–917.

Yanovsky, M.J., Casal, J.J., and Luppi, J.P. 1997. The *VLF* loci, polymorphic between ecotypes of Landsberg *erecta* and Columbia dissect two branches of phytochrome A signalling pathways that correspond to the very-low fluence and high-irradiance responses of phytochrome. Plant J. **12**: 659–667.

Yanovsky, M.J., Casal, J.J., and Whitelam, G.C. 1995. Phytochrome A, phytochrome B and HY4 are involved in hypocotyl growth responses to natural radiation in *Arabidopsis*: weak deetiolation of the *phyA* mutant under dense canopies. Plant Cell Environ. **18**: 788–794.

Yanovsky, M.J., Luppi, J.P., Kirchbauer, D., Ogorodnikova, O.B., Sineshchekov, V.A., Adam, E., Kircher, S., Staneloni, R.J., Schäfer, E., Nagy, F., and Casal, J.J. 2002. Missense mutation in the PAS2 domain of phytochrome A impairs subnuclear localization and a subset of responses. Plant Cell **14**: 1591–1603.

Yokoyama, S. 1997. Molecular genetic basis of adaptive selection: Examples from color vision in vertebrates. Annu. Rev. Genet. **31**: 315–336.

Zanis, M.J., Soltis, D.E., Soltis, P.S., Mathews, S., and Donoghue M.J. 2002. The root of the angiosperms revisited. Proc. Natl. Acad. Sci. U.S.A. **99**: 6848–6853.

19
The variable nature of herbivore defense: evidence for a rapidly diverging Kunitz trypsin inhibitor gene in *Populus*

Manoela Miranda, Mary Christopher, and C. Peter Constabel

Abstract: Plant defense against insect herbivores is a very active process, and involves rapid up-regulation of defense and signaling genes. As part of our ongoing studies on wound- and herbivore-induced genes in hybrid poplar (*Populus trichocarpa* × *P. deltoides*), we identified four inducible genes encoding Kunitz-type trypsin inhibitors. These are potential anti-nutritive defense proteins against insect herbivores, and the presence of four distinct trypsin inhibitor (TI) genes suggests a multiplicity of defenses in *Populus*. Here we make use of EST data sets to compare TIs from different *Populus* species and hybrids. Our results provide preliminary evidence for the rapid evolution of at least one of the TI genes.

Introduction

Plant defense against pests and pathogens is a dynamic process acting at several levels, and generally includes the induction of a suite of defense-related genes. To date, many herbivore- and wound-induced defense genes have been identified in a number of plants, mostly annual weeds or crop species. Many of these genes encode enzymes required for the synthesis of toxic secondary metabolites or anti-nutritive proteins such as the protease inhibitors (PIs). PIs are small proteins that inhibit proteolytic enzymes, and when ingested by insects they can inhibit digestive enzymes in the gut, reduce insect growth, and have other direct toxic effects (Ryan 1990). Protease inhibitors include many structurally diverse proteins, and can belong to one of eight distinct protein families. In tomato, wounding induces synthesis of at least four distinct PI types (Bergey et al. 1996). This diversity can be explained by the multiplicity of digestive enzymes found in insect guts, and by the fact that PIs are generally effective only against a subset of target enzymes (Ryan 1990). Furthermore, some lepidopteran herbivores can switch their digestive enzymes if they encounter a diet high in PIs (Jongsma et al. 1995; Broadway 1996). One of the most common types of PIs are the serine protease inhibitors, which include the Bowman-Birk and Kunitz trypsin inhibitor (TI) types. The diversity of PIs found in the plant kingdom hints at selection favoring the rapid evolution of PIs. This was recently shown directly in a genomic analysis of the tomato *pin2* gene (Barta et al. 2002).

Populus species are common throughout the northern hemisphere, and occupy diverse niches (Stettler et al. 1996). Due to its small genome size, the genus *Populus* is an attractive model for molecular genetics, and has become an important resource for tree molecular biology research (Sterky et al. 1998). Large collections of expressed sequence tags (ESTs) originating from hybrid aspen (*P. tremula* × *P. tremuloides*), European aspen (*P. tremula*), trembling aspen (*P. tremuloides*) and black cottonwood (*P. trichocarpa*) have been deposited in the GenBank database. Furthermore, sequencing of the entire genome of *P. trichocarpa* is expected to be completed soon (Wullschleger et al. 2002). Strong inducible defenses have been observed in hybrid poplar (*P. trichocarpa* × *P. deltoides*) (Parsons et al. 1989; Constabel et al. 2000), and projects are under way to identify additional defense-related inducible genes. In this paper, we report our analysis of four genes with strong similarity to Kunitz trypsin inhibitor (TI) genes. The objective of this study was to compare putative orthologous TI sequences from different *Populus* species and hybrids using available EST data. This has provided preliminary evidence for the rapid evolution of at least one of the TI genes.

Manoela Miranda, Mary Christopher, and C. Peter Constabel.[1] Centre for Forest Biology and Department of Biology, PO Box 3020, STN CSC, Victoria, BC V8W 3N5, Canada.
Correct citation: Miranda, M., Christopher, M., and Constabel, C.P. 2004. The variable nature of herbivore defense: evidence for a rapidly diverging Kunitz trypsin inhibitor gene in *Populus*. *In* Plant Adaptation: Molecular Genetics and Ecology. Proceedings of an International Workshop held December 11–13, 2002, in Vancouver, British Columbia, Canada. *Edited by* Q.C.B. Cronk, J. Whitton, R.H. Ree, and I.E.P. Taylor. NRC Research Press, Ottawa, Ontario. pp. 153–158.

[1]Corresponding author e-mail address: cpc@uvic.ca

Materials and methods

Plant material and wounding experiments

Hybrid poplar (*P. trichocarpa* × *P. deltoides*, clone H11-11) and other *Populus* species and hybrids were propagated and maintained as described by Constabel et al. (2000). For wounding experiments, leaves were wounded by crushing with pliers and samples harvested 48 h after wounding. Protein extractions, SDS-PAGE, and western analyses were performed with an antibody raised against the trembling aspen PtTI2 product (Haruta et al. 2001) following standard procedures (Harlow and Lane 1988).

Computational analyses

A collection of 1,747 ESTs from hybrid poplar was previously generated in our laboratory through single-pass sequencing of a cDNA library constructed from wound-induced leaves (Christopher et al. 2004; M. Miranda and P. Constabel, unpublished data). From these data, a non-redundant set of sequences was produced using StackPack (Miller et al. 1999). To identify sequence clusters containing Kunitz trypsin inhibitor homologs, BLAST searches were performed against the GenBank databases at NCBI (http://www.ncbi.nlm.nih. gov/BLAST/; Altschul et al. 1990). For the longest non-redundant clones with similarity to TIs, the entire coding region of these genes was sequenced. Multiple sequence analyses and sequence identities were calculated using the ClustalV algorithm using MegAlign (DNAStar, London, UK). The cluster analysis was carried out using SeqMan (DNAStar). For analysis of ESTs obtained from GenBank, the database was screened for *Populus* sequences in December, 2002. These sequences were downloaded and formatted into individual databases for each species or hybrid. Comparisons between the hybrid poplar ESTs and these databases were performed locally by a standalone BLASTN program (ftp://ftp.ncbi.nih.gov/blast/). Data were extracted from the BLAST results by PERL scripts and loaded into a relational database (MySQL; http://www.mysql.com). All EST-based sequences used for further analyses were obtained by assembling single ESTs into clusters which helps to detect possible sequencing errors in the EST data. The resulting contigs, i.e., the consensus sequences obtained by clustering shorter overlapping ESTs, were then used for further analyses.

Results and discussion

Characterization of four Kunitz-type TIs from hybrid poplar

Ten clones with sequence similarity to Kunitz TI genes were identified among ESTs from a small-scale sequencing project. The clones of interest were sequenced completely so that the entire predicted coding sequences were obtained. Sequence clustering analyses grouped these cDNAs into four distinct gene clusters with a low level of similarity between them, i.e., the contigs of these clusters shared 25–62% sequence identity at the nucleotide level (Table 1). Despite the low similarity among themselves, all cDNA sequences showed significant sequence similarity with other Kunitz-type TIs in GenBank, suggesting that we had identified four distinct TI-like genes from hybrid poplar. Sequence comparisons against the NCBI databases revealed that one of the contigs was identical to the previously identified wound inducible TI gene *win3* from hybrid poplar (Bradshaw et al. 1990). A second contig, named *TI-3* (accession AY378088), was very similar to a trembling aspen TI cloned in this laboratory, *PtTI3* (Haruta et al. 2001). The two other contigs formed by the TI-homologs were clearly distinct from these previously described TI genes, and were named *TI-4* (AY378089) and *TI-5* (AY378090). All four TI genes were strongly up-regulated following wounding of poplar leaves, in both wounded and systemically wounded leaves (Christopher et al. 2004), suggesting that they are all involved in plant defense. The parallel wound-induction of four distinct TI-like genes suggests a multiplicity of poplar herbivore defenses, perhaps driven by the evolution of resistant herbivore digestive enzymes (Jongsma et al. 1995). If TIs are under strong positive selection, one would predict that TIs genes should be diverging rapidly, which can be observed by comparing the degree of sequence similarity of the same TI genes among *Populus* species and hybrids. We made use of available *Populus* ESTs in GenBank to test this hypothesis.

Table 1. Percent identities among the different types of TIs of hybrid poplar (*P. trichocarpa* × *P. deltoides*).

	TI-3	TI-4	TI-5	win3
TI-3	—	29.4[a]	24.9	61.7
TI-4	22.8[b]	—	34.3	27.5
TI-5	15.8	25.0	—	28.5
win3	49.5	21.5	19.6	—

[a]Alignment of nucleotide sequences.
[b]Alignment of predicted protein sequences (ClustalV method, MegAlign, DNAStar, London, UK).

Comparative analysis of TI-like sequences in *Populus* EST databases

To compare putative TI orthologs in EST collections from different *Populus* species, we required a baseline for analysis. Therefore, we performed a global homology comparison among our hybrid poplar sequence collection and the publicly available NCBI GenBank sequences from hybrid aspen (56,186 sequence entries), European aspen (14,100 entries) and black cottonwood (24,067 entries). For all three database searches, the highest BLAST hits were 78–100% identical to the hybrid poplar query sequences (when homologous regions longer than 100 bp were considered). The median values for the comparisons of our hybrid poplar ESTs with the hybrid aspen, European aspen or black cottonwood databases was 95%, 94%, and 96%, respectively (Fig. 1). This confirmed that the transcriptome of poplars and aspens share very high sequence similarity, and provided the baseline on which specific comparisons could be based.

To identify putative orthologs of the four TI-like genes in other species and hybrids, we performed local BLASTN searches against the hybrid, European and trembling aspen, as well as black cottonwood EST sets in GenBank. Based on the level of similarity of the entire sequence sets (Fig. 1), we stipulated that only sequences sharing >90% identity over a region longer than 100 bp would be considered as possible orthologs. Following these criteria, potential candidates for all four TI orthologs were found only in the hybrid aspen EST set (Fig. 2). In addition, all TIs except the putative *TI-5* ortholog were identified among the black cottonwood ESTs. In the European and trembling aspen EST sets, however, only some of the TI types were found. The different representation in the databases is likely due to the varying number of ESTs in these sets (see above). As expected, the four TIs form discrete clusters across species, confirming the distinct nature of the four Kunitz TI-like genes (Fig. 2).

Next, we compared our hybrid poplar TI sequences with the aspen and black cottonwood ESTs in more detail. Multi-

Fig. 1. Percent identity among *Populus* nucleotide sequences in publicly available databases. Nucleotide sequence similarity comparisons between 1,747 hybrid poplar (*P. trichocarpa* × *P. deltoides*) ESTs and publicly available EST sequences originating from hybrid aspen (*P. tremula* × *P. tremuloides*), European aspen (*P. tremula*) and black cottonwood (*P. trichocarpa*) were performed with a standalone BLASTN program (Altschul et al. 1990). Only values from homologous regions longer than 100 bp were used. * shows the median values for each of the databases.

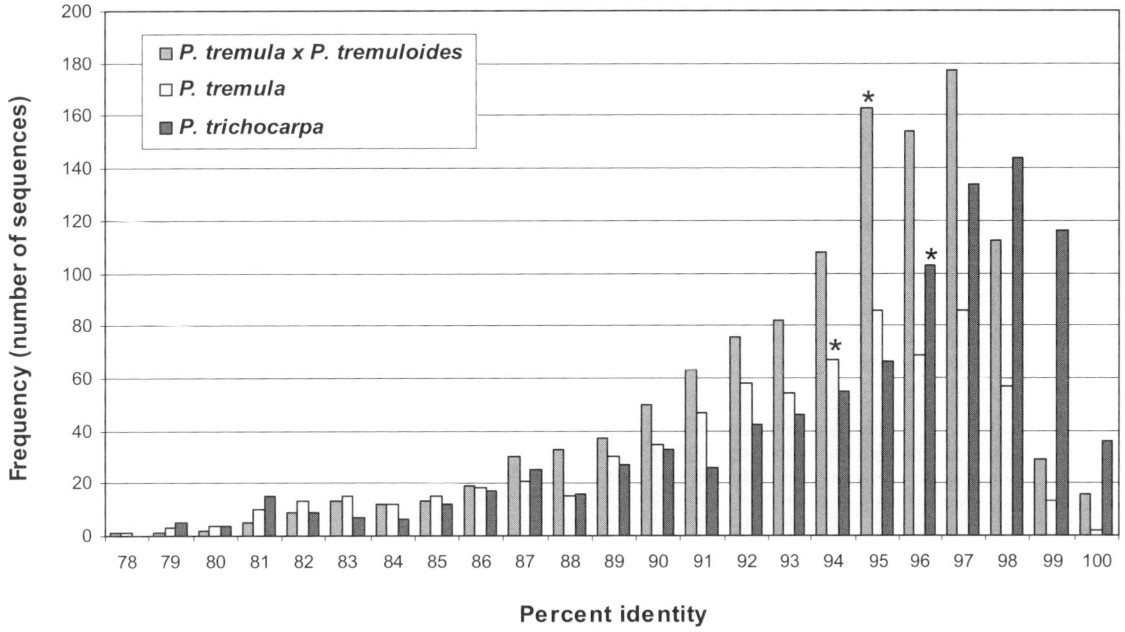

Fig. 2. Phylogenetic tree of predicted TI protein sequences from poplars and aspens. The predicted protein sequences were aligned using the ClustalV method (MegAlign, DNAStar, London, UK), and grouped into four distinct groups (*win3*, *TI-3*, *TI-4* and *TI-5*). The TI-homologs from hybrid poplar are shown in bold and originate from our own EST collection. The remaining ESTs are deposited in public databases. For each TI homolog, clusters of at least ten EST sequences were constructed and the resulting contig sequences were used for the analyses. GenBank accession numbers of the longest EST from each cluster appear in brackets. Hybrid poplar (*P. trichocarpa* × *P. deltoides*); hybrid aspen (*P. tremula* × *P. tremuloides*); European aspen (*P. tremula*); black cottonwood (*P. trichocarpa*).

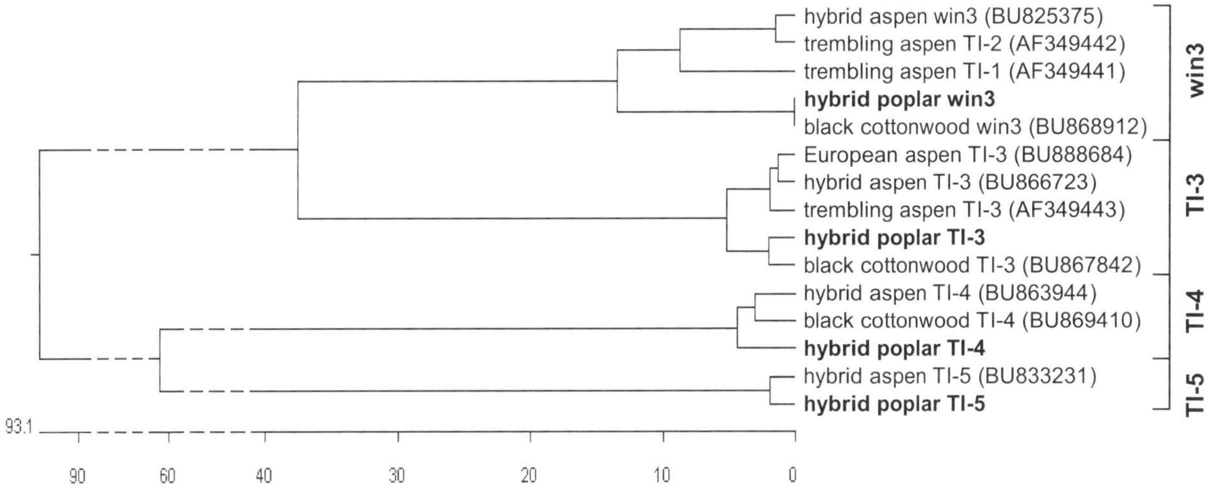

ple ESTs belonging to the same gene were clustered together and contigs built to extend the length of the TI sequences, as well as to detect possible sequencing errors. Alignments over the entire available coding regions were carried out, and the percent identities calculated. As a comparison, a similar analysis was also performed for the Rubisco small subunit (*rbcS*), a gene that has been previously used as a marker for assessing phylogenetic distances of closely related species (Meagher et al. 1989), and of caffeoyl CoA O-methyltransferase (*CCoAMT*), a gene involved in phenylpropanoid metabolism. Identities between the hybrid poplar TI sequences and black cottonwood ESTs was very high, though in some cases less than 100% (Table 2). This may be due to differences in genotypes used for the different projects, or may reflect the fact that we cannot at this point distinguish the *P. deltoides* and *P. trichocarpa* alleles in hybrid poplar.

Table 2. Sequence comparisons of hybrid poplar (*P. trichocarpa* × *P. deltoides*) TIs and their putative orthologs in black cottonwood (*P. trichocarpa*), hybrid aspen (*P. tremula* × *P. tremuloides)* and trembling aspen (*P. tremuloides*).

	Black cottonwood			Hybrid aspen			Trembling aspen		
Hybrid poplar	*nt*	aa	dS/dN	nt	aa	dS/dN	nt	aa	dS/dN
TI-3	97.4[a]	95.5	1.09	92.9	88.1	1.79	92.1	86.1	1.39
TI-4	94.5	91.5	1.98	93.6	91.9	4.62	N.A.	N.A.	N.A.
TI-5	N.A.	N.A.	N.A.	96.2	94.8	4.39	N.A.	N.A.	N.A.
win3	100	100	—	81.8	72.0	0.58	80.3	70.8	0.63
rbcS	100	100	—	96.7	94.5	1.28	88.2	91.3	6.81
CCoAMT	99.6	100	—	98.0	97.6	4.40	91.2	95.1	18.18

[a] Percent identity was calculated from alignments of nucleotide (nt) or predicted amino acid (aa) sequences using the ClustalV method (MegAlign, DNAStar, London, UK). Substitution rates were calculated by the Synonymous Nonsynonymous Analysis Program (SNAP; Nei and Gojobori 1986). Rubisco small subunit (*rbcS*) and caffeoyl CoA O-methyltransferase (*CCoAMT*) cDNAs were used as controls.

dS: synonymous substitution; dN: non-synonymous substitution; — : undefined value for dN/dS when dS is equal to zero; N.A.: putative orthologous sequences are not available in the database.

The percent amino acid identity between hybrid poplar and aspens TIs was lower, in the range of 70.8%–97.6%.

While the comparison of putative orthologs from hybrid poplar and hybrid aspen indicated that the percent identities of *TI-3*, *TI-4* and *TI-5* were all greater than 92% at the nucleotide level, identity levels of *win3* between hybrid poplar and trembling aspen or hybrid aspen were substantially lower (Table 2). This suggests that the win3 protein is evolving rapidly, and therefore we also calculated synomous and non-synonomous substitution rates (dS and dN, respectively) for *win3* and other cDNAs for which we had sequence information (Table 2). These analyses revealed that in *TI-3, TI-4,* and *TI-5*, as well as *rbcS* and *CCoAMT,* synonymous substitutions are more common than non-synonymous, as evidenced by dS/dN > 1. This was also true for a number of other control genes analyzed (data not shown). On the other hand, differences among poplar and aspen *win3* originate mostly from non-synonymous substitutions (dS/dN < 1), indicating a faster rate of amino acid substitution relative to neutral mutations in this protein (Table 2). Although other explanations are possible, we speculate that positive selection is occurring in *win3*, resulting in more variable trypsin inhibitors. Earlier comparisons of the *win3* genes within hybrid poplar by Hollick and Gordon (1993) identified the 60 C-terminal amino acid residues of the win3 protein as hypervariable, and the authors also suggest this could indicate positive selection for sequence divergence. The rapid adaptation of insect gut proteases has been demonstrated in several insects (Jongsma et al. 1995; Broadway 1996), and this provides a rationale for selection for rapid evolution of trypsin inhibitor proteins.

In order to further visualize the differences between the likely *win3* orthologs, an alignment of the predicted protein sequences from both hybrids as well as trembling aspen and black cottonwood was carried out (Fig. 3). This alignment demonstrated that the first 50 amino acid residues of the N-terminus of these predicted proteins are highly conserved. A signal peptide, with the most probable cleavage site between the 28th and the 29th amino acid residues, was predicted in this region (SignalP, Nielsen et al. 1997). A variable region and an insertion/deletion of 7–8 amino acids were observed at position 55 of the alignment (Fig. 3). This insertion/deletion clearly separates the poplar and aspen *win3* sequences, since only the trembling and hybrid aspens contain these additional amino acids. In addition, an insertion was apparent at position 171 of the predicted proteins in the hybrid aspen sequence. Similar to that shown by Hollick and Gordon (1993), the C-terminal regions of the win3 predicted proteins are hypervariable, and show deletions/insertions in the aspen protein sequences (Fig. 3). A comparison among proteins in public databases containing the Kunitz TI conserved domain showed that both insertions/deletions correspond to variable regions of the protein alignment, interspersed between more conserved domains (protein family ID: PF00197; http://pfam.wustl.edu).

Western blot analysis of wound-induced win3 TI proteins

We further extended our analysis of TIs to the protein level using western blots of control and wounded leaf extracts from the different *Populus* genotypes, and developed with an antibody raised against the trembling aspen *win3* ortholog PtTI2 (Haruta et al. 2001). As expected, the win3 protein was detected in wounded leaves of all *Populus* genotypes tested, but not in non-wounded controls (Fig. 4). Both hybrid poplar and black cottonwood showed similar protein patterns with strong TI signals. By contrast, the trembling and hybrid aspens revealed bands of distinctly different sizes (Fig 4). In general, we observed a surprising degree of size polymorphisms among the different genotypes, not seen with other defense proteins such as polyphenol oxidase (data not shown). These win3 protein-size polymorphisms provide further evidence for the high variability of the win3 trypsin inhibitor in poplars and aspens, consistent with our analysis of sequence data and further supporting the idea of rapid evolution of these defense genes. We note that at the population level, work on trembling aspen PtTI2 in this laboratory also demonstrated high levels of restriction fragment polymorphisms in native clones of *P. tremuloides* on Southern blots (Haruta et al. 2001).

Several reports have noted rapid evolution of protease inhibitor genes in other taxa. For example, in tomato, the *Pin2*

Fig. 3. Alignment of the predicted win3 protein sequences from poplars and aspens. *Win3* contig sequences were obtained as described for Fig. 2, and the predicted proteins aligned using the ClustalV algorithm of MegAlign (DNAStar, London, UK). The arrowhead indicates the predicted signal peptide cleavage site (SignalP, Nielsen et al. 1997).

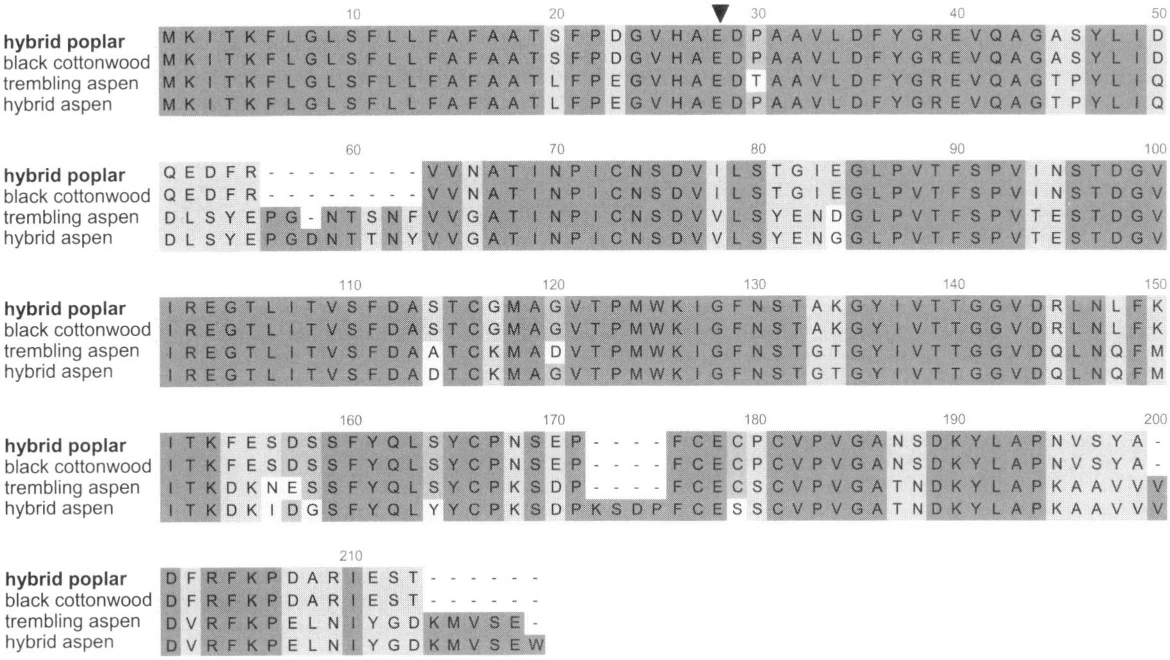

Fig. 4. Western blot analysis of win3 protein in leaves of different *Populus* species and hybrids. C and W indicate control and wounded leaves, respectively.

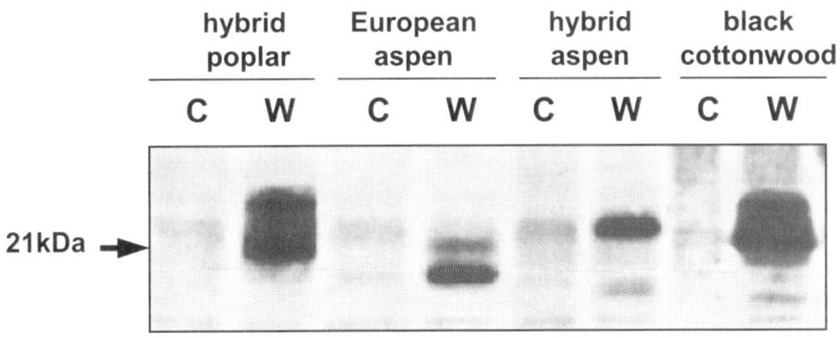

family of inhibitors from tomato has diverged to an extent that it is not easily detected outside of the Solanaceae (Van der Hoeven et al. 2002; Barta et al. 2002). Tiffin and Gaut (2001) investigated *wip1*, a wound inducible Bowman-Birk PI in *Zea mays*, and found evidence for an accelerated evolutionary rate of this PI in *Zea* relative to other monocotyledoneous genera. Likewise, rapid evolution of animal PIs was deduced from the existence of hypervariable reactive centers of homologous proteins isolated from related species (Hill and Hastie 1987; Laskowski et al. 1987). In addition, positive selection acts on several plant resistance and defense-related genes causing rapid evolution of these genes (reviewed by Stahl and Bishop 2000). The availability of *Populus* genome sequences will allow such evolutionary analyses to be expanded to other defense related genes. Ultimately, our aim is to understand the extent to which defense responses vary among *Populus* species and genotypes, in the context of their ecological adaptations.

Acknowledgments

The authors gratefully acknowledge the support of the Natural Sciences and Engineering Research Council (NSERC) of Canada and the Alberta Agricultural Research Institute in the form of research grants to CPC, and the technical assistance of Charlotte Yates.

References

Altschul, S.F., Gish, W., Miller, W., Myers, E.W, and Lipman, D.J 1990. Basic local alignment search tool. J. Mol. Biol. **215**: 403–410.

Barta, E., Pintar, A., and Pongor, S. 2002. Repeats with variations: accelerated evolution of the *Pin2* family of proteinase inhibitors. Trends Genet. **18**: 600–603.

Bergey, D.R., Howe, G.A., and Ryan, C.A. 1996. Polypeptide signaling for plant defensive genes exhibits analogies to defense signaling in animals. Proc. Natl. Acad. Sci. U.S.A. **93**: 12053–12058.

Bradshaw, H.D., Jr., Hollick, J.B., Parsons, T.J., Clarke, H.R., and Gordon, M.P. 1990. Systemically wound-responsive genes in poplar trees encode proteins similar to sweet potato sporamins and legume Kunitz trypsin inhibitors. Plant Mol. Biol. **14**: 51–59.

Broadway, R.M. 1996. Dietary proteinase inhibitors alter complement of midgut proteases. Arch. Insect Biochem. Physiol. **32**: 39–53.

Christopher, M.E., Miranda, M., Major, I.T., and Constabel, C.P. 2004. Gene expression profiling of systemically wound-induced defenses in hybrid poplar. Planta (in press).

Constabel, C.P., Yip, L., Patton, J.J., and Christopher, M.E. 2000. Polyphenol oxidase from hybrid poplar. Cloning and expression in response to wounding and herbivory. Plant Physiol. **124**: 285–295.

Harlow, E., and Lane, D. 1988. Antibodies. A laboratory manual. Cold Spring Harbor Laboratory Press, Cold Spring Harbor.

Haruta, M., Major, I.T., Christopher, M.E., Patton, J.J., and Constabel, C.P. 2001. A Kunitz trypsin inhibitor gene family from trembling aspen (*Populus tremuloides* Michx.): cloning, functional expression, and induction by wound and herbivory. Plant Mol. Biol. **46**: 347–359.

Hill, R.E., and Hastie, N.D. 1987. Accelerated evolution in the reactive centre regions of serine protease inhibitors. Nature **326**: 96–99.

Hollick, J.B., and Gordon, M.P. 1993. A poplar tree proteinase inhibitor-like gene promoter is responsive to wounding in transgenic tobacco. Plant Mol. Biol. **22**: 561–572.

Jongsma, M., Bakker, P., Peters, J., Bosch, D., and Stiekema, W. 1995. Adaptation of *Spodoptera exigua* larvae to plant proteinase inhibitors by induction of gut proteinase activity insensitive to inhibition. Proc. Natl. Acad. Sci. U.S.A. **92**: 8041–8045.

Laskowski, M., Jr., Kato, I., Ardelt, W., Cook, J., Denton, A., Empie, M.W., Kohr, W.J., Park, S.J., Parks, K., Schatzley, B.L., Schoenberger, O.L., Tashiro, M., Vichot, G., Whatley, H.E., Wieczorek, A., and Wieczorek, M. 1987. Ovomucoid third domains from 100 avian species: isolation, sequences, and hypervariability of enzyme-inhibitor contact residues. Biochemistry **26**: 202–221.

Meagher, R.B., Berry-Lowe, S., and Rice, K. 1989. Molecular evolution of the small subunit of ribulose bisphosphate carboxylase: nucleotide substitution and gene conversion. Genetics **123**: 845–863.

Miller, R.T., Christoffels, A.G., Gopalakrishnan, C., Burke, J., Ptitsyn, A.A., Broveak, T.R., and Hide, W.A. 1999. A comprehensive approach to clustering of expressed human gene sequence: The Sequence Tag Alignment and Consensus Knowledge base. Genome Res. **9**: 1143–1155.

Nei, M., and Gojobori, T. 1986. Simple methods for estimating the numbers of synonymous and nonsynonymous nucleotide substitutions. Mol. Biol. Evol. **5**: 418–426

Nielsen, H., Engelbrecht, J., Brunak, S., and von Heijne, G. 1997. Identification of prokaryotic and eukaryotic signal peptides and prediction of their cleavage sites. Prot. Eng. **10**: 1–6.

Parsons, T.J, Bradshaw, H.D., Jr., and Gordon, M.P. 1989. Systemic accumulation of specific mRNAs in response to wounding in poplar trees. Proc. Natl. Acad. Sci. U.S.A. **86**: 7895–7899.

Ryan, C.A. 1990. Protease inhibitors in plants: genes for improving defenses against insects and pathogens. Annu. Rev. Phytopath. **28**: 425–449.

Stahl, E.A., Bishop, J.G. 2000. Plant-pathogen arms races at the molecular level. Curr. Opin. Plant Biol. **3**: 299–304.

Sterky, F., Regan, S., Karlsson, J., Hertzberg, M., Rohde, A., Holmberg, A., Amini, B., Bhalerao, R., Larsson, M., and Villarroel, R. 1998. Gene discovery in the wood-forming tissues of poplar: analysis of 5,692 expressed sequence tags. Proc. Natl. Acad. Sci. U.S.A. **95**: 13330–13335.

Stettler, R.F., Bradshaw, H.D., Jr., Heilman, P.E., and Hinckley, T.M. 1996. Biology of *Populus* and its Implications for Management and Conservation. NRC Research Press, Ottawa.

Tiffin, P., and Gaut, B.S. 2001. Molecular evolution of the wound-induced serine protease inhibitor *wip1* in *Zea* and related genera. Mol. Biol. Evol. **18**: 2092–2101.

Van der Hoeven, R., Ronning, C., Giovannoni, J., Martin, G., and Tanksley, S. 2002. Deductions about the number, organization, and evolution of genes in the tomato genome based on analysis of a large expressed sequence tag collection and selective genomic sequencing. Plant Cell **14**: 1441–1456.

Wullschleger, S.D., Jansson, S., and Taylor, G. 2002. Genomics and forest biology: *Populus* emerges as the perennial favorite. Plant Cell **14**: 2651–2655.

List of participants

Aitken, Sally. *University of British Columbia*
Allen, Geraldine. *University of Victoria*
Awadalla, Phillip. *University of California, Davis*

Baucom, Regina S. *University of Georgia*
Beland, Jaclyn. *University of British Columbia*
Bell, Charles. *Yale University*
Berube, Yanik. *University of British Columbia*
Binns, Shannon. *University of British Columbia*
Black, Scott T.A. *University of Victoria*
Bohlmann, Jorg. *University of British Columbia*
Borevitz, Justin O. *Salk Institute, San Diego*
Bouck, Amy C. *University of Georgia*
Bradshaw, Toby. *University of Washington*
Breen Carroll, Amy. *University of Alaska*
Byers, Diane L. *Illinois State University*

Carlson, Michael R. *British Columbia Forest Service Research*
Chen, Charles. *University of British Columbia*
Cheung, Dorothy. *University of British Columbia*
Constabel, C. Peter. *University of Victoria*
Cronk, Quentin C.B. *University of British Columbia*
Cruzan, Mitchell B. *Portland State University*

Douglas, Carl J. *University of British Columbia*
Douglas, Diane. *British Columbia Ministry of Forests*

Elle, Elizabeth. *Simon Fraser University*
Elmhirst, Janice F. *Elmhirst Diagnostics & Research, BC*

Fulop, Daniel. *Harvard University*

Gilchrist, Erin J. *University of British Columbia*
Givnish, Thomas. *University of Wisconsin-Madison*
Good-Avila, Sara. *Acadia University*
Guest, Heidi. *University of Victoria*
Guy, Robert D. *University of British Columbia*

Hahn, Matthew. *Duke University*
Hodges, Scott A. *University of California, Santa Barbara*
Hu, Xinsheng. *University of British Columbia*
Husband, Brian. *University of Guelph*

James, Shelley. *Bishop Museum, Honolulu*
Jones, Carolyn. *Van Dusen Botanical Garden, Vancouver*

Kentner, Edward K. *University of Georgia*
Kumar, Anshuman. *University of British Columbia*

Levin, Donald. *University of Texas*
Lewis, Gary. *University of British Columbia*
Liewlaksaneeyanawin, Cherdsak. *University of British Columbia*
Lund, Stephen. *University of British Columbia, Wine Research Centre*

Mable, Barbara. *University of Guelph*
Mathews, Sarah. *University of Missouri*
McKhann, Heather. *INRA-CNG, France*
Montgomery, Rebecca A. *University of Wisconsin-Madison*
Moore, Richard. *North Carolina State University*
Moyle, Leonie. *University of California, Davis*

Neel, Maile C. *University of Massachusetts*
Nepokroeff, Molly. *University of South Dakota*

O'Connell, Lisa. *University of British Columbia*
Olmstead, Richard. *University of Washington*
Olson, Matthew. *University of Alaska*
Otto, Sally. *University of British Columbia*
Oyama, Ryan K. *John Innes Centre, UK*

Piggott, Nina. *Pacific Forestry Centre, Victoria*
Purugganan, Michael D. *North Carolina State University*

Rajakaruna, Nishanta. *University of British Columbia*
Rausher, Mark D. *Duke University*
Ree, Rickard H. *University of British Columbia*
Rhode, Jennifer M. *Portland State University*
Ritland, Kermit. *University of British Columbia*
Rogers, Sean. *Paprican, Vancouver*
Rungis, Dainis. *Canadian Tilling Initiative, Vancouver*

Sargent, Risa. *University of British Columbia*
Sayre, Cindy M. *Van Dusen Botanical Garden, Vancouver*
Schluter, Dolph. *University of British Columbia*
Schlichting, Carl D. *University of Connecticut*
Shi, Suhua. *Zhongshan University, PRC*
Sumampong, Grace. *Simon Fraser University*

Tang, Jihong. *Paprican, Vancouver*
Taylor, D. Lee. *University of Alaska*
Taylor, Iain E.P. *University of British Columbia*
Tiffin, Peter. *University of Minnesota*
Thompson, Stacey. *University of British Columbia*

van Kleunen, Mark. *University of British Columbia*

Waldron, Anthony. *University of British Columbia*
Wheeler, Erica. *University of Victoria*
White, Gemma M. *Rothamsted Research, UK*
Whittle, Carol-Ann. *Dalhousie University*
Whitton, Jeannette. *University of British Columbia*
Wilkin, Jennifer. *University of British Columbia*
Willis, John. *Duke University*

Yanchuk, Alvin. *British Columbia Forest Service Research*
Ying, Cheng C. *British Columbia Ministry of Forests*

Index

A

abiotic factors, 7, 83, 111, 113, 116
Acacia, 4, 5
Achillea, 86
adaptive convergence, 22
adaptive divergence, 38, 39, 119, 120
adaptive landscape, 38, 39, 61 et seq., 66 (Fig.)
additive genetic model, 23, 61, 62
Aegilops, 31, 34
age-dependent effects, 21
agriculture and silviculture, 9, 77
Agrobacterium, 96, 127
Agropyron, 34
Agrostis, 87, 104, 105
Alaska, 127
algae, green, 21
alkaline flats, 103, 105
allele, 14, 15, 61, 62, 142
allopatry, 84, 103, 105
alloplasmic lines, 31–33
allopolyploidy, 30, 31, 34, 130
allozyme, 106, 112, 130
Amborella, 147, 149
amplified fragment length polymorphisms (AFLP), 56, 128
ancestral factors, 98, 100, 146
ancient adaptation, 143, 148, 149
Antirrhinum, 75, 76
Anthoxanthum, 87, 104, 105
Ants, 77
Aquilegia, 19, 84, 111, 112
Arabidopsis halleri (formerly *Arabis halleri*), 129
Arabidopsis lyrata, 48, 95, 96, 99, 127 et seq., 130 (Fig.)
Arabidopsis suecica (formerly *Arabis suecica*), 129
Arabidopsis thaliana, 5, 7, 8, 10–12, 15, 20–24, 46, 48–50, 53 et seq., 95–97, 99, 116, 127, 128, 129 (Fig.), 131, 135–137, 141–144, 146–149
Arabidopsis core collection, 135, 137, 138 (Table), 141, 142 (Fig.)
Arabis, 129
archaebacterium, 20
Archaefructus, 147
Arenaria, 87
Armeria, 4, 105
arrays, 14, 55, 56, 58, 94, 97
Artemisia, 86
artemisinin, 70
arthropod feeding, 74, 153 et seq.
Asparagus, 4
aspen, hybrid, 153, 154, 156, 157 (Fig.)
aspen, European, 153, 154
aspen, trembling, 153, 154, 156, 157 (Fig.)
Asteraceae, 9, 10, 32
Atropa, 31, 32
Australopyrum, 34
Austrobaileyaceae, 147
autogamy, 85
autopolyploidy, 119, 120, 122, 130
Avena, 33, 34

B

bacteria, 20, 24
bacterial artificial chromosome (BAC), 58, 96
bark beetle, 72, 76, 77
barley, 127
Bataillon, T., 135
bean, 127
Berberis, 4
Beta, 33, 112
biosynthesis, terpenoids, 70, 72, 76, 77
biotic selection, 112, 113
Bohlmann, J., 70
boreal forest, 120
Borevitz, J., 53
bottlenecks, 13, 16, 83
Brassicaceae/*Brassica*, 54, 56, 99, 127–129, 144
breadwheat, 31, 32, 34
broccoli, 127
Brunel, D., 135
buckwheat, 32

C

C4 photosynthesis, 38
Ca^{2+}/ Mg^{2+} uptake, 103
Cabomba, 147
Cacti, 4
cadmium, 104
Caenorhabditis, 20, 24
caffeoyl CoA O-methyltransferase (*CCoAMT*), 155, 156
Calluna, 4
Camilleri, C., 135
canola, 127
Cardamine, 129
Cardaminopsis, 129
Casuarina, 4
cDNA, 8, 53, 57, 58, 94, 97, 154
cedar, 16
cell fusion lines, 30
cell wall composition, 128
Chamaecrista, 35
Chamerion, 119 et seq., 121 (Fig.)
character displacement, 87
chasmogamy, 112, 114
chemical ecology, 46, 72
Chlamydomonas, 24
chloroplast, 11, 15, 16, 30, 31–34, 72
Christopher, M., 153
chromosome, 5, 30, 53, 119, 120
Chrysanthemum, 87

circadian rhythm, 53, 55, 75
clades, 18, 24, 25, 99
Clarkia, 75, 84, 85, 112, 114
cleistogamy, 114
clone, 3, 11, 12, 16, 130
cluster trees, 141 (Fig.)
co-adaptation, 11, 12, 31, 85
coastal bluffs, 105
coffee, 127
cold, 22, 135–137, 139
Collinsia, 95, 111, 112 et seq.
columbine, 19
common garden, 64, 116, 122
common goldfields, 105
competition, 83, 87, 89, 135
conifers, 4, 16, 144, 146, 148
Constabel, C.P., 153
copper, 86, 89, 104, 105
corn, 15
cotton, 127
cottonwood, black, 153–157 (Fig.)
Crepis, 32, 86, 87
Cronk, Q.C.B., 3
Cruzan, M.B., 61
cryptic adaptation, 23, 41
cybrids, 31, 32
Cycas, 144, 146
Cytisus, 4
cytochrome oxidase, 31
cytoplasm, 30, 31, 33, 35, 87, 99
cytotypes, 119, 120, 123

D

Dalechampia, 89
Datura, 86
David, J.L., 135
de-etiolation, 143, 147
defense against insects, 72, 75, 153
Delphinium, 86
Dendroctonus, 77
development, 5, 22, 31, 84, 116
dimethylallyl diphosphate (DMAPP), 70, 72
dioecy, 85
diploid-polyploid clines, 119
Discussion Report, 7, 38, 94
disease resistance, 48
disjunction, abnormal 30
diterpenes, 23, 62 (Fig.), 70, 72, 73
diversification, 19, 84, 103, 104, 136
diversity mapping, 11, 13
Dobzhansky-Muller model, 62
domestication, 9, 45, 54, 70
Drosophila, 20, 23, 24, 55
drought, 83, 99, 107
drugs, 70

E

eco-evo-devo framework, 8, 9
ecology, 3, 11, 30, 34, 83, 84, 86, 91, 98, 112, 116, 119, 120 (Fig.), 121, 136, 141, 147
ecomolecular synthesis, 5, 7, 8, 10, 39, 94, 96
EcoTILLING, 5
ecotypes, 99, 127, 131
edaphic factors, 9, 82, 83, 99, 103 et seq., 111
eggplant, 46
Elle, E., 111
Elymus, 34
Empetrum, 4
environment, 20–23, 53, 86, 127, 138, 142, 143
Ephedra, 146
epigenetic phenomena, 18, 21, 38
Epilobium, 32–34
epistasis, 11, 21, 53, 61–63, 65–67, 96, 98
equilibrium drift theory, 13
Eryngium, 4
Escherichia coli, 20
Eucalyptus, 4, 127
evolutionary synthesis, 38, 121
evolvability, 18, 19 (Table), 23, 24, 25
expressed sequence tags (ESTs), 14, 58, 96, 99, 153–155
extreme array mapping (XAM), 56

F

Fabaceae, 9
farnesyl diphosphate (FPP), 70, 72
Festuca, 4
Fisher, 5
fitness advantages, 6, 21, 40, 61, 62, 63, 64, 65, 136, 139
fitness landscape (surface) see adaptive landscape
flavonoids, 105, 106
fleshy fruits, 8, 9
flow cytometry, 122, 130
flower, 4, 30, 33, 46, 48, 53–55, 61, 67, 70, 75, 76, 82, 84, 85, 88, 90, 105, 111–113,
116 (Fig.), 122, 128, 130, 135–140, 142, 143, 148
forest tent caterpillar, 75
Formica, 77
founder events, 83, 119, 123
fruit, 54, 137–139
functional genomics, 5, 127
fungi, 21, 22, 72

G

gamete, 34, 85, 122, 124
gametophyte, 11, 16, 85
gender dimorphism, 111
gene, 7, 11, 12 (Fig.), 21, 23, 30, 57, 97, 120, 123, 124, 143, 146
gene flow, 14, 84, 85, 103, 123
gene genealogy, 7
gene trees, 30
gene × environment interaction/(g × e), 95
genealogical relationships, 49, 56
86, 114, 124, 128, 132
genetic/genetics, 11, 12, 13, 14, 16, 18, 20, 22, 39, 41, 45, 56, 58, 72, 82, 96, 111, 116, 120 (Fig.), 121, 124, 135
genetic variability, 18, 20, 119, 120, 138, 137
genome, 5, 9, 11, 12, 15, 16, 13, 38, 41, 48, 53, 56, 66, 95, 96, 97, 98, 119, 121, 123, 124, 127, 128 (Table), 129, 153

genomics, 45
genotype × environment interaction (G × E) , 21, 104, 107, 137
genotype-phenotype (G → P), 8, 18, 19, 21–25, 21 (Table), 39, 50 (Fig.), 135
geographic range, 119, 121 (Fig.), 122, 127, 129
geographic variation, 111, 112
Geonoma, 85
geranyl diphosphate (GPP), 70, 72, 77
geranyl geranyl diphosphate (GGPP), 70, 72
germacrene D, 75, 76 (Fig.)
germination, 123, 143
Gilia, 33, 34, 85, 87
Ginkgo, 5, 144, 146, 147
Gnetum/Gnetales, 146
Gossypium, 86
Gould objection, 3
guano deposits, 103

H

Hakea, 4
Haldane, 5
haplotype,12, 16, 33, 47, 49 (Fig.), 53, 55, 96
Hawaiian silverswords, 24, 99
heavy metal, 83, 88
Helianthus, 31, 35, 46
herbivore/herbivory, 24, 54, 70, 75, 76, 77, 83, 89, 99, 153 attack
heritability, 23, 135, 137, 138, 139, 142
herkogamy, 111, 112
heterochrony, 5, 21, 116
heterotopy, 5, 21
hill climbing, 61
Hippophae, 4
hitchhiking, 12
holey adaptive landscape, 61 et seq.
homeosis, 5
homeostasis, 21
homoplasy, 9, 40
Hordeum, 34
Huber, D.P.W., 70
Husband, B.C., 38
hybrid, 30–34, 63, 82, 84–86, 88, 89, 105, 106
hybridization, 33, 57, 62, 86, 113, 120
Hyocyamus, 32

I

Ilex, 4
Illiciaceae, 147
inbreeding, 11, 12, 16, 32, 83, 112, 116
incompatibility, 30, 85
inflorescence, 7, 46, 53, 128, 136, 139
insect behavior, 21, 22, 70, 72, 75–77, 153, 156
ionic stress, 105, 107
*Ipomoe*a, 99
Ipomopsis, 34, 86, 89
Ips pini, 72, 77
ipsdienol, 77
Iris, 4, 32, 34, 86
island radiations, rapid, 6
island, 6, 9, 40, 83, 85, 112
isolation, 32, 82, 83, 84, 89, 103
isopentenyl diphosphate (IPP), 70, 72
isozyme variation, 14, 104

J

Jamaican click beetles, 3
jasmonate/jasmonic acid, 74
juvenile hormone, 77

K

Kermode bear, 15
knock out, 5, 8, 53, 57
Kranz anatomy, 38
Kunitz trypsin inhibitor (TI), 153, 154, 156

L

Lasthenia, 103 et seq.
Laurus, 4
Le Corre, V., 135
lead, 105
Leavenworthia, 84, 85
Levin, D.A., 30
life history, 12, 53, 95, 135, 136
Liliales, 8
limonene, 72
linalool synthase, 75
Lingula, 5
linkage disequilibrium (LD), 11–13, 16, 33, 45–47 (Fig.), 48, 50 (Fig.), 55, 96–99, 135
linkage, 16, 45, 56, 99, 103, 105
Lobelia, 111
loblolly pine, 13
lodgepole pine, 11, 13 (Fig.)
Lotus, 15, 58
Louisiana irises, 46, 99
lupins, 4
Lycium, 32
Lycopersicon, 48, 86, 87
Lycopodium, 4

M

Mable, B.K., 127
MacDonald-Kreitman test, 4
maize, 5, 7, 12, 32, 33, 45, 46, 49, 50, 54, 58
Malacosoma disstria, 75
maladaptation, 39, 40, 86, 123
male sterility, 31, 32, 87
mangrove, 7, 99
Mathews, S., 143
mating system, 15, 84, 123, 129
McKhann, H., 135
Medicago, 58
MEP (2-C-methyl-D-erythritol 4-phosphate) pathway, 70, 71 (Fig.), 72
Mesembryanthemum, 4
metal-contaminated sites, 104
methyl jasmonate (MeJA), 73, 74, 75 (Fig.)
methylthioalkylmalate synthase, 54

mevalonate, 70, 71 (Fig.), 72, 77
micoarrays, 8, 22–24, 39, 53, 55, 96, 97, 99
Mimulus, 8, 32, 34, 46, 58, 85–89, 95, 97, 99, 105
mine tailings, 103, 104, 111
Miranda, M., 153
mis-hybridization, 57
mitochondria, 30–32
mixed mating systems, 112
models, 66, 72, 83, 141
modern synthesis of evolution, 5
modularity, 21, 22
molecular paleontology, 98
monoterpene, 70, 72–74 (Fig.), 75, 76
morphology, 19, 140 (Table)
morphotypes, 63
mountain monkeyflower, 14
Moyle, L.C., 82
Murren, C.J., 18
mutant collections, 53
mutation, 5, 16, 20, 21, 57, 61, 62, 65, 143, 148, 149
myrcene, 75, 77

N

Na^+ tolerance, 107
Na^+/H^+ antiporter, 106
nectar spur, 19, 84, 103
neopolyploids, 121, 122, 124
Nerium, 4
nickel, 105
Nicotiana, 31, 32, 87
nodulation genetic systems, 9
Nolina, 32
nucleocytoplasmic interactions, 30, 31, 32, 33, 35
nutriceuticals, 70
nutrients, 83
Nymphaeales, 147

O

oak-woodlands, 105
ocimene synthase, 75
octadecanoids, 74, 75
oligonucleotide microarrays, 53, 57, 97
Olsen, Kenneth M., 45
Onagraceae, 119, 120
Ophyrs, 85
organellar divergence, 31
orthology, 98
Oryza, 48, 49
outbreeding depression, 86
oxylipins, 75

P

paedomorphosis, 114
palmetto, 63
parallel evolution, 7, 40, 103, 104, 105, 106, 108
parapatry, 105
parasitoids, 70, 75
pathogens, 31, 83, 89, 128
PCR, 5, 16, 131
pearl millet, 33, 48
Pedicularis, 84
pepper, 46
Peronospora, 21
pest management, 70
Petunia, 84
Phaseolus, 86, 87
phenology, 30, 84, 105, 111
phenotype, 18, 19, 21, 23, 38, 39, 49, 53, 55, 62, 65, 94, 95, 113 (Table), 115 (Table), 119, 121, 139
phenotypic plasticity, 3, 5, 21, 22, 38, 61, 65
phenotypic variation, 18, 19 (Fig.), 20, 23, 38, 39, 41, 64, 95, 111, 122, 114
phenylpropanoid metabolism, 155
pheromones, 76, 77
Phlox, 34, 87
photomorphogenesis, 54, 143, 145 (Fig.), 148
photosynthesis, 30, 32
phylogenetic tree, *Arabidopsis* phytochromes, 146 (Fig.)
phylogenetic tree, poplar and aspen TI proteins, 155 (Fig.)
phylogenetic tree, terpenoid synthases, 73 (Fig.)
phylogeny, 7, 12 (Fig.), 15 (Fig.), 18, 19, 23, 33, 40, 99, 100, 104, 106, 129, 144 (Fig.), 147, 155
Physochlaine, 32
phytochrome, 67, 9, 54, 143 et seq.
Picea, 72, 73, 145
Pinaceae, 144, 146, 147
pinene, 72
Pinus, 13, 16, 146, 127
Piper, 22
Piriqueta, 61, 63–67, 63 (Fig.)
Pissodes strobi, 72
plant defense, 75
plant volatiles, terpenoid, 76
plant-insect interactions, 70
pleiotropy, 21, 53, 84, 85, 86, 98, 103, 105, 123
ploidy, 121, 124 (Fig.), 127
Poaceae, 32
Polemonium, 89
pollen, 16, 34, 87, 105, 108, 112, 123
pollination, 30, 40, 61, 70, 76, 82, 84, 90, 111, 113, 108 (Fig.)
pollinator, 19, 46, 76, 77, 83–87, 99, 103, 111–113, 116, 123
polymorphism, 14, 56, 57, 94–97, 99, 120, 127, 156
polyphenisms, 22
polyphenol oxidase, 156
polyploidy, 11, 40, 84, 90, 99, 106, 119, 122, 123, 124, 130, 131
polyterpenes, 70
poplar, 15, 96, 98, 127, 154, 156, 156 (Table)
poplar, hybrid, 75, 153, 154, 155, 156, 157 (Fig.)
populations, 6, 11, 14, 47, 48, 56, 62, 98, 103, 113, 119, 130
Populus, 13, 75, 95, 153 et seq., 155 (Fig.)
positive selection, 7, 8, 143, 145, 147, 157
post-mating barriers, 82, 85, 87, 88, 105
post-zygotic barriers, 30, 31, 35, 82, 86, 88
potato, 46, 127
Potentilla, 34, 86, 95, 99
predators, 75, 83, 89
pre-mating barriers, 82, 84, 85, 87, 88
prenyl transferases, 72, 73

pre-zygotic barriers, 90, 123
proteomics, 97, 127
Prunella, 112
Pseudoroegneria, 34
Purugganan, Michael D., 45

Q

quantitative trait loci (QTL), 9, 14, 23, 24, 39, 40, 45, 46, 53, 54, 58 (Fig.), 66, 83, 89, 95, 97–99, 111, 124, 135, 141, 142
Quercus, 33

R

Rajakaruna, N., 103
RAPD variation, 106
rapeseed, 127
Reboud, X., 135
recombinants, 12, 20, 21, 53, 61, 65, 66
redwoods, 32
Ree, R., 7
regulation, 5, 9, 96
reinforcement, 87, 90, 103
reproduction/reproductive, 19, 30, 46, 62, 65, 67, 76, 82, 84, 85, 87, 88 (Table), 95, 99, 103, 105, 106, 108, 111–113, 119, 123, 124, 148
restriction fragment length polymorphism (RFLP), 56
reverse genetics, 57
Rhode, J.M., 61
Rhododendron, 85
ribulose-1,5-bisphosphate carboxylase (Rubisco), 32, 33, 155, 156
rice, 16, 46, 48, 49, 54, 55, 58, 127, 146,
Ritland, K., 11
Robinia, 4
Rocky Mountains, 120
root, 22, 96
Rosa, 75
Roux, F., 135
Rubus, 87
Ruscus, 4

S

Saccharomyces, 22
salt tolerance, 22, 83, 107
Salvia, 84, 86
Saxifraga, 4
Scarcelli, N., 135
scents, 70
Schiedea, 85
Schisandraceae, 147
Schlichting, C.D., 18
Scopolia, 32
Scots pine, 14
Scrophulariaceae, 112
Secale, 31
Sedum, 4
seed, 53, 82, 85, 86, 89, 112, 123, 145
seed, production, 33, 108 (Fig.), 113, 114 (Fig.), 135, 136
seedling, 86, 145, 148
selection, 3, 4, 14, 24, 38, 40, 72, 82, 83, 84, 91, 99, 105, 119, 123, 144
self-compatible, 16, 105, 123, 131
self-incompatibility (SI) , 127, 129–131, 148
selfing, 14, 84, 85, 111, 112, 128, 136
Sempervivum, 4
serine protease inhibitors, 153
serpentine, 83, 84, 103, 105, 111
sesquiterpenes, 70, 72, 75
Sewell-Wright, 5
sexual selection, 82, 91
shade, 143, 147, 148
shifting balance theory, 40
signaling, 39, 75, 156
Silene, 33, 104
single feature polymorphisms (SFPs), 56, 57, 97, 99
single nucleotide polymorphisms (SNPs), 11, 13 (Fig.), 14 (Fig.), 15, 16, 45, 46, 48, 49 (Table), 55, 97, 99, 135, 137
Sinorhizobiumi, 9
slash pine, 63
snapdragon, 75
sodium, 103, 105, 106, 107 (Fig.)
soil, 10, 83, 86, 103
Solanaceae, 31, 32, 46, 157
Solanum, 32, 34
somatic mutation rates, 16
Sordaria, 21
sorghum, 33, 46, 127
soybean, 127
Spearman coefficient, 135, 138
speciation, 6, 19, 20, 25, 30–33, 46, 62, 82, 83, 90, 99, 103, 119
spinach, 15
sporophytic self-incompatibility (SSI), 127
spruce, 16, 98
spruce, Norway, 73, 74
Stachys sp., 4
starch biosynthesis, 48
sterility, 30, 86
stochasty, 119, 123
stomata, 75
Streptocarpus, 4, 33
Stylidium, 84, 87
sugar cane, 127
sulfated flavonoid, 105, 106, 107
Sulfolobus, 20
sympatry, 84, 86
synapomorphy, 9, 10, 19
synonymy, 16, 39, 156
synteny, 95

T

taxol, 70
taxonomy, *Arabadopsis lyrata*, 129
taxonomy, *Collinsia*, 112
tent caterpillars, 75
teosinte, 5, 54
termites, 77
terpene synthases (TPS), 70 et seq.

terpenoids, 70 et seq., 71 (Fig.)
tetraploidy, 34, 120, 123, 130, 131
tetraterpenes, 70
theoretical/analytical tools, 94, 97
TILLING, 5
tolerance, 86, 104, 105, 106
Tolmiea, 120
tomato, 39, 46, 54, 58, 127, 146, 153, 156
toxicity, 72, 75, 153
Tragopogon, 34, 87
transcription factors, 5, 22 (Table), 53, 57, 96, 154
transformation techniques, 94, 96
transplant garden, 64, 121
traumatic resin ducts, 72
TREESNIPS project, 14
trichome, 137
Trifolium, 111
Trimeniaceae, 147
triploidy, 122, 123, 130, 148
triterpenes, 70
Triticum, 31, 32, 34
tritrophic interactions, 75
trypsin inhibitor protein, 153 et seq., 154 (Table), 156 (Table)
turkey oak 63, 64
Turneraceae, 63

U

Ulex, 4

V

Vaccinium, 4
variability, 18, 20 (Table), 23 (Table)
variation, 10, 11, 24, 39, 55, 96, 98, 99, 104, 112, 127, 131, 132
vegetative trait, 33, 90, 65 (Fig.), 136,
vernal pools, 103, 105
vernalization, 53, 135, 137, 139
vessels, 148
viability selection, 86, 87
Viola, 112
vivipary, 99
volatile signaling, 70

W

Waxy gene, 49
weevil damage, 72, 73
Western blot analysis, 156, 157 (Fig.)
wheat leaf rust, 31
wheat, 33, 105, 127
white pine weevil, 72
Whitton, J., 103
wild barley, 48
wild hybrids, 34
wind-pollination, 85, 90
winter annuals, 114
wounding, 22, 74, 153 et seq.
Wright, rugged landscape model, 66
Wright, shifting balance theory, 39

X

xeromorphs, 3, 4 (Table), 85

Y

yeast, 24, 56

Z

Zamia, 146
Zea, 32, 48, 157
zinc, 104, 105